Contents

AN INTRODUCTION TO

Geographical Information Systems

Second edition

IAN HEYWOOD,

SARAH CORNELIUS

AND STEVE CARVER

Harlow, England • London • New York • Boston • San Francisco • Toronto • Sydney • Singapore • Hong Kong
Tokyo • Seoul • Taipei • New Delhi • Cape Town • Madrid • Mexico City • Amsterdam • Munich • Paris • Milan

Pearson Education Limited
Edinburgh Gate
Harlow
Essex CM20 2JE
England

and Associated Companies throughout the world

Visit us on the World Wide Web at:
www.pearsoneduc.com

First published 1998
Second edition published 2002

ISBN 0130 611980

British Library Cataloguing-in-Publication Data
A catalogue record for this book is available from the British Library

Library of Congress Cataloging-in-Publication Data
Heywood, D. Ian.
An introduction to geographical information systems/Ian Heywood, Sarah Cornelius, and Steve Carter. -- 2nd ed.
p.cm
Includes bibliographical references (p.).
ISBN 0-13-061198-0 (pbk.)
1. Geographica Information systems. I. Title: Geographical information systems. II. Cornelius, Sarah. III. Carver, Steve. IV. Title.

G70.212 .H49 2002
910'.285--dc21

2002075994

10 9 8 7 6 5 4 3
07 06 05 04 03

Typeset in Sabon 10/12 pt by 30
Printed by Ashford Colour Press Ltd., Gosport

Preface

How this book is written

Your encounters with GIS to date may be similar to those of a Martian arriving on Earth and being faced with a motor car. Imagine a Martian coming to Earth and stumbling across a motor car showroom. Very soon he (or she) has heard of a 'car' and may even have seen a few glossy brochures. Perhaps you are in the same position. You have heard of the term GIS, maybe even seen one or two demonstrations or the paper output they produce.

Developing the analogy of the Martian and the car leads us to a dilemma. There are two approaches to explaining to the Martian what the car is and how it works. The first method is a bottom-up approach. This involves taking the car apart into its component pieces and explaining what each part does. Gradually, we put the pieces back together so that by the time we have reassembled the car our extraterrestrial friend has a good appreciation of how the car works in theory. However, he may still have little idea about how to use it or what to do with it in practice.

The second method, the top-down approach, starts by providing the Martian with several examples of what the car is used for and why. Perhaps we take him for a test run, and then explore how the different components of the car work together to produce an end result. If this approach is adopted, the Martian will never be able to build a car engine, but he will have a clear appreciation of how, when, why and where a car could be used. In addition, if we explore the subject in sufficient technical detail he will know how to choose one car in preference to another or when to switch on the lights rather than the indicator.

We feel that the same two methods can be used to inform you about GIS. Since we believe you are

reading this book not because you want to write your own GIS software, but because you wish to develop a better appreciation of GIS, the approach adopted is similar to the top-down method. We focus on the practical application of GIS technology and where necessary and appropriate take a more detailed look at how it works. In a book of this size it is impossible for us to explain and describe every aspect of GIS. Therefore, we concentrate on those areas that enable you to make sense of the application of GIS, understand the theories upon which it is based and appreciate how to set up and implement your own GIS project.

The second edition

The Martian has returned for a second visit to Earth to find out how things have changed in the car market. And what a difference he has found in the few years since his first visit. New models come with satellite navigation systems, mobile phone and portable computer docking stations, DVD screens in the headrests to keep the kids entertained and all kinds of other innovations. These are the visible changes – the safety features have also improved – there are twin and side airbags, automatic tyre pressure checking and other improved technological features. There are a few more electric cars around, and other fuels are becoming more popular – low sulphur for example. The range of models has increased – and now you buy a car to go with your lifestyle.

So too with GIS – what a change since the date of the first edition of this book! Now we have true mobile GIS, Internet-based systems, and real multimedia. There have been technological advances in data models, in the representation of 3D and application of the object-oriented model. There are new sources of data – from recent population censuses, from the de-restriction of GPS, and completely new sources such as LIDAR. The range of GIS software has increased dramatically, and there has been a specialization of products towards different types of users and applications.

The second edition has been expanded and updated to reflect these developments. In addition, there have been various changes and corrections made in response to suggestions made by reviewers, and from lecturers and teachers using the book. Many, many thanks to all of you who have commented on the first edition.

How to use this book

There are three ways to use this book:

1. Browse through the text, consider the chapter headings, and review the general questions and issues at the start of each chapter. Supplemented by reading Chapter 1 this will give you a quick tour through the world of GIS and highlight some of the important issues and concepts.
2. A better way to approach the text is to read the book chapter by chapter. This will allow you to discover the ideas presented and develop an understanding at your own pace. To assist you we have included revision questions and pointers for further study at the end of each chapter. The revision questions are provided to encourage you to examine and revisit the major themes and issues introduced in each chapter.
3. The third way to use this book is as a reference source. You might wish to delve into those sections that you feel are relevant at a particular time. Perhaps you can read the appropriate chapter to supplement a course you are following. There is a comprehensive index and glossary at the back of the text and pointers to further sources of information. We have tried to reference only readily available published material. References to other additional information sources on the Internet can be found on the website for the book. These lists of further sources are by no means comprehensive, but are offered as a starting point. We suggest that you also consult CD-ROM indexes, abstracts and on-line sources for more up-to-date materials.

Before you start

The Martian could not drive his car without an understanding of the road network or road signs. Similarly, a full understanding of GIS requires some computing background, particularly in topics like operating systems and file management. This book assumes basic familiarity with

PC computing. Anyone who has any experience of word processing, using spreadsheets, databases or mapping packages, should be able to apply that knowledge in a GIS context. The typical first-year undergraduate IT course offered in subject areas like geography, biology, business studies or geology is sufficient for you to cope with the ideas presented in this book. We assume that you are familiar with terms such as hardware and software, and the major components of a computer: for example, monitor, keyboard, hard disk drive, floppy disk drive, CD-ROM drive, processor and memory. We make no other assumptions – this book is written to be accessible to students of GIS from any professional or academic background: from archaeology, through biology, business studies, computing, demography, environmental management, forestry, geography, history ... and on to zoology.

If you want to become a GIS expert, you will need to be comfortable with more advanced computing issues, and will have to expand your computing background to include skills in such areas as programming and networks. These issues are beyond the scope of this book.

A book cannot substitute for hands-on experience in a subject area as practical as GIS. Therefore, for a fuller appreciation of GIS we encourage you to enroll on a course that offers practical experience of GIS, or to find a system to use in your own time. There are a number of excellent GIS 'courses' now available electronically and reference has been made to these on the website for the book.

How this book is structured

The book is organized into 13 chapters in two parts. The first part of the book deals with GIS theory and concepts. After an introductory scene-setting chapter, which also introduces the case studies used throughout the book, important topics like spatial data, database theory, analysis operations and output are all examined. In the second part of the book we have focused on a selection of GIS 'issues' – the development of GIS, data quality, organizational issues and project management. There are, of course, other issues we could have considered, but those selected reflect many of the key themes of current interest, and areas important for anyone undertaking practical work in GIS. The book ends with a chapter on the future of GIS in which a reflective look is taken at current GIS, and some predictions are made for the future of the technology. In summary, the chapters are:

Part 1 Fundamentals of GIS

Chapter 1: What is GIS?

This chapter provides an overview of GIS. It examines what GIS is, what it can do and, in brief, how it works. The chapter starts by looking at the types of generic questions GIS can answer and expands on these with reference to three case studies. GIS is then defined, and a range of issues and ideas associated with its use identified. Much of the material introduced in this chapter will be covered in more detail later in the book.

Chapter 2: Spatial data

This chapter looks at the distinction between data and information and identifies the three main dimensions of data (temporal, thematic and spatial). The main characteristics of spatial data are described. A review of how the traditional map-making process shapes these characteristics is presented. The three basic spatial entity types (points, lines and areas), which form the basic building blocks of most GIS applications, are introduced. Maps and a range of other sources of spatial data are reviewed.

Chapter 3: Spatial data modelling

How do you model spatial form in the computer? This chapter considers in detail how the basic spatial entities (points, lines and areas) can be represented using two different approaches: raster and vector. Two other entity types that allow the modelling of more complex spatial features (networks and surfaces) are introduced. Finally, modelling of three- and four-dimensional spatial data is reviewed.

Chapter 4: Attribute data management

This chapter introduces the methods available for handling attribute data in GIS. The need for formal methods for database management is discussed. The principles and implementation of a relational database model are considered in detail, since this model is the most frequently used in current GIS. Database options for large-scale users are presented, including the use of centralized and distributed database systems. Finally, a brief introduction to the object-oriented approach to database management is reviewed.

Chapter 5: Data input and editing

This chapter gives an overview of the process of creating an integrated GIS. It takes us from source data through data encoding, to editing and on to manipulatory operations such as re-projection, transformation, and rubber sheeting. The chapter provides examples of how these processes are carried out, and highlights issues pertinent to the successful implementation of a GIS application.

Chapter 6: Data analysis

Methods for making measurements and performing queries in GIS are introduced in this chapter. Proximity, neighbourhood and reclassification functions are outlined, then methods for integrating data using overlay functions explained. Interpolation techniques (used for the prediction of data at unknown locations) are introduced and the analysis of surfaces and networks considered. Finally, analysis of quantitative data is reviewed.

Chapter 7: Analytical Modelling in GIS

This chapter provides a summary of process models before considering how they can be implemented in GIS. These models are then approached from an applications perspective, and three examples are examined: physical process models; human process models and decision-making models. To conclude, the chapter considers some of the advantages and disadvantages of using GIS to construct spatial process models.

Chapter 8: from new maps to enhanced decisions

An understanding of the basic principles of map design is essential for the effective communication of information and ideas in map form. In addition, an understanding of the complexity of the map design process helps appreciation of the power of maps as a visualization tool. This chapter considers the advantages and disadvantages of cartographic and non-cartographic output. In the conclusion to this chapter there is a brief discussion of the role of GIS output in supporting decision-making.

Part 2 Issues in GIS

Chapter 9: The development of computer methods for the handling of spatial data

This chapter considers how GIS have developed to their current state. The methods of handling spatial data that were used before computers were available are examined. These give an insight into what we require computers to do, and how they can help (or hinder) existing practice. Computer methods for handling spatial data existed before GIS, so these are reviewed, then developments in GIS are discussed together with developments in a selection of complementary disciplines. To conclude we examine reasons for different rates of growth in different countries and the role of policy makers in the development of GIS. The chapter does not attempt to present a comprehensive history of GIS but aims to give some context for the systems and concepts we work with today.

Chapter 10: Data quality issues

The terms used for data errors and quality are explained at the beginning of this chapter, since the first step to solving problems is to be able to recognise and describe them. The remainder of the chapter outlines the types and sources of errors in GIS to help you identify and deal with problems at the appropriate stage of a GIS project. Techniques for modelling and managing errors are also considered.

Chapter 11: Human and organizational issues

This chapter takes us away from relatively small-scale research-type applications of GIS, where one user can be in control from start to finish, and takes a look at some of the issues surrounding larger scale commercial and business applications. In utilities, local government, commerce and consultancy, GIS must serve the needs of a wide range of users, and should fit seamlessly and effectively into the information technology strategy and decision-making culture of the organisation. In this chapter we examine the users of GIS and their needs; how to justify investment in GIS; how to select and implement a system, and the organizational changes that may result.

Chapter 12: GIS project design and management

How do we understand a problem for which a GIS solution is being sought? Two methods are introduced in this chapter: constructing a rich picture and a root definition. The method for constructing a GIS data model is then discussed. Here a distinction is made between the conceptual data model and its physical implementation in the computer. A closer look at the various project management approaches and techniques and tools available for the implementation of a GIS project follow. Potential implementation problems and tips for project evaluation are also considered. To conclude, a checklist is provided to help with the design and implementation of your own GIS project. A case study is used to illustrate the approach throughout the chapter.

Chapter 13: The future of GIS

GIS of the late 1990s is reviewed here and some of the predictions for GIS in the 1990s are revisited. The problems and limitations of current GIS are considered, and from this research and development issues for the future are identified. The predictions of other authors and the impact of GIS on our daily lives are considered. Finally, the question of whether GIS will still exist in a few years time is addressed.

In summary

We hope that after you have read this book, you will have the knowledge and enthusiasm to start applying GIS in the context of your own course, discipline or organization. While the text will not have taught you how to drive a specific GIS product we hope that it will give you an appreciation of the concepts on which GIS are based, the methods they use and the applications to which they can be put. In addition, the book should give you an appreciation of some of the difficulties and considerations associated with setting up any GIS project. We hope you enjoy the book!

Note

In common with Bonham-Carter (1995) we have found it convenient to use the abbreviation GIS in a number of ways:

- to refer to a single system (e.g. a GIS software package that illustrates this is...)
- to refer to several or all geographical information systems (e.g. there are many GIS that...)
- to refer to the field of geographical information systems (e.g. GIS has a long history), and
- to refer to the technology (e.g. the GIS answer would be...).

Trade marks

Mention of commercial products does not constitute an endorsement, or, indeed, a refutation of these products. The products are simply referred to as illustrations of different approaches or concepts, and to help the reader relate to the practical world of GIS.

A Companion Web Site accompanies *An Introduction to Geographical Information Systems*, 2nd edition by Heywood, Cornelius and Carver

Visit the ***Introduction to Geographical Information Systems*** Companion Website at www.booksites.net/heywood to find valuable teaching and learning material including:

For students:

- Study material designed to help you improve your results
- Extensive links to valuable resources on the web
- Revision questions to help you check your understanding
- Search for specific information on the site

For lecturers:

- A secure, password protected site with teaching material
- A downloadable Instructor's Manual
- Powerpoints of key artwork from the book, to assist in lecturing
- Links to articles and resources on the web
- A syllabus manager that will build and host a course web page.

Publisher's Acknowledgements

We are grateful to the following for permission to reproduce copyright material:

Ordnance Survey for the Ordnance Survey National Grid System map, Figure 2.6.

In some instances we have been unable to trace the owners of copyright material and we would appreciate any information that would enable us to do so.

Acknowledgements

It took a long time for this book to get from initial idea into print! Along the way many, many people helped with support, advice and ideas both directly and indirectly. We have taken the opportunity here to thank some of those who have helped, but we recognise that we have forgotten many others who deserve appreciation.

For help with the text we should thank Steve Wise and an anonymous reviewer who made very constructive comments on drafts of the early chapters of the book. Tony Hernandez, Chris Higgins, Derek Reeve, David Medyckyj-Scott and Tony Heywood are also thanked for reading drafts of various sections. Help with the diagrams, plates and cartoons came from Gustav Dobrinski, Bruce Carlisle and Simon Kenton. The cartoons were drawn by Vincent Silcock after original cartoons by Steve Carver. Tracey McKenna provided emergency help with word processing when required.

All of the case studies used in the book are the result of the work of teams of researchers and academics. Stan Openshaw and Martin Charlton were involved with the original radioactive waste siting study. Jim Petch, Eva Pauknerova, Ian Downey, Mike Beswick and Christine Warr worked on the Zdarske Vrchy study. The initial ideas for the House Hunting game came from a team that included James Oliver and Steve Tomlinson. This last case study is available as a final product, an excellent multimedia CAL resource produced by the GeographyCAL initiative based at the University of Leicester in the UK. For this version Roy Alexander provided additional input, and John McKewan undertook programming. John Castleford has been instrumental in seeing the product to its final form as a marketable and useful teaching resource.

Many other individuals have supported our work on this book – Helen Carver, Thomas Blaschke and Manuela Bruckler deserve special mention. In addition, the Universities of Leeds and Salzburg, the Manchester Metropolitan University and the Vrije Universiteit Amsterdam have allowed us to use their resources to help compile the materials.

At Addison Wesley Longman and Prentice Hall we have worked with many individuals. Particularly we should thank Vanessa Lawrence for help and advice at the start of the project, Sally Wilkinson, who took over during the middle stages, and Tina Cadle, Matthew Smith, Shuet-Kei Cheung and Patrick Bonham, who steered the book towards publication. For their work on the second edition thanks are due to Matthew Smith, Bridget Allen, Morten Funglevand, Paula Parish and Nicola Chilvers.

And lastly, with over 30 years of teaching experience between us we should thank all our students and colleagues. We started this project to help our students understand the mysteries of GIS. Some have been an inspiration, and gone on to exciting and rewarding careers in the GIS field. Others struggled with basic concepts and ideas, and helped us to realise just what we needed to take time to explain. Most importantly, writing for distance learning students has taught us all tremendous lessons about the English language. Some of our colleagues on the UNIGIS distance learning course deserve particular mention – Jim Petch, Derek Reeve, Josef Strobl, Nigel Trodd, Christine Warr, Steve Tomlinson, and Bev Heyworth – for supporting the development of a rich resource of GIS teaching materials. We apologise to them unreservedly if they see ideas and examples inspired by their course materials within the book!

We hope that the experience we have gained between the initial idea for the book and the date of publication has helped to make this a useful and readable introduction to GIS.

Ian Heywood, Sarah Cornelius, Steve Carver
September 2001

Abbreviations and acronyms

1:1	One-to-one relationship
1:M	One-to-many relationship
2.5D	Two-and-a-half-dimensional
2D	Two-dimensional
3D	Three-dimensional
4D	Four-dimensional
AGI	Association for Geographic Information (UK)
AI	Artificial Intelligence
AM/FM	Automated Mapping/Facilities Management
BGS	British Geological Survey
BS	British Standard
BSU	Basic Spatial Unit
CAD	Computer-aided Design
CAL	Computer-aided Learning
CASE	Computer-assisted Software Engineering
CCTV	Closed-circuit Television
CD-ROM	Compact Disc – Read Only Memory
CEN	Comité Européen de Normalisation (European Standards Commission)
CGIS	Canadian Geographic Information System
CORINE	Coordinated Information on the European Environment
CRT	Cathode Ray Tube
CUS	Census Use Study (USA)
DBMS	Database Management System
DIGEST	Digital Geographic Information Exchange Standards
DNF	Digital National Framework
DoE	Department of the Environment (UK)
DOS	Disk Operating System
dpi	Dots Per Inch
DSS	Decision Support System
DTM	Digital Terrain Model
DXF	Data Exchange Format
EAM	Entity Attribute Modelling
ED	Enumeration District (UK)
EDA	Exploratory Data Analysis
EDM	Electronic Distance Metering
ESMI	European Spatial Meta-information system
ESRC	Economic and Social Research Council (UK)
ESRI	Environmental Systems Research Institute
ETHICS	Effective Technical and Human Implementation of Computer-based Systems
GAM	Geographical Analysis Machine
GBF-DIME	Geographic Base File, Dual Independent Map Encoding
GDF	Geographic Data File
GeoTIFF	Geographic Tagged Image File Format
GIS	Geographical Information System
GISc	Geographical Information Science
GLONASS	Global Orbiting Navigation Satellite System
GML	Geography Markup Language
GPS	Global Positioning Systems
GUI	Graphical User Interface
HTTP	Hypertext Transfer Protocol
IAEA	International Atomic Energy Authority
ID	Identifier
IS	Information System
ISO	International Organization for Standardization
IT	Information Technology
LAN	Local Area Network
LBS	Location-based Services
LCD	Liquid Crystal Display
LIDAR	Light Detection and Ranging
M:N	Many-to-many relationship
MAUP	Modifiable Areal Unit Problem
MCE	Multi-criteria Evaluation
MEGRIN	Multipurpose European Ground Related Information Network

MSS	Multispectral Scanner
NAVSTAR	Navigation System with Time and Ranging
NCDCDS	National Committee on Digital Cartographic Data Standards (USA)
NCGIA	National Center for Geographic Information and Analysis (USA)
NERC	Natural Environmental Research Council (UK)
NII	Nuclear Installations Inspectorate (UK)
NIREX	Nuclear Industry Radioactive Waste Executive (UK)
NGDC	National Geospatial Data Clearinghouse
NJUG	National Joint Utilities Group (UK)
NTF	National Transfer Format (UK)
NTF	Neutral Transfer Format
NWW	North-west Water
OCR	Optical Character Reader
OGC	Open GIS Consortium
OGIS	Open Geodata Interoperability Specification
OO	Object-oriented
OS	Ordnance Survey (UK)
PC	Personal Computer
PDA	Personal Digital Assistant
PERT	Program Evaluation and Review Technique
PLSS	Public Land Survey System (USA)
PPGIS	Public Participation GIS
REGIS	Regional Geographic Information Systems Project (Australia)
RINEX	Receiver Independent Exchange Format
RRL	Regional Research Laboratory (UK)
SDSS	Spatial Decision Support System
SDTS	Spatial Data Transfer Standard
SLC	System Life Cycle
SPOT	Système pour l'Observation de la Terre
SQL	Standard Query Language (or Structured Query Language)
SSA	Soft Systems Analysis
SSADM	Structured Systems Analysis and Design
SWOT	Strengths, Weaknesses, Opportunities, Threats
TCP/IP	Transmission Control Protocol/ Internet Protocol
TIGER	Topologically Integrated Geographic Encoding Reference File (USA)
TIN	Triangulated Irregular Network
TM	Thematic Mapper
TV	Television
URL	Uniform Resource Locator
UK	United Kingdom
URISA	Urban and Regional Information Systems Association (USA)
USA	United States of America
UTM	Universal Transverse Mercator
VIP	Very Important Point
VR	Virtual Reality
VRML	Virtual Reality Modelling Language
WALTER	Terrestrial database for Wales
WAP	Wireless Application Protocol
WWW	World Wide Web
XML	Extensible Markup Language

Part 1

Fundamentals of GIS

1 What is GIS?

KEY QUESTIONS AND ISSUES

- What is GIS?
- What are the applications of GIS?
- What are the characteristics of GIS?
- How is the real world represented in GIS?
- What analysis can GIS perform?
- Where can I find more information about GIS?

Introduction

This chapter provides an overview of GIS. It examines what GIS is, what it can do and, in brief, how it works. We begin with a look at the types of generic questions GIS can answer and expand on these with reference to three case studies. GIS is then defined, and a range of issues and ideas associated with its use identified. Much of the material introduced in this chapter will be covered in more detail later in the book, and pointers to the appropriate sections are provided. If we return to the analogy of the Martian and the car, introduced in the Preface, this is where the Martian finds out what exactly a car is, why it is useful and what he needs to know to make it work.

One of the best ways to introduce GIS is to consider the generic types of questions they have been designed to answer. These include questions about location, patterns, trends and conditions:

- Where are particular features found?
- What geographical patterns exist?
- Where have changes occurred over a given period?
- Where do certain conditions apply?
- What will the spatial implications be if an organization takes certain action?

Box 1.1 provides examples of the different kinds of problems that can be addressed by asking these types of questions. An imaginary ski resort, Happy Valley, provides the context.

If you have a geographical background you may be asking what is new about these questions. Are these not the questions that geographers have been contemplating and answering for centuries? In part they are, though in many cases geographers and others using spatial data have been unable to find answers to their questions because of the volume of data required and a lack of time and techniques available to process these data. The following examples of three GIS applications are used to illustrate the capabilities of GIS as a tool for geographical analysis. All three involve the manipulation of data in ways that would be difficult or impossible by hand, and each illustrates different issues associated with the application of GIS.

BOX 1.1

Questions GIS could answer in Happy Valley

The Happy Valley GIS has been established to help the ski resort's managers improve the quality of the ski experience for visitors. The examples below are only a few of the situations in which asking a spatial question can help with the management of Happy Valley. You will find many other examples throughout the rest of this book.

1. *Where are particular features found?* When skiers visit Happy Valley they need to know where all the visitor facilities are located. To help the Happy Valley GIS is used to produce maps of the ski area. In addition, visitors can ask direct questions about the location of facilities using 'touch screen' computerized information points located in shops and cafes throughout the ski resort. These information points provide skiers with a customized map showing them how to find the facilities they require.
2. *What geographical patterns exist?* Over the last two ski seasons there have been a number of accidents involving skiers. All these incidents have been located and entered into the GIS. The Happy Valley management team is trying to establish whether there is any spatial pattern to the accidents. Do accidents of a certain type occur only on specific ski pistes, at certain points on a ski piste such as the lift stations or at particular times of day? So far one accident black spot has been identified where an advanced ski run cuts across a slope used by beginners, just below a mountain restaurant.
3. *Where have changes occurred over a given time period?* In Happy Valley avalanches present a danger to those skiers who wish to venture off the groomed ski pistes. The management team and the ski patrol use the GIS to build up a picture of snow cover throughout the area. This is done by regularly recording snow depth, surface temperature, snow water content and snow strength at a number of locations. A study of the geographical changes in these parameters helps the management team prepare avalanche forecasts for different locations in Happy Valley.
4. *Where do certain conditions apply?* Every day, during the winter season, the Happy Valley management team provides information on which ski pistes are open. Since this depends on the snow cover, avalanche danger and wind strength, data on these factors are regularly added to the GIS from reports provided by the ski patrols and local weather service. The warden can use the GIS to help identify which runs should be opened or closed.
5. *What will the spatial implications be if an organization takes certain action?* The access road to Happy Valley is now too narrow for the number of skiers visiting the area. A plan is being prepared for widening the road. However, any road-widening scheme will have impacts on a local nature reserve as well as surrounding farm land. The Happy Valley GIS is being used to establish the amount of land that is likely to be affected under different road-widening schemes.

Searching for sites

Searching for the optimum location to put something is a task performed by individuals and organizations on a regular basis. The task may be to find a site for a new retail outlet, a new oil terminal or a new airport. Sometimes the task is more demanding than others, involving searches through large numbers of maps and related documents.

Over the last 15 years finding a suitable site for the disposal of radioactive waste in the UK has become a sensitive and important issue. NIREX (Nuclear Industry Radioactive Waste Executive) is the UK company with responsibility for the identification of suitable radioactive waste disposal sites. It has the task of interpreting government radioactive waste policy and siting guidelines and presenting possible sites at public inquiries. One of the problems for NIREX is the lack of comprehensive and coherent guidelines for the identification of suitable sites. However, NIREX is expected to show that it has followed a rational procedure for site identification (Department of the Environment, 1985). Hydrology, population distribution and accessibility are examples of important siting factors, but how such factors should be interpreted is left up to NIREX. Where, therefore, should NIREX site a nuclear waste repository?

In the past, NIREX used a pen and paper approach to sieve through large numbers of paper maps containing data about geology, land use, land ownership, protected areas, population and other relevant factors. Areas of interest were traced from these maps by hand, then the tracings overlaid to identify areas where conditions overlapped. NIREX is not the only organization that has searched for sites using these techniques. This was a standard approach employed in the siting of a wide range of activities including shopping centres, roads and offices. The method is time-consuming and means that it is impossible to perform the analysis for more than a few different siting criteria. Inevitably the best sites are often missed. GIS techniques offer an alternative approach, allowing quick remodelling for slight changes in siting criteria, and produce results as maps eminently suitable for presentation at public inquiries.

Openshaw *et al.* (1989) demonstrated the use of GIS for this application, and their method is summarized in Figure 1.1. First, a number of data layers were established, each containing data for a separate siting criterion (for example, geology, transport networks, nature conservation areas and population statistics). These were converted from paper to digital format by digitizing (this technique is explained in Chapter 5), or acquired from existing digital sources such as the UK Census of Population. These data layers were then processed so that they represented specific siting criteria. The geology layer was refined so that only those areas with suitable geology remained; the transport layer altered so that only those areas close to major routes were identified; and the nature conservation layer processed to show protected areas where no development is permitted. The population layer

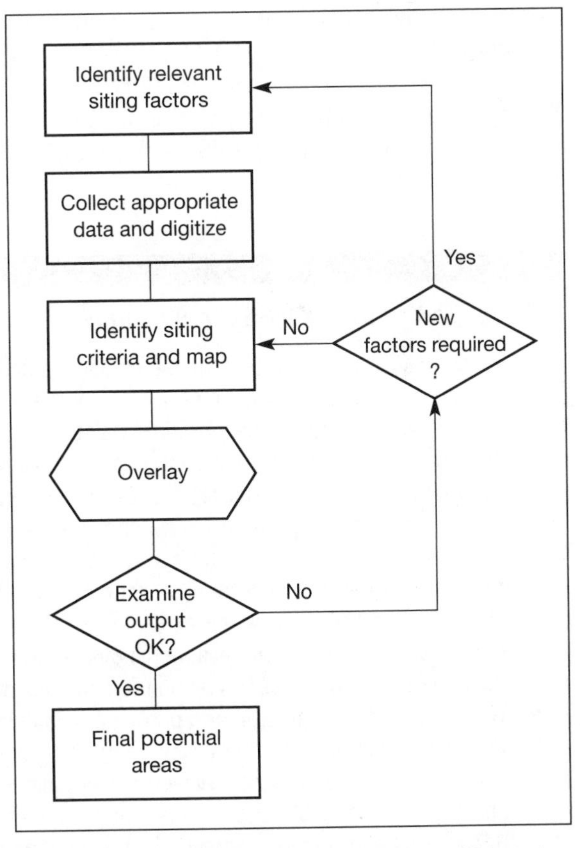

Figure 1.1 Using GIS for siting a NIREX waste site (*Adapted from Openshaw et al.*, 1989)

was analysed so that areas with high population densities were removed. GIS software was then used to combine these new data layers with additional layers of information representing other siting criteria. The final result was a map showing the locations where all the specified siting criteria were satisfied and, thus, a number of locations suitable for the siting of a nuclear waste repository. The advantage of using the GIS to perform this task was that the siting criteria could be altered and the procedure repeated. Examples for several siting scenarios are shown in Plate 5. These illustrate how changes in siting criteria influence the geographical distribution of potential sites.

This example shows how a GIS approach allows comparative re-evaluation and testing of data and conditions. In this way the decision maker can evaluate options in a detailed and scientific manner. The work of Openshaw *et al.* (1989) also illustrates three other important issues associated with the use of GIS: the problem of errors in spatial data sets; the difficulty in establishing criteria for abstract spatial concepts; and the potential value of using GIS to communicate ideas (Box 1.2).

Evaluating land use planning

Virtually every country in the world has areas of natural beauty and conservation value that are managed and protected in the public interest. Those managing these areas face the problem of balancing human activities (such as farming, industry and tourism) with the natural elements of the landscape (such as climate, flora and fauna) in order to maintain the special landscape character without exploitation or stagnation.

The protected area of Zdarske Vrchy, in the Bohemian–Moravian highlands of the Czech Republic, is an example of an area that has suffered as a result of ill-considered state control. Unregulated farming, tourism and industrial activities have placed the landscape under severe pressure. Czech scientists and environmental managers have relied on traditional mapping and statistical techniques to monitor, evaluate and predict the consequences of this exploitation. However, with the change in the political administration of the country there are scientists and managers who are looking not only into a policy of sustainable development for Zdarske Vrchy, but also at GIS as a tool to help with policy formulation (Petch *et al.*, 1995).

BOX 1.2

Issues raised by the NIREX case study

- *Errors in source data*, such as those introduced during the conversion of data to digital form, may have a significant effect on the GIS site-searching process. Mistakes in capturing areas of appropriate geology from paper maps may lead to inappropriate waste repository sites being identified, because areas on the ground will have different geological properties from those recorded in the GIS. Errors in spatial data sets and the associated issues of data quality are discussed in detail in Chapter 10.
- The GIS site-searching process relies on the translation of *abstract concepts* such as 'near to' and 'far from' into precise conditions that can be mapped. This can be a problem. How do you create a map that shows all the geographical zones 'far from' a centre of population? The only method is to make an arbitrary decision about what sort of distance (for example, 10, 20 or 30 km) 'far from' represents. In some cases rules may be applied to guide this process, but in others the numerical representation of a criterion may depend upon the preferences of the person responsible for choosing and implementing the criterion.
- *GIS output* can be used to inform public participation in the decision-making process. A series of maps could be used to illustrate why a particular geographical location has been identified as a suitable site for the disposal of radioactive waste. However, the issues raised above – data quality and the problems of creating spatial criteria for abstract concepts – suggest that output from GIS should be viewed with caution. Just because a map is computer-generated does not mean that the picture it presents is correct. More on this issue can be found in Chapter 8.

GIS is seen as a tool to bring together disparate data and information about the character and activities that take place in the Zdarske Vrchy region. Data from maps, aerial photographs, satellite images, ecological field projects, pollution monitoring programmes, socio-economic surveys and tourism studies have been mapped and overlaid to identify areas of compatibility and conflict.

In this context GIS permits scientists and managers in Zdarske Vrchy to interact with their data and ask questions such as (after Downey *et al.*, 1991, Downey *et al.*, 1992 and Petch *et al.*, 1995):

- What will be the long-term consequences of continuing recreational activity for the landscape? (Plate 6)
- Where will damage from acid rain occur if a particular industrial plant continues to operate?
- Where is the best location for the reintroduction of certain bird species?
- Where should landscape conservation zones be established?

One application of GIS in Zdarske Vrchy has been to identify areas of the landscape for conservation. Traditionally, water storage in the region has relied on the use of natural water reservoirs such as peat wetlands and old river meanders (Petch *et al.*, 1995). Current land use practices, in particular forestry and farming, are resulting in the gradual disappearance of these features. The consequences of these changes have been localized droughts and floods in areas further downstream. In turn, these changes have brought about a reduction in plant species and wildlife habitats. Managers in the Zdarske Vrchy region wanted to identify conservation zones to protect the remaining natural water reservoirs as well as to identify those areas where it may be possible to restore the water retention character of the landscape. To do this they needed to establish the characteristics of the landscape that determine whether or not a particular location is likely to retain water. Specialists in hydrology, geology and ecology were consulted to identify a range of important criteria describing:

- the type of soil and its water retention ability;
- the character of the topography (for example, presence or absence of hollows or hills);
- the type of land use, as certain agricultural practices exploit the water retention capacity of the landscape; and
- the presence or absence of human-enhanced water drainage channels.

GIS professionals were then asked to find appropriate sources of spatial data that could be used to represent these criteria. A range of sources was identified including:

- paper maps (for soil type and geology);
- contour maps (for topography);
- ecological field maps (for drainage conditions); and
- remote sensing (for land use).

These data were acquired and entered into the GIS. The hydrologists, ecologists and geologists were then asked how each of the criteria (land use, topography, soil type and drainage) might interact to influence the retention capacity of the landscape at a particular location. First, the scientists used the GIS to look at the relationship of the criteria they had identified in areas where natural water reservoirs were still in existence. This involved adding more data to the GIS about the location of existing water retention zones. These data came from the regional water authorities as paper maps. The relationship between the geographical distribution of the various landscape characteristics at these locations was then used to develop a model. This model allowed the managers of the Zdarske Vrchy region to predict which other areas could be restored as natural water reservoirs. The next stage was to check which of these areas were located in existing conservation zones, as it was easier to change existing conservation regulations than to set up new conservation areas. Figure 1.2 summarizes the method used and shows how the GIS was used to integrate data from a range of different sources. Figure 1.2 also shows how these data were overlain with additional data about existing water retention zones and existing conservation areas to identify potential new conservation sites.

The Zdarske Vrchy project shows how GIS can be used to bring together data from a wide variety of sources to help address a range of environmental management problems. This use of GIS is not unique to environmental planning in

Eastern Europe and is being practised by environmental managers all over the world. In addition, the Zdarske Vrchy project reveals a number of other important issues associated with the use of GIS. These include the problem of data sources being in different map projections, the value of GIS as a modelling tool and the role for GIS as a participatory problem-solving tool (Box 1.3).

Finding a new home

At some stage in our lives, most of us will need to look for a new home. Perhaps because of a new job, or a change in family circumstances, our accommodation requirements will change and we will have to look for a new place to live. This can be a time-consuming and frustrating task. The requirements of individual family members need to be considered.

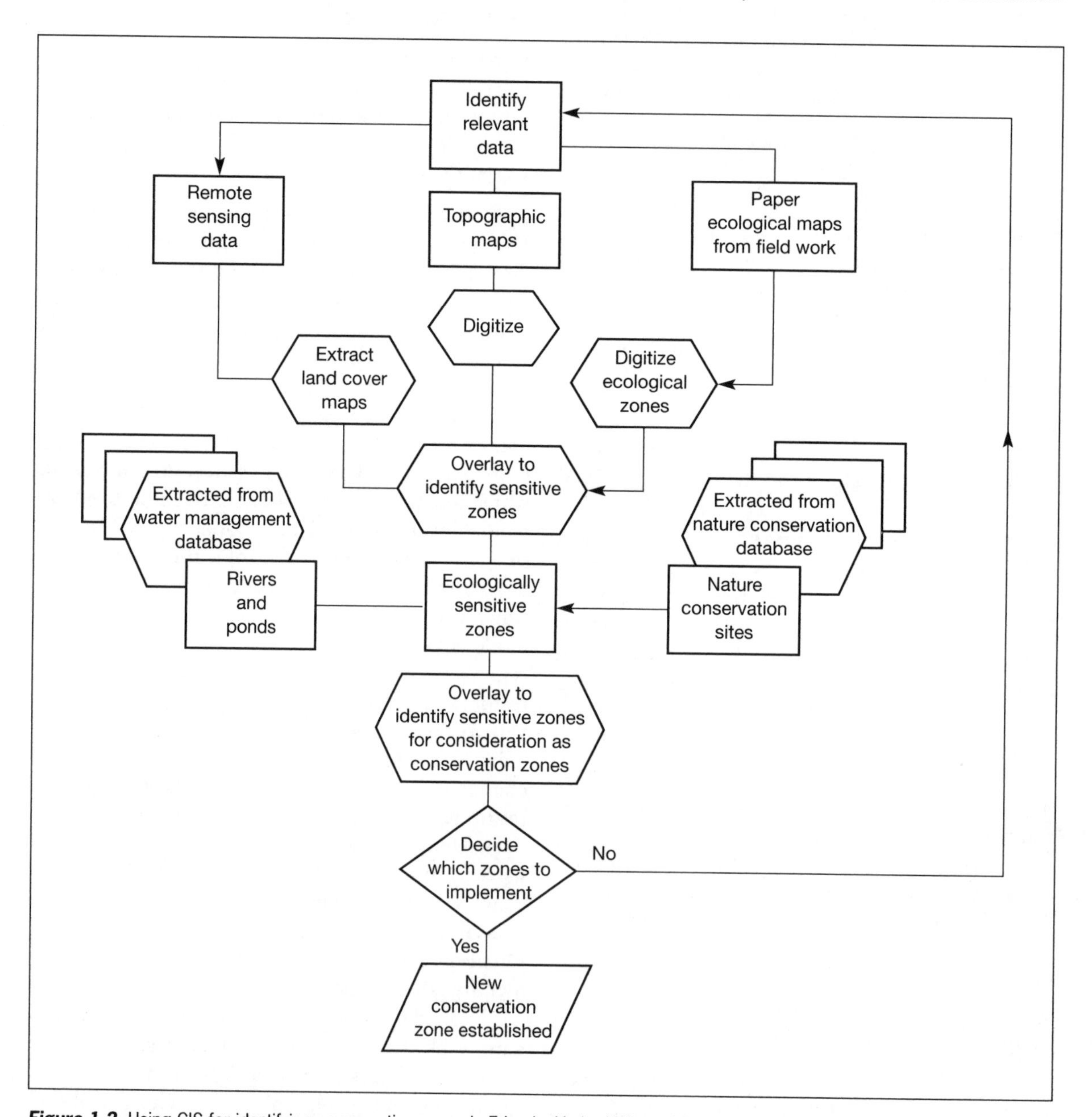

Figure 1.2 Using GIS for identifying conservation zones in Zdarske Vrchy (*Adapted from Petch et al.,* 1995)

Do they need to be close to schools, major roads or a railway station? Perhaps they would prefer to be in an area where insurance costs are lower. Maybe they want to be in an urban area to be close to shops and their place of work. To find a new home acceptable to all the family a decision support system (software to help you make a decision) may be appropriate. GIS can act in this role.

Much of the data needed to help answer the questions posed above can be gathered and converted into a format for integration in GIS. Heywood *et al.* (1995) have successfully completed this exercise and created a house-hunting decision support system. To find a suitable home participants were first required to decide which of a series of factors (insurance costs, proximity to schools, railways and roads, urban areas) were important in their decision-making. These factors were allocated weights and scores reflecting their importance. Constraints, areas where a new home would not be suitable under any conditions, were also identified. Constraints excluded certain areas from the analysis altogether; for example, participants could decide that they did not wish to live within 500 m of a major road.

Once the weighting process had been completed, the data selected were combined in a GIS using a multi-criteria modelling technique. This technique will be explored in more detail in Chapter 7, but, in brief, the data layers were combined using weightings, so that the layer with the highest weight had the most influence on the result. The resulting maps were used to help target the house-hunting process. The method is summarized in Figure 1.3 and Plate 9.

Locations of houses that were for sale were plotted over the top of the suitable areas identified, and ranked according to the number of criteria which they meet. If further details of these houses were available in computerized form, they were accessed by pointing at the map to find out, for instance, how many bedrooms they had, or whether they had a garden. To achieve this the information on the map (locations of properties) was linked with a database of house features. This type of data is often referred to as attribute

BOX 1.3

Issues raised by the Zdarske Vrchy case study

- The greatest problem associated with bringing data together for the creation of the Zdarske Vrchy GIS was deciding which *map projection* to adopt as the common frame of reference. Several different projection systems were used by the source maps. A projection system is the method of transformation of data about the surface of the earth on to a flat piece of paper. Because many methods exist which can be used to perform this task, maps drawn for different purposes (and maybe even for the same purpose but at different points in time) may use different projection systems. This does not present any problems as long as the maps are used independently. However, when the user wishes to overlay the data in a GIS the result can be confusing. Features that exist at the same location on the ground may appear to lie at different geographical positions when viewed on the computer screen. This problem became apparent in the Zdarske Vrchy project when the road network, present on two of the maps, was compared. Map projections are explained in more detail in Chapter 2.
- The Zdarske Vrchy case study also shows how GIS can be used to create models of environmental processes with maps used as the building blocks for the model. The topic of *modelling and GIS* is returned to in Chapters 6 and 7.
- Bringing people together to search for a solution to a common problem is often difficult. Different specialists will have different ideas about the problem. For example, an ecologist might recommend one approach, an engineer a second and an economist a third. The Zdarske Vrchy project showed how, through the use of GIS, the common medium of the map could be used as a tool to help experts from different backgrounds exchange ideas and compare possible solutions. The idea that GIS can be used as a *participatory problem solving tool* has also received considerable attention from the GIS research community (Carver *et al.*, 1997). Chapter 7 considers this topic in more detail.

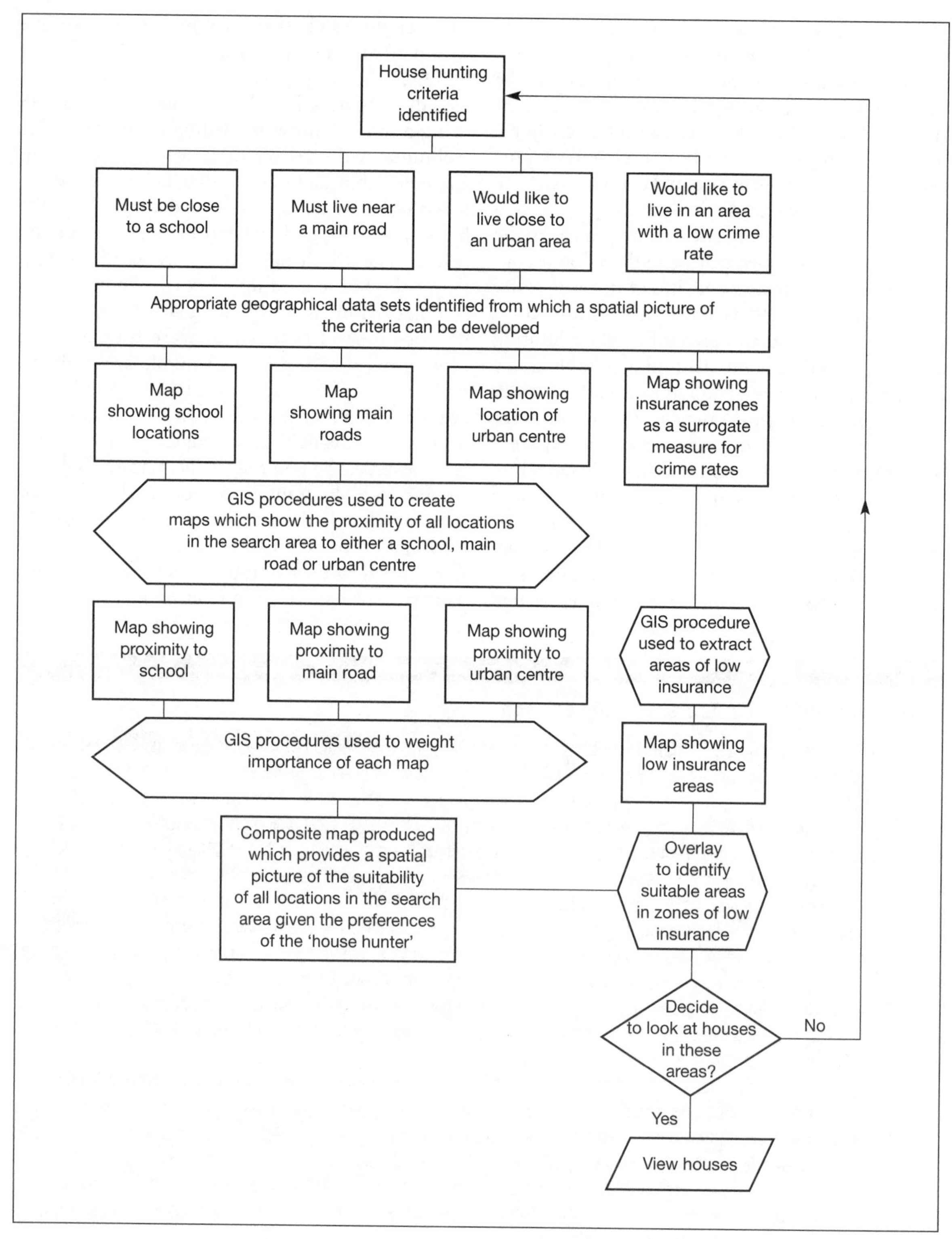

Figure 1.3 Using GIS to assist in house hunting

data. Attribute data and database concepts are considered in detail in Chapter 4.

There are examples of software already developed to perform similar tasks, including Wigwam in the UK (Anon., no date) and GeoData in the USA (ESRI, 1995). These systems have been designed to make it possible for a home buyer to visit an estate agent, explain the type of house and neighbourhood they prefer, and come away with a map showing the locations of houses for sale which meet their requirements. These products bring help to the home buyer deciding where to look for a new home in an unfamiliar area. GIS in this context is a decision support system.

The house hunting example shows how GIS can be used to link databases with similar types of data. This improves the speed and efficiency with which an appropriate location can be found. In this respect, it is a similar application to the NIREX case study discussed earlier. However, it differs from the NIREX example in that a large part of the search process may be carried out using the attributes associated with a spatial feature. In the case of the house hunting example, these may be the number of bedrooms a property may have or its price. Heywood *et al.* (1995) raise other issues associated with the use of GIS as a decision support tool. These are the problem of different GIS software products giving different results, the problem of defining search criteria and the human constraints on the decision-making process (Box 1.4).

These three examples are not enough to illustrate the range of applications and problems that GIS can be used to address. Even when a problem cannot be solved entirely using GIS, there may be the potential for some GIS input to aid the decision-making process. Table 1.1 offers additional pointers to further examples of GIS use by local government, defence agencies, utility companies, commerce and business.

Defining GIS

There have been so many attempts to define GIS that it is difficult to select one definitive definition. Maguire (1991) offers a list of 11 different definitions. This variety can be explained, as Pickles (1995) suggests, by the fact that any definition of GIS will depend on who is giving it, and their background and viewpoint. Pickles also considers

BOX 1.4

Issues raised by the house hunting case study

- Heywood *et al.* (1995) compared the results they obtained using two GIS software products to identify appropriate neighbourhoods for one house buyer. The results were different. The differences were in part explained by small differences in the search methods used by each GIS. Different GIS will implement similar methods in slightly different ways, and there will also be variations due to the way data are stored in the GIS. Therefore, a clear understanding of the way *GIS software* works is crucial if you are to be able to understand and explain your results. Chapters 2 to 8 expand on these issues.
- Heywood *et al.* (1995) also considered that there could be dificulties with results due to the way the problem was defined at the outset. For example, the limited selection of data suggested above may be available to help with the site selection, but in reality it may be that you wish to be in close proximity to a swimming pool and a sports club. If data are not available for these factors, they cannot be included in the analysis. So, defining the problem, and identifying all relevant criteria, are crucial steps in the *design of GIS projects*. The result you obtain will be influenced by the questions you ask. If you do not ask the right questions, you will not get the right answer. Therefore, good project design is an essential component of using GIS. In Chapter 12 we provide you with a methodology to help you plan your own GIS project.
- *Human factors* such as awareness and training also influence the effectiveness of GIS as a decision support system as they will help the user formulate appropriate questions. Chapter 11 looks at these issues in more detail.

that definitions of GIS are likely to change quickly as technology and applications develop further. Some of the shorter definitions give an idea of what a GIS is, albeit in a superficial way. For example, Rhind (1989: 28) proposes that GIS is 'a computer system that can hold and use data describing places on the Earth's surface'. Fuller definitions give more idea of what GIS can do, as well as what they are. Those provided by Burrough (1986; 6): 'a set of tools for collecting, storing, retrieving at will, transforming, and displaying spatial data from the real world for a particular set of purposes', and the Department of the Environment (1987: 132): 'a system for capturing, storing, checking, integrating, manipulating, analysing and displaying data which are spatially referenced to the Earth', fall into this category.

Table 1.1 Application areas for GIS

Activity	*Application*
Socio-economic/ government	Health Local government Transport planning Service planning Urban management
Defence agencies	Target site identification Tactical support planning Mobile command modelling Intelligence data integration
Commerce and business	Market share analysis Insurance Fleet management Direct marketing Target marketing Retail site location
Utilities	Network management Service provision Telecommunications Emergency repairs
Environmental management	Landfill site selection and mineral mapping potential Pollution monitoring Natural hazard assessment Resource management Environmental impact assessment

In general, the definitions of GIS cover three main components. They reveal that GIS is a *computer system*. This implies more than just a series of computer boxes sitting on a desk, but includes hardware (the physical parts of the computer itself and associated peripherals – plotters and printers), software (the computer programs that run on the computer) and appropriate procedures (or techniques and orders for task implementation). They also tell us that GIS uses *spatially referenced* or *geographical data*, and that GIS carries out various *management and analysis tasks* on these data, including their input and output. The Department of the Environment (1987) lists the capabilities that a 'well-designed GIS' should be able to provide:

1 Quick and easy access to large volumes of data.
2 The ability to:
   - select detail by area or theme;
   - link or merge one data set with another;
   - analyse spatial characteristics of data;
   - search for particular characteristics or features in an area;
   - update data quickly and cheaply; and
   - model data and assess alternatives.
3 Output capabilities (maps, graphs, address lists and summary statistics) tailored to meet particular needs.

In short, GIS can be used to add value to spatial data. By allowing data to be organized and viewed efficiently, by integrating them with other data, by analysis and by the creation of new data that can be operated on in turn, GIS creates useful information to help decision making. A GIS can, as was alluded to in the house hunting case study, be described as a form of spatial decision support system.

Some authors consider that there are important elements of a GIS in addition to those common to the definitions above. For example, Burrough (1986) suggests that GIS have three main elements: 'computer hardware, application software modules, and a proper organizational context'. Others, such as Maguire (1989), stress that data are the most important part of GIS. In practice, none of the main elements (the computer system, data or processing tools) will function as a GIS in isolation, so all might be considered of equal importance. However, it is perhaps the nature of the data used, and the attention given to the processing and interpretation of these data, that should lie at the centre of any definition of GIS.

GIS draws on concepts and ideas from many different disciplines. The term Geographic Information Science has been adopted to refer to the science behind the systems. Geographic Information Science draws on disciplines as diverse as cartography, cognitive science, computer science, engineering, environmental sciences, geodesy, landscape architecture, law, photogrammetry, public policy, remote sensing, statistics and surveying. Geographic Information Science involves the study of the fundamental issues arising from the creation, handling, storage and use of geographic information (Longley *et al.*, 2001), but it also examines the impacts of GIS on individuals and society and the influences of society on GIS (Goodchild, 1997).

Goodchild (1997) offers a useful summary of key concepts that help with the definition of GIS:

- Geographical information is information about places on the Earth's surface.
- Geographic information technologies include global positioning systems (GPS), remote sensing and geographic information systems.
- Geographical information systems are both computer systems and software.
- GIS can have many different manifestations.
- GIS is used for a great variety of applications.
- Geographic Information Science is the science behind GIS technology.

Components of a GIS

There is almost as much debate over the components of a GIS as there is about its definition. At the simplest level, a GIS can be viewed as a software package, the components being the various tools used to enter, manipulate, analyse and output data. At the other extreme, the components of a GIS include: the *computer system* (hardware and operating system), the *software, spatial data, data management and analysis procedures* and the people to operate the GIS. In addition, a GIS cannot operate in isolation from an application area, which has its own tradition of ideas and procedures. It is this more comprehensive perspective that is adopted here.

Computer systems and software

GIS run on the whole spectrum of computer systems ranging from portable personal computers (PCs) to multi-user supercomputers, and are programmed in a wide variety of software languages. Systems are available that use dedicated and expensive workstations, with monitors and digitizing tables built in; others will run on bottom-of-the-range PCs. In all cases, there are a number of elements that are essential for effective GIS operation. These include (after Burrough, 1986):

- the presence of a processor with sufficient power to run the software;
- sufficient memory for the storage of large volumes of data;
- a good quality, high-resolution colour graphics screen; and
- data input and output devices (for example, digitizers, scanners, keyboard, printers and plotters).

Likewise, there are a number of essential software elements that must allow the user to input, store, manage, transform, analyse and output data. Discussion of these issues follows in Chapters 4 to 8. However, although GIS generally fit all these requirements, their on-screen appearance (user interface) may be very different. Some systems require instructions to be typed at a command line, while others have 'point and click' menus operated using a mouse. The type of interface individual users find easier to operate is largely a matter of personal preference and experience.

Unlike the issue of software functionality there is limited discussion of hardware and interface technology in this book. This is because we consider these technologies to be changing so rapidly that any discussion would soon be out of date. If the reader is interested in the latest technical advances in GIS hardware and interfaces then World Wide Web (WWW) addresses are provided for a number of leading GIS developers and product directories on the book website (see p. xii).

Spatial data

All GIS software has been designed to handle spatial data (also referred to as geographical data). Spatial data are characterized by information about position, connections with other features and details of non-spatial characteristics (Burrough, 1986; Department of the Environment, 1987). For example, spatial data about one of Happy Valley's weather stations may include:

- latitude and longitude as a geographical reference. This reference can be used to deduce relationships with nearby features of interest. If the latitude and longitude of a weather station are known, the relative position of other weather stations can be deduced, along with proximity to ski slopes and avalanche areas;
- connection details such as which service roads, lifts and ski trails would allow the meteorologist access to the weather station;
- non-spatial (or attribute) data, for instance details of the amount of snowfall, temperature, wind speed and direction.

In a similar way spatial data about a ski piste may include:

- a series of spatial references to describe position;
- details of other runs that cross or join the ski piste;
- attribute data such as the number of skiers using the piste and its standard of difficulty.

The spatial referencing of spatial data is important and should be considered at the outset of any GIS project. If an inappropriate referencing system is used, this may restrict future use of the GIS (Openshaw, 1990). The challenge is to adopt a flexible and lasting referencing system, since a GIS may be intended to last many years. A full discussion of spatial referencing can be found in Chapter 2.

The traditional method of representing the geographic space occupied by spatial data is as a series of thematic layers. Consider, for example, traditional cartographic maps that may be available for an individual area. There may be a map for geology, one for soils and a topographic map showing cultural and environmental features on the surface. Computer models of space frequently use a similar approach. For example, the house hunting GIS example discussed earlier contained layers of data including insurance, transport, schools, and urban–rural land use. This was the first method of modelling space to be developed. This method, known as the layer-based approach, is still used by most GIS.

An alternative method of representing reality in a computer is to consider that space is populated by discrete 'objects' (Goodchild, 1995). Goodchild (1995) presents the example of a utility company which needs to map and manage a vast array of telegraph poles, connection boxes and cables. Each of these may be regarded as a discrete object, and there is empty space between the objects. This method is known as the object-oriented approach. Chapter 3 looks at both these approaches to modelling spatial data in more detail and considers their advantages and disadvantages.

Spatial data, represented as either layers or objects, must be simplified before they can be stored in the computer. A common way of doing this is to break down all geographic features into three basic entity types (an entity is a component or building block used to help data organization). These are points, lines and areas (Figure 1.4). If we return to the example of Happy Valley, points may be used to represent the location of features such as restaurants, lift pylons or rescue stations. Lines can be used to represent features such as roads, rivers and ski lifts. Area features are used to represent geographical zones, which may be observable in the real world (such as the Happy Valley car park) or may be artificial constructs (such as administrative areas). Points, lines and areas can all be used to represent surfaces. For example, spot heights or contour lines can be used to create a surface model of the Happy Valley landscape. In addition, points (representing junctions) and lines (representing roads) can be used to create a network model of Happy Valley's roads. These representations of real-world phenomena are normally held in a GIS according to one of two models – raster (sometimes referred to as grid or tesseral) or vector (Dale and McLaughlin, 1988; Peuquet, 1990). These two different approaches are compared in Figure 1.4. The raster model is particularly applicable where remotely sensed images are used (since the data are

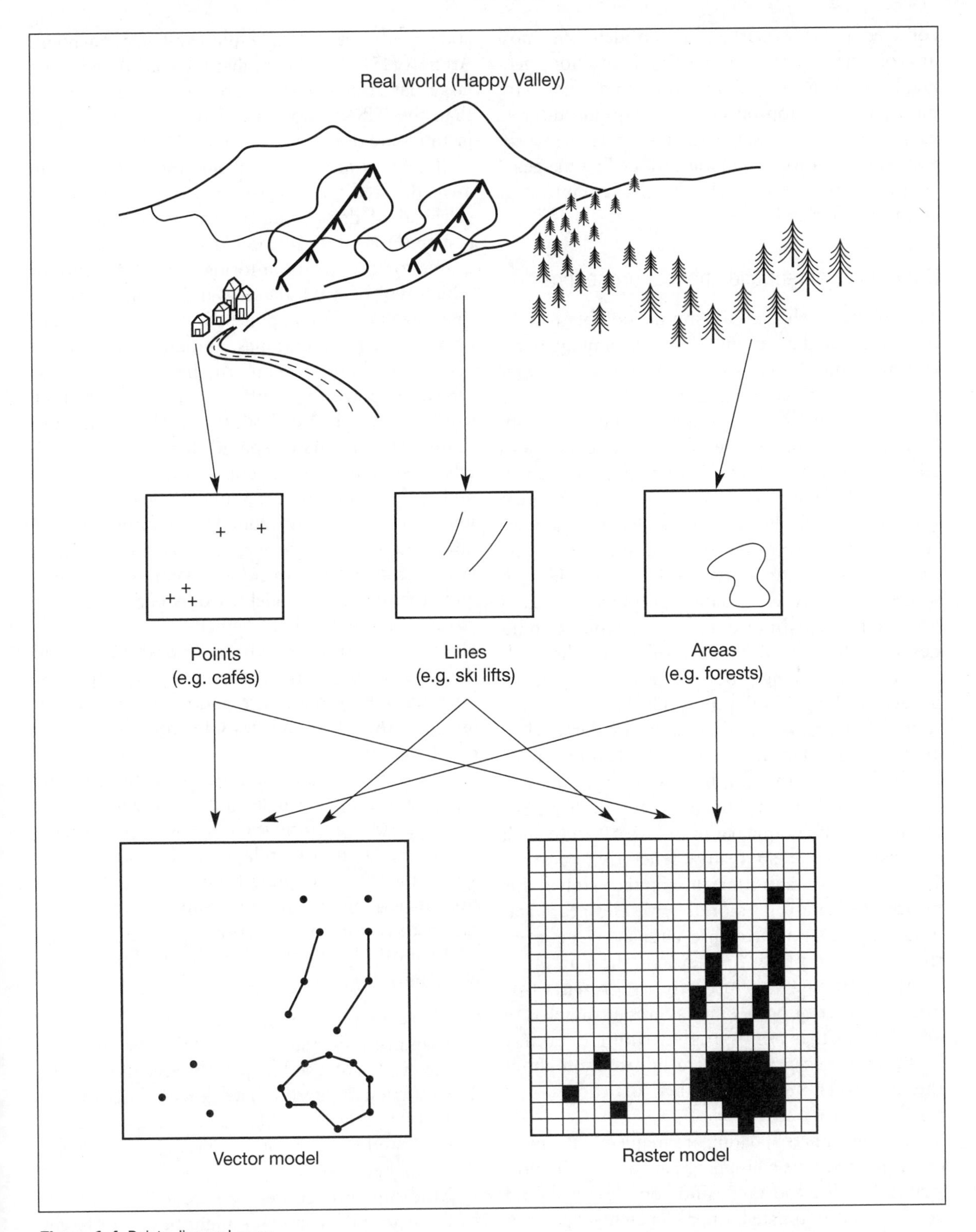

Figure 1.4 Points, line and areas

collected in this format) and is considered the most appropriate choice for modelling continuous geographic phenomena such as snow depth. The vector model is more appropriate for mapping discrete geographic entities such as road and river networks and administrative boundaries. Modelling the geography of the real world in the computer is considered further in Chapter 3.

Data management and analysis procedures

The functions that a GIS should be able to perform include data input, storage, management, transformation, analysis and output. *Data input* is the process of converting data from its existing form to one that can be used by the GIS (Aronoff, 1989). It is the procedure of encoding data into a computer-readable form and writing the data to the GIS database. This process should include verification procedures to check that the data are correct and transformation procedures to allow data from different sources to be used. GIS need to handle two types of data – graphical data and non-spatial attribute data. The graphical data describe the spatial characteristics of the real-world feature being modelled. For example, the hotels in Happy Valley may be described by a series of points. In some cases, particularly when area and line features are used to model real-world features, the graphical data may include information about the linkages between them. For example, if the boundary of an area feature such as a car park is also a snow fence that prevents skiers from overshooting the nursery slopes this information may be stored with the graphical data. Non-spatial attribute data describe what the features represent. They tell the computer what a particular set of entities represents (for instance, a set of points may represent hotels). In addition, further non-spatial attribute data may be stored which provide extra information about the hotels (standard, number of rooms and restaurant facilities).

Data input and updating are frequently the most expensive and time-consuming part of any GIS project and their importance and complexity should never be underestimated. Approximately 80 per cent of the duration of many large-scale GIS projects is concerned with data input and management. Aronoff (1989) estimates that the construction of a large database could cost five to ten times more than the GIS software and hardware. Data input methods are discussed in Chapter 5.

The *data management* functions necessary in any GIS facilitate the storage, organization and retrieval of data using a database management system (DBMS). A DBMS is a set of computer programs for organizing information, at the core of which will be a database. Database applications that have no GIS component include management of payrolls, bibliographies, and airline and travel agency booking systems. In the same way that DBMS organize these different types of data they can be used to handle both the graphical and non-graphical elements of spatial data. An ideal GIS DBMS should provide support for multiple users and multiple databases, allow efficient updating, minimize repeated (or redundant) information and allow data independence, security and integrity (Smith *et al.*, 1987). Relational databases, flat files and other database models used by GIS will be discussed in more detail in Chapter 4.

It is the ability of GIS to *transform* spatial data, for example from one entity type (points, lines and areas) to another, and to perform *spatial analysis*, that distinguishes GIS from other types of information systems.

Transformation is the process of changing the representation of a single entity, or a whole set of data. In GIS, transformation may involve changing the projection of a map layer or the correction of systematic errors resulting from digitizing. In addition, it may be necessary to convert data held as rasters to vectors or vice versa.

Aronoff (1989) classifies GIS analysis procedures into three types:

1 Those used for storage and retrieval. For example, presentation capabilities may allow the display of a soil map of the area of interest.
2 Constrained queries that allow the user to look at patterns in their data. Using queries, only sandy soils could be selected for viewing or further analysis.
3 Modelling procedures, or functions, for the prediction of what data might be at a different time and place. Predictions could be made

about which soils would be highly vulnerable to erosion in high winds or during flooding or the type of soil present in an unmapped area.

Transformation and analysis procedures can also be classified based on the amount of data analysed. Data in GIS are normally held in a series of layers. For instance, a 1:50,000 topographic map might be digitized to create a series of layers – one layer for road data, one for buildings, one for recreational interest (parking, picnic sites and youth hostels) and additional layers for soils and population data. Data layers normally contain data of only one entity type, that is point, or line, or area data. Analysis can be carried out either on one layer at a time, or on two or more layers in combination. The techniques available in GIS for manipulating and correcting data in preparation for analysis, and the analysis methods available, are discussed in greater depth in Chapters 5, 6 and 7.

The form of *data output* used will depend on cost constraints, the audience to whom the results are directed and the output facilities available. A local authority may produce simple tables, graphs and maps for the communication of important points to councillors, whilst professional map makers may produce detailed plots for publication. In other cases, data may be output in digital form for transfer to another software package for statistical analysis, desktop publishing or further analysis. However, most GIS output is in the form of maps. These may be displayed on-screen for immediate communication to individuals or small groups, photographed, stored digitally or plotted to produce permanent hard copy. Most GIS provide the ability to design screen formats and forms for plotting and these help to ensure that all maps have titles, keys, north arrows and scales, just as with traditional cartographic output. The facilities available for map design can be very extensive, incorporating a myriad of different colours, symbols and line styles, and it is often possible to design additional symbols for your own use. This can make map design very time-consuming, but effective. The audience for any GIS product is an important consideration when designing output, but generally it is best to keep products clear and simple. Options for data output and communication of results from GIS are discussed further in Chapter 8.

People and GIS

Most definitions of GIS focus on the hardware, software, data and analysis components. However, no GIS exists in isolation from the organizational context, and there must always be people to plan, implement and operate the system as well as make decisions based on the output. GIS projects range from small research applications where one user is responsible for design and implementation and output, to international corporate distributed systems, where teams of staff interact with the GIS in many different ways. In most organizations the introduction of GIS is an important event, a major change bringing with it the need for internal restructuring, retraining of staff and improved information flows. Research has been undertaken to highlight the factors that promote successful GIS and it has been suggested that in certain business sectors, innovative flexible organizations with adequate resources and straightforward applications are more likely to succeed (Campbell and Masser, 1995). However, not all GIS are successful. There is evidence that many systems fail, and more are under-used (Cornelius and Medyckyj-Scott, 1991). So the issues surrounding how to choose a system and how to implement it successfully require examination. These topics are covered in Chapter 11.

Conclusions

GIS technology is now well-established and, as we will see in Chapter 9, has been in use since the 1960s. The growth in application areas and products through the later years of the twentieth century has helped GIS to become an accepted tool for the management and analysis of spatial data. This trend is set to continue as computer technology continues to improve with faster and more powerful machines and as more data become available in digital formats directly compatible with GIS. In addition, the striking advances in related technologies such as surveying and field data collection, visualization and database management technology are likely to influence this growth. Further comments on the future of GIS can be found in Chapter 13.

There have been some notable failures in GIS. Sometimes data difficulties or other technical problems have set back system developments and applications; however, there are also human and organizational problems at the root of GIS failures. Before we can begin to appreciate these fully, to ensure that our GIS applications are successful, it is important to have a good understanding of what a GIS can do and the data it is working with (Chapter 2).

REVISION QUESTIONS

- What is GIS? Write your own definition of GIS and consider why GIS can be difficult to define.
- GIS is routinely used in applications such as siting new industrial developments. What are the advantages of using GIS in this context?
- What type of questions could GIS help environmental managers address?
- What are the components of GIS?
- Explain the following terms and phrases:
  1. Spatial data.
  2. Attribute data.
  3. Spatial referencing.
  4. Spatial entities.
- What is the difference between a GIS and geographical information science?

Further study

There are many sources of further reading that complement the material presented in this chapter. Aronoff (1989) offers a management perspective on GIS and is a well-written and readable general introduction. Chrisman (1997) and DeMers (1997) take a more technical approach but also offer a comprehensive introduction to GIS. Maguire *et al.* (1991) and Longley *et al.* (1999) are comprehensive reference works on GIS, considering technical, theoretical and applied issues. Longley *et al.* (2001) is a shorter work for use in conjunction with these two books. In Maguire *et al.* (1991) the overview chapters (Maguire, 1991; Goodchild, 1991b and Dangermond, 1991) offer interesting views of the important elements of the field.

For an overview of GIS from a more applied perspective there are a range of texts available. Korte (2000) provides a 'smart managers guide' to GIS that offers comment on industry trends and particular software systems. Martin (1996) looks at GIS with a focus on socio-economic applications in a UK setting; while Grimshaw (1994) considers GIS from an information management perspective and offers some useful business case studies. Burrough and McDonnell (1998) and Bonham-Carter (1995) consider applications in land management and the geosciences respectively. Both of these books offer general introductory chapters, then considerable discussion of GIS data models and analysis methods that will be covered in later chapters of this text. Harder (1997), Mitchell (1997), Lang (1998) Davis (1999) and Hanna (1999) also cover applications of GIS at an introductory level.

Like many GIS texts, Martin (1996), Burrough and McDonnell (1988) and Korte (2000) contain glossaries of commonly used GIS terms and acronyms. Another useful reference work is the *International GIS Dictionary* by McDonnell and Kemp (1998).

For up-to-date accounts of applications, technological developments and GIS issues there are several magazines and journals available. These include *GeoWorld, GeoInfoSystems, GeoEurope* and *GINews*. More academic journals include the *International Journal of Geographical Information Science* and *Transactions in GIS*. Journals and magazines from other disciplines (for example, surveying, computer-aided design (CAD) and computing) also contain introductory articles on GIS from time to time.

Electronic sources of information on GIS on the World Wide Web should also be considered. Because these change rapidly we have developed a website for this book, which includes links to online sources. The site can be can be found at www.booksites.net/heywood and contains information about the book itself, appropriate computer-aided learning (CAL) materials and links to a range of other GIS sites.

Aronoff S (1989) *Geographic Information Systems: a Management Perspective*. WDL Publications, Ottawa
Bonham-Carter G F (1995) *Geographic Information Systems for Geoscientists: Modelling with GIS*. Pergamon Press, New York

Burrough P A, McDonnell R (1998) *Principles of Geographical Information Systems*. Oxford University Press, Oxford

Chrisman N R (1997) *Exploring Geographic Information Systems*. Wiley, New York

Dangermond J (1991) The commercial setting of GIS. In: Maguire D J, Goodchild M F, Rhind D W (eds.) *Geographical Information Systems: Principles and Applications*. Longman, London, pp. 55–65, Vol. 1

Davis D (1999) *GIS for Everyone*. ESRI Press, Redlands

DeMers M N (1997) *Fundamentals of Geographic Information Systems*. Wiley, New York

Goodchild M F (1991b) The technological setting of GIS. In: Maguire D J, Goodchild M F, Rhind D W (eds.) *Geographical Information Systems: Principles and Applications*. Longman, London, pp. 45–54, Vol. 1

Goodchild M F (1997) What is Geographic Information Science? *NCGIA Core Curriculum in GIScience*. http://www.ncgia.ucsb.edu/giscc/units/u002/u002.html, posted October 7, 1997.

Grimshaw D J (1994) *Bringing Geographical Information Systems into Business*. Longman, London

Hanna K C (1999) *GIS for Landscape Architects*. ESRI Press, Redlands

Harder C (1997) *ArcView GIS means Business: Geographic Information System Solutions for Business*. ESRI Press, Redlands

Harder C (1998) *Serving Maps on the Internet: Geographic Information and the World Wide Web*. ESRI Press, Redlands

Korte G B (2000) *The GIS Book*. 5th edn. OnWord Press, USA

Lang L (1998) *Managing Natural Resources with GIS*. ESRI Press, Redlands

Longley P A, Goodchild M F, Maguire D J, Rhind D W (eds.) (1999) *Geographical Information Systems: Principles, Techniques, Management and Applications*. John Wiley, New York

Longley P A, Goodchild M F, Maguire D J, Rhind D W (2001) *Geographic Information Systems and Science*. Wiley, Chichester

Maguire D J (1991) An overview and definition of GIS. In: Maguire D J, Goodchild M F, Rhind D W (eds.) *Geographical Information Systems: Principles and Applications*. Longman, London, pp. 9–20, Vol. 1

Maguire D J, Goodchild M F, Rhind D W (eds.) (1991) *Geographical Information Systems: Principles and Applications*. Longman, London

Martin D (1996) *Geographical Information Systems and their Socio-economic Applications*, 2nd edn. Routledge, London

McDonnell R, Kemp K (1998) *International GIS Dictionary*. 2nd edn. GeoInformation International, London.

Mitchell A (1997) *Zeroing In: Geographic Information Systems at Work in the Community*. ESRI Press, Redlands

2 Spatial data

KEY QUESTIONS AND ISSUES

- What is a model?
- What is the difference between data and information?
- What are the main characteristics of spatial data?
- What are map projections and why are they important?
- What are the different methods of spatial referencing?
- What is topology?
- What are the thematic characteristics of spatial data?
- What are the main sources of spatial data?
- Why are data standards an important issue in GIS?

Introduction

All Geographical Information Systems are computer representations of some aspect of the real world. GIS present a simplified view of the world as it would be impossible to represent reality in its entirety in a computer. As the case studies introduced in Chapter 1 showed, this simplified view contains only the data the GIS designer considers necessary to solve a particular problem. Thus, the GIS for Zdarske Vrchy did not contain data on the location of houses for sale, nor did the house hunting GIS contain data on the distribution of flora and fauna. The simplified view of the real world adopted by GIS is often termed a model. A model is 'a synthesis of data' (Haggett and Chorley, 1967) which is used as a 'means of "getting to grips" with systems whose spatial scale or complexity might otherwise put them beyond our mental grasp' (Hardisty *et al.*, 1993). GIS is used to help build models where it would be impossible to synthesize the data by any other means. Models also contain our ideas about how or why elements of the real world interact in a particular way. Therefore, a GIS populated with data and ideas about how these data interact is a spatial model. Haggett and Chorley (1967) point out that a spatial model places emphasis on reasoning about the real world by means of translation in space. This is exactly the reason why GIS is used to solve geographical problems.

Before looking at how spatial models are constructed using a GIS it is necessary to consider the character of the spatial data they use as their raw material. In the context of our Martian and the car analogy, this is where we explain to the Martian about the fuel a car requires to make it run. First, however, it is necessary to review our understanding of the term 'data' and take a closer look at the distinction between data and information.

Data are observations we make from monitoring the real world. Data are collected as facts or evidence that may be processed to give them meaning and turn them into information. There is a clear distinction between data and information, although the two terms are often used interchangeably. To help appreciate the distinction it is perhaps easiest to think of data as raw numbers, such as those you might see listed in a table. Field notes made by the Happy Valley ski patrol containing snow depth measurements, the printout from the Happy Valley automatic weather station or a table of responses from a survey of skiers are examples. All you see are numbers that have no particular meaning. To make the numbers useful you need to add context. For somebody else to interpret your tables or lists of figures they would need to know to what the data refer and which scale or unit of measurement has been used for

recording the data. With these details the data become information. Therefore, information is data with meaning and context added (Hanold, 1972).

There are a wide variety of data sources, though all data fall into one of two categories: primary or secondary. Counts of skiers using a particular ski run are an example of primary data collected through first-hand observation. Secondary data will have been collected by another individual or organization, for example consumer surveys of customers buying ski equipment. Many secondary data sources are published and include maps, population census details and meteorological data (Griffith and Amrhein, 1991). Box 2.1 provides examples of primary and secondary data sources used in Happy Valley.

All primary and secondary data have three modes or dimensions: temporal, thematic and spatial. For all data it should be possible to identify each of these three modes. For example, for data about an avalanche accident which took place in Three Pines Valley on 14 February 1995, the three modes are:

- *temporal* – 14 February 1995;
- *thematic* – avalanche accident; and
- *spatial* – Three Pines Valley.

The temporal dimension provides a record of when the data were collected and the thematic dimension describes the character of the real-world feature to which the data refer. Additional thematic data for the avalanche accident might relate to the size and type of the avalanche. In GIS the thematic data are often referred to as non-spatial or attribute data.

BOX 2.1

Sources of primary and secondary data used in Happy Valley

Primary data sources

- *Daily meteorological records* collected by the Happy Valley ski patrols. These data are used to help make decisions about which runs to open and which to close.
- *Number of lift passes* purchased each day. These data are used to monitor the demand for skiing on different days of the week.
- *The number of skiers* using a specific lift on a particular day. The automatic turnstiles at the entry points to all the lifts collect these data. They are used to monitor lift usage.
- *The number of avalanches* recorded by the ski patrols. These data are used to help predict avalanche risk.

Secondary data sources

- Published *meteorological maps* for the Happy Valley area. These data are used to assist with avalanche forecasting.
- *Local topographic maps.* Back-country ski trail maps are prepared using local topographic maps.
- National and regional *lifestyle* data derived from market research surveys are used to estimate the demand for skiing and target the marketing of Happy Valley.

The spatial dimension of data can be regarded as the values, character strings or symbols that convey to the user information about the location of the feature being observed. In the case of the avalanche accident we know that the incident occurred in Three Pines Valley. In this case, the spatial reference used is a textual description that would only be of use to those who are familiar with the area. However, because GIS have no 'local knowledge' all spatial data used in GIS must be given a mathematical spatial reference. One of the most common is a map co-ordinate. Here, a co-ordinate pair (x,y) is used to locate the position of a feature on a uniform grid placed on a map. Spatial referencing is considered in more detail later in this chapter.

It is common to find the term temporal data used to describe data organized and analysed according to time, thematic data used for data organized and analysed by theme, and spatial data for data organised and analysed by location. However, even though one dimension may be used to organize data, the other dimensions will still be present.

GIS place great emphasis on the use of the spatial dimension for turning data into information, which, in turn, assists our understanding of geographic phenomena. Therefore, we consider next the characteristics of spatial data in detail and examine how the map metaphor has shaped these characteristics. This is followed by a review of the thematic dimension of spatial data and a review of a range of sources of spatial data, including surveys, aerial photographs, satellite images and field data sources.

Maps and their influence on the character of spatial data

The traditional method for storing, analysing and presenting spatial data is the map. The map is of fundamental importance in GIS as a source of data, a structure for storing data and a device for analysis and display. Perhaps more importantly, maps have shaped the way most of us think about space in two dimensions. Therefore, understanding maps and how they are produced is an essential starting point for exploring the characteristics of spatial data.

Maps take many different forms and come at a range of different scales. Examples range from simple sketch maps, such as those used to show colleagues and friends how to get to a party, to the more complex topographic and thematic maps that can be found in national atlases.

It is common to make a distinction between thematic and topographic maps. Thematic maps show data relating to a particular theme or topic, such as soil, geology, geomorphology, land use, population or transport. Topographic maps contain a diverse set of data on different themes. Thus, land use, relief and cultural features may all appear on the same topographic map. Unwin (1981) argues that the topographic map is simply a 'composite of many different kinds of maps'.

Even though there are many different types of maps the mapping process is of a general nature. During this process the cartographer must (after Robinson *et al.*, 1995):

- establish the *purpose* the map is to serve;
- define the *scale* at which the map is to be produced;
- select the *features* (spatial entities) from the real world which must be portrayed on the map;
- choose a method for the *representation* of these features (points, lines and areas);
- *generalize* these features for representation in two dimensions;
- adopt a *map projection* for placing these features onto a flat piece of paper;
- apply a *spatial referencing system* to locate these features relative to each other; and
- *annotate* the map with keys, legends and text to facilitate use of the map.

When a cartographer produces a map, an underlying geometric structure is created which allows the user to describe the relationships between features. It will be clear, for instance, that islands lie within lakes, that fields are adjacent and that if you follow a particular road you will reach a certain destination. This structure, known as *topology*, is based on the geometric relationships of objects (see Chapter 3). Following the map-making process outlined above helps the cartographer shape the character of the final map. The key characteristics which result from this process are described in more detail below.

Purpose

All maps, and other sources of spatial data, are generated with a purpose in mind. In most cases that purpose is to turn data into information which will be communicated to a third party. Every year the managers of Happy Valley produce a map of the ski area for use by visitors. The map shows the location of ski trails, car parks, hotels, emergency shelters and ski lifts. Its purpose is to help visitors orient themselves and decide how to spend their time. Naturally, such a map can have a strong influence over the user. For example, visitors are unlikely to dine at a restaurant if they cannot find it on the map. However, there are restaurants in the area that are not shown on the official Happy Valley map. The ski company wishes to encourage visitors to use its facilities, not those owned by competitors. This simple example illustrates how purpose can influence the character and quality of a spatial data set. Clearly, you would not use this map on its own if you were trying to compile a data set of the restaurants in Happy Valley. However, you may not know that the map was incomplete.

In some cases maps may have a single purpose. The propaganda maps produced by the Allies during the Second World War were designed to convince the general public that the war effort was going well (Monmonier, 1991). The true geography of Europe was distorted to emphasize the area occupied by the allied forces. This was effective in boosting the morale of the Allies, but produced maps of limited use in other circumstances. Other maps, such as the topographic maps produced by national mapping agencies, aim to meet the needs of a wide range of users, ranging from utility companies to outdoor enthusiasts. In this case, the maps will be more geographically accurate than the propaganda maps described above, but they will still contain generalized data to enable them to be of wide generic use. These generalizations will limit their use for certain applications. A utility company is unlikely to use national maps, on their own, to plan the detailed installation of a new set of electricity cables, because these maps do not contain details about the location of the existing cable network.

Whilst not strictly a spatial characteristic in its own right, the purpose for which a spatial data set has been created will influence the quality and spatial detail provided by the data set. An appreciation of the purpose behind the production of a data set is, therefore, an essential prerequisite for judging whether or not the data are appropriate for use in a particular situation. Linked closely to purpose is the idea of scale.

Scale

Virtually all sources of spatial data, including maps, are smaller than the reality they represent (Monmonier, 1991; Keates, 1982). Scale gives an indication of how much smaller than reality a map is. Scale can be defined as the ratio of a distance on the map to the corresponding distance on the ground (Martin, 1996). An alternative definition is offered by Laurini and Thompson (1992) as the order of magnitude or level of generalization at which phenomena exist or are perceived or observed. Scale can be expressed in one of three ways: as a ratio scale, a verbal scale or a graphical scale (Figure 2.1).

Ratio	1:5,000	1:1,000,000
Verbal (nominal)	1 cm represents 50 m	1 cm represents 10 km
Graphical	0 100 200 km	0 10 20 30 40 km

Figure 2.1 Expressions of scale

Examples of ratio scales are 1:5000 and 1:5,000,000. At a scale of 1:5000 a 1 mm line on the map represents a 5000 mm line on the ground. In the same fashion a line of 1 m on the map represents a line of 5000 m on the ground; the units do not matter as long as they are the same. A verbal scale would express the scale in words, for example '1 cm represents 50 m'. Finally, a graphic scale (or scale bar) is usually drawn on the map to illustrate the distances represented visually. Graphic scales are frequently used on computer maps. They are useful where changes to the scale are implemented quickly and interactively by the user. In such cases, recalculating scale could be time-consuming, and the ratios produced (which may not be whole numbers) may be difficult to interpret. Redrawing a graphic scale in proportion to the map is relatively straightforward and simple to understand. However, it is often possible in GIS to specify the scale at which you require your maps using a ratio representation.

Standard topographic maps contain examples of verbal, ratio and graphical scales. It should be remembered that small-scale maps (for example, 1:250,000 or 1:1,000,000) are those which cover large areas. Conversely, large-scale maps (for example, 1:10,000 or 1:25,000) cover small areas and contain large amounts of detail. With some data used in GIS, such as aerial photographs or satellite imagery, the scale is not immediately obvious and may have to be calculated by the user. Scale is also important when using spatial entities (points, lines and areas) to represent generalized two-dimensional versions of real-world features.

Spatial entities

Traditionally, maps have used symbols to represent real-world features. Examination of a map will reveal three basic symbol types: points, lines and areas (Monmonier, 1991). These were introduced in Chapter 1 (Figure. 1.4) and are the basic spatial entities. Each is a simple two-dimensional model that can be used to represent a feature in the real world. These simple models have been developed by cartographers to allow them to portray three-dimensional features in two dimensions on a piece of paper (Laurini and Thompson, 1992; Martin, 1996). Box 2.2 provides more details on the types of features that points, lines and areas can be used to represent.

The representation of real-world features using the point, line and area entity types appears relatively straightforward. However, the method chosen to represent a spatial feature will depend on the scale used. Consider the way cities are represented on maps of different scales. On a world map a point would be the most appropriate method of representation, given the number of cities to be included. However, at national and regional scales a point could provide an oversimplified view of the extent of the geographical area covered by a city. A point used here would tell us nothing about the relative size of cities, so it is more likely that the cartographer would choose to represent the cities using areas. At the local scale even the area spatial entity may be considered too simplistic and the cartographer may choose to build up a representation of the city using a mixture of point, line and area entities. Points may be used for the representation of features such as telephone boxes, areas for residential blocks and parks, and lines for road networks. Choosing the appropriate entity to represent real-world features is often difficult.

Generalization

All spatial data are a generalization or simplification of real-world features. In some instances generalization is needed because data are required at a particular scale. In other cases generalization is introduced by the limitations of the technical procedures used to produce data. The grain size of photographic film, or the resolution of a remote sensing device, will determine the level of detail discernible in the resulting air photo or satellite image. Generalization may also be introduced directly by human intervention in order to improve the clarity of an image or to enhance its major theme.

All the data sources used in GIS – aerial photographs, satellite images, census data and particularly maps – contain inherent generalizations. This simplification of detail is necessary in order to maintain clarity. If a cartographer wishes to depict the course of a river on a map, decisions

BOX 2.2

Basic spatial entities

Points

Points are used to represent features that are too small to be represented as areas. An example is a postbox. The data stored for a postbox will include geographic location and details of what the feature is. Latitude and longitude, or a co-ordinate reference, could be given together with details which explain that this is a postbox in current use. Of course, features that are represented by points are not fully described by a two-dimensional geographical reference. There is always a height component since the postbox is located at some height above sea level. If three dimensions are important to a GIS application this may also be recorded, usually by adding a *z* value representing height to give an (*x*,*y*,*z*) co-ordinate.

Lines

Lines are used to represent features that are linear in nature, for example roads or rivers. They can also be used to represent linear features that do not exist in reality, such as administrative boundaries or international borders. It can be difficult for a GIS user to decide when a feature should be represented by a line. Should a road be represented by a single line along its centre, or are two lines required, one for each side of the road?

A line is simply an ordered set of points. It is a string of (*x*,*y*) co-ordinates joined together in order and usually connected with straight lines. Lines may be isolated, such as geological fault lines, or connected together in networks, such as road, pipeline or river networks. Networks are sometimes regarded as a separate data type but are really an extension of the line type. More will be said about networks and their analysis in later chapters. Like points, lines are in reality three-dimensional. For instance, a hydrogeologist may be interested in underground as well as surface drainage. Adding a *z* co-ordinate (representing depth or height) to the points making up the line representing a stream allows an accurate three-dimensional representation of the feature.

Areas

Areas are represented by a closed set of lines and are used to define features such as fields, buildings or administrative areas. Area entities are often referred to as polygons. As with line features, some of these polygons exist on the ground, while others are imaginary.

Two types of polygons can be identified: island polygons and adjacent polygons. Island polygons occur in a variety of situations, not just in the case of real islands. For example, a woodland area may appear as an island within a field, or an industrial estate as an island within the boundary of an urban area. A special type of island polygon, often referred to as a nested polygon, is created by contour lines. If you imagine a small conical hill represented by contour lines, this will be represented in polygon form as a set of concentric rings. Adjacent polygons are more common. Here, boundaries are shared between adjacent areas. Examples include fields, postcode areas and property boundaries.

A three-dimensional area is a surface. Surfaces can be used to represent topography or non-topographical variables such as pollutant levels or population densities. Some authors (for example Martin, 1996; Laurini and Thompson, 1992) consider surfaces to be a separate fourth entity type. This issue will be considered in more detail in Chapter 3.

need to be taken regarding the amount of detail to be included. These decisions are largely governed by the scale of the map – the level of detail shown is proportional to scale. At very small scales (for example, 1:20,000,000), a large river like the Mississippi in the USA may appear as a single blue line. This line will show the approximate course of the river from its source in the northern state of Minnesota to its exit into the Gulf of Mexico. All bends and turns in the river will have been smoothed to create a simple, easy to understand map. At larger scales (for example, 1:5,000,000) it is possible to show something of the meandering nature of this great river and its network of tributaries: the Red, Arkansas, Missouri and Ohio. At even larger scales (for example, 1:250,000) it becomes possible to indicate width, river banks, small bends and meander cut-offs. At larger scales still (for example, 1:50,000) it may be possible to indicate depth and the positions of sandbanks and shoals that might be important for navigation purposes. As the scale increases, the cartographer has greater scope for including more detail. The relationship between scale and detail is referred to as *scale-related generalization* and is illustrated in Figure 2.2.

Decisions regarding what features to include on the final map and which to leave out also need to be made by the cartographer. If the cartographer were to include every single tributary of the Mississippi river network on the 1:20,000,000 scale map, the map would be covered by dense

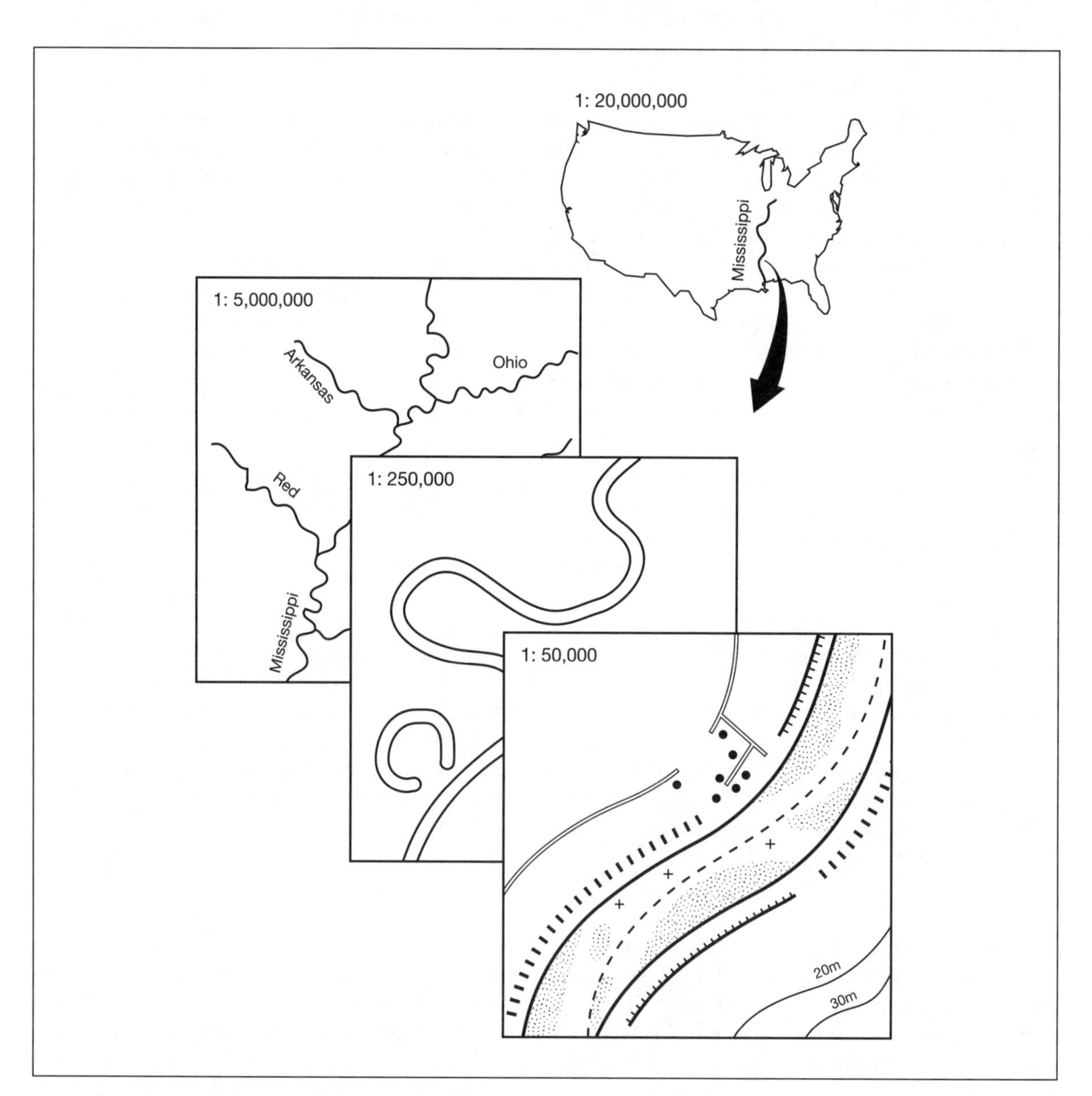

Figure 2.2 Scale related generalization

blue linework and impossible to read. For the sake of clarity, the cartographer has to be selective about drawing map features.

Another problem facing the cartographer is how to depict features in proportion to their size on the ground. If a river is drawn as a line 0.5 mm thick on a 1:1,000,000 scale map, this would imply that the river is 500 m wide. In reality it may be only 50 m wide. If the width of line features drawn on the map were determined rigidly by the map scale, then most features on small-scale maps could not be seen with the naked eye. For example, a 50 m wide river on the 1:1,000,000 scale map would have to be drawn using a line only 0.05 mm wide. Similarly, if a road running along the banks of the river were to be depicted accurately in this fashion, it would need to be drawn on top of the river on the map. In order to make the road distinguishable from the river, the cartographer has to displace the road to leave a gap between it and the river. To cope with these and other problems relating to the necessary generalization of map features, cartographers have adopted a broad code of practice relating to selection, simplification, displacement and smoothing. This is summarized in Box 2.3. A good understanding of the processes of cartographic generalization is important if data from paper maps and other spatial data sources are to be used effectively within GIS.

Projections

For the GIS analyst to make use of simple spatial entities (points, lines and areas) it is necessary to locate them in two dimensions. The analyst, like the cartographer, must treat the world as a flat surface to achieve this. While this is a gross generalization, for most purposes it works well. Moreover, current technology provides us with no other realistic choice (though future developments in three-dimensional modelling and virtual reality may change this). The method by which the 'world is laid flat' is to use a map projection.

Map projections transfer the spherical Earth onto a two-dimensional surface. In doing so they approximate the true shape of the Earth. This process introduces errors into spatial data, the character of which will vary depending on the projection method chosen. Some projections will cause distance between spatial entities to be preserved while direction is distorted. In other cases, shape may be preserved at the expense of accurate area estimates. One way to visualize the problem of representing a spherical world in two dimensions is to imagine a plastic beach ball overprinted with a map of the world showing lines of latitude and longitude. The inflated ball is a globe, with the countries in their correct locations, and shown as area entities with correct relative shapes and sizes. Imagine that you have to deflate the ball

BOX 2.3

Cartographic generalization: code of practice

1 *Selection.* First, the map feature for generalization is selected. If more than one source is available to the cartographer this may involve choosing the most appropriate representation of the feature or a blending of the two.
2 *Simplification.* Next, a decision will be taken to simplify the feature. For the example of the river this may involve the removal of some minor bends. The aim of generalization will usually be to simplify the image but maintain the overall trend and impression of the feature.
3 *Displacement.* If there are features that are located side by side in the real world, or that lie on top of one another, the cartographer may choose to displace them by a small degree so that they are both visible on the map image. This may have the effect of displacing a feature several hundred metres depending on the map scale used.
4 *Smoothing* and *enhancement.* If the source data from which a cartographer is working are very angular, because they have been collected from a series of sampling points, a smoothing technique may be used to apply shape and form to the feature. This will give a better representation.

(adapted from Robinson *et al.*, 1995)

and lay it flat on a table while still displaying all the countries. The only way to do this is to cut the beach ball into pieces. In doing this you would find that the distances between countries will be altered and their shape distorted. The principle is the same with map projections.

If you imagine the beach ball has a hole at the 'north pole' large enough for a light bulb to be inserted, it is transformed into a light fitting. When the light is switched on, an image of the surface is projected onto the walls of the room. Careful examination of the images on the walls reveals that the centre of the image reflects the globe most accurately. It is on this simple concept that the whole range of map projections is based.

Today there are a wide range of map projections in use, and there were even more used in the past. Different map projections are used in different parts of the world for mapping different sized areas and for different applications. Think again of the globe as a light fitting. The picture of the Earth from our 'globe-light' will vary depending upon the shape of the room in which the light is placed.

In a circular room, assuming our globe is hanging from the north pole, there will be a continuous picture of the Earth. Countries nearest the equator will appear in their true relative geographical positions. The equator is the line of latitude nearest the wall and so represents the *line of true scale*, along which distances (and consequently the map scale) are not distorted. However, at the top or bottom of the wall the location of the countries is distorted, with the distance between countries increased. Our view of the poles will be very distorted, or missing altogether. In fact, if the poles are included on our projected map the points representing the north and south poles become so distorted as to be projected as a straight line equal in length to that of the equator.

In a square room with flat walls, only a part of the Earth's surface will be visible on any one wall. This will depend on the position of the light. The view will be of half the globe or less. Distortion will be similar to that in the circular room except that, since the single wall is straight and not curved, the image of the world will be distorted at all four edges of the wall and not just the top and bottom.

If the room is shaped like a tepee (with circular walls tapering towards a point at the apex of the structure) the line of true scale is no longer the equator, as in the circular room, but some line of latitude nearer the north pole that lies at a tangent to the sloping walls of the tepee. The exact line of latitude will depend on the angle of the tepee walls: the steeper the walls, the lower the latitude (nearer the equator); the shallower the walls; the higher the latitude (nearer the pole). Not all projections of this kind have the north pole uppermost; projections can have the south pole uppermost or may have the apex of the tepee centred over a different point altogether.

The circular room is equivalent to the family of *cylindrical* projections (which includes the Mercator projection), where the surface of the Earth is projected onto a cylinder which encompasses the globe (Figure 2.3a). This projection is very suitable for making maps of an area which have only a small extent in longitude. It has been chosen as the basic projection for use by the Ordnance Survey to map the UK. The transverse Mercator projection has the advantage of maintaining scale, shape, area and bearings for small areas. This explains why it has become a popular projection for mapping small areas of the globe. The single wall illustrates the *azimuthal* family of projections (Figure 2.3b) and the tepee is equivalent to the *conic* family (Figure 2.3c).

Many of the map-based spatial data sources used in GIS have a projection associated with them. To undertake meaningful analysis it is necessary to know something about the projections being used. The results of analyses will be affected in different ways by different map projections. If a GIS application requires the accurate calculation of areas, then using a projection which distorts areas is obviously not suitable. When using data at large scales (covering small areas) the effects may be slight, but at small scales (covering large areas) the effects can be substantial. Finally, since one of the functions of a GIS is to allow the integration of data from different sources, the ability to alter projections is a fundamental ability of many GIS. There are hundreds of different map projections and some GIS seem to offer the capability to re-project data for most of these. Only the most common projections have been considered above.

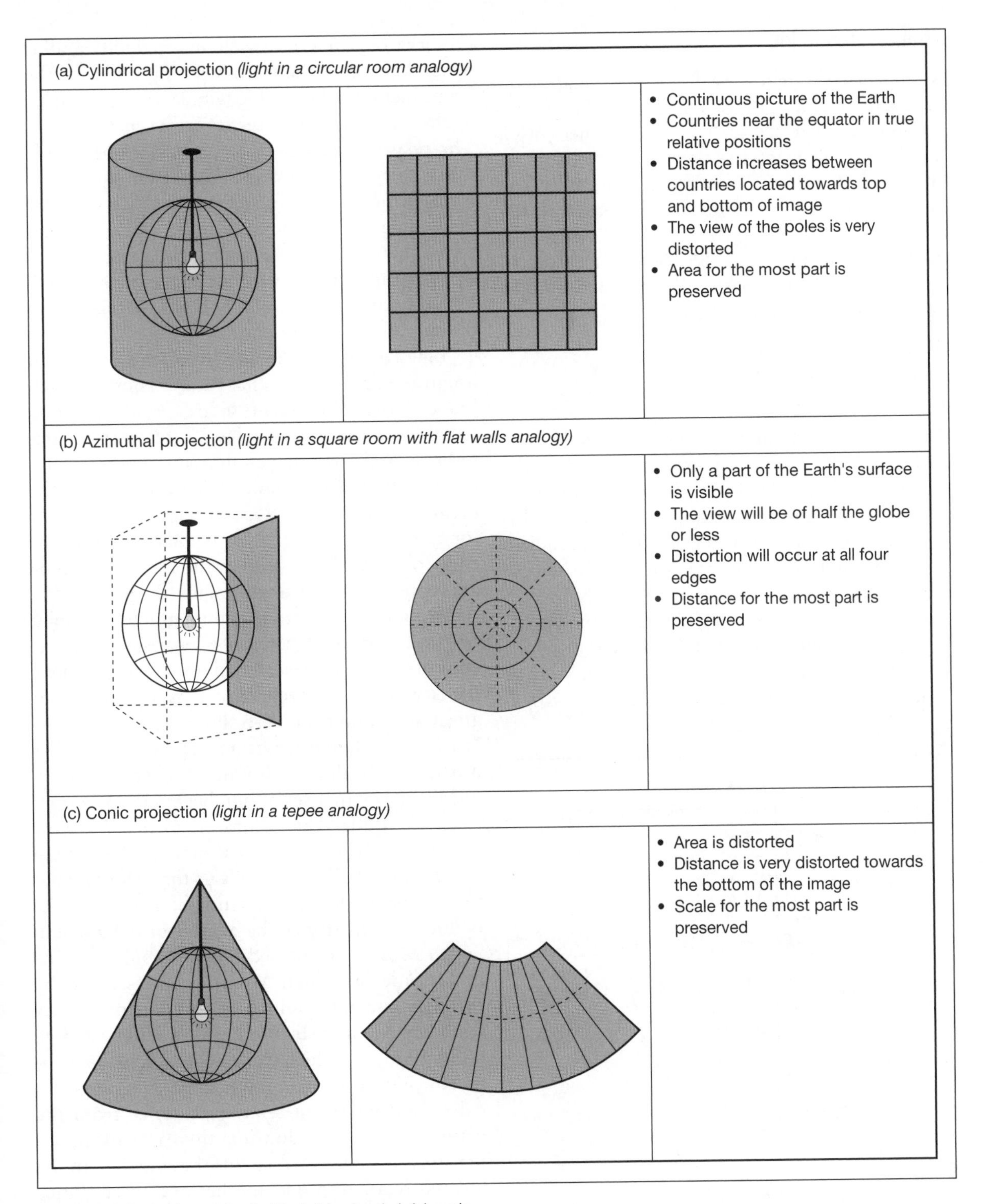

Figure 2.3 Projections: (a) cylindrical; (b) azimuthal; (c) conic

Spatial referencing

A referencing system is used to locate a feature on the Earth's surface or a two-dimensional representation of this surface such as a map. There are a number of characteristics that a referencing system should have. These include stability, the ability to show points, lines and areas, and the ability to measure length, size (area) and shape (Dale and McLaughlin, 1988). Several methods of spatial referencing exist, all of which can be grouped into three categories:

- geographic co-ordinate systems;
- rectangular co-ordinate systems; and
- non co-ordinate systems.

The only true *geographic co-ordinates* are latitude and longitude. The location of any point on the Earth's surface can be defined by a reference using latitude and longitude. Lines of longitude (also known as meridians) start at one pole and radiate outwards until they converge at the opposite pole (Figure 2.4). Conceptually they can be thought of as semicircles. If you slice a globe along two opposing lines of longitude you will always cut the globe in half. The arbitrary choice for a central line of longitude is that which runs through the Royal Observatory in Greenwich in England, and is hence known as the Greenwich meridian or the prime meridian. Lines of longitude are widest apart at the equator and closest together at the poles. The relative distance between lines of longitude where they intersect lines of latitude (or parallels) is always equal. However, the real distance will vary depending on the line of latitude which is intersected. For example, the distance between the lines of longitude intersecting the same parallel will increase towards the equator, with the maximum distance existing at the equator itself.

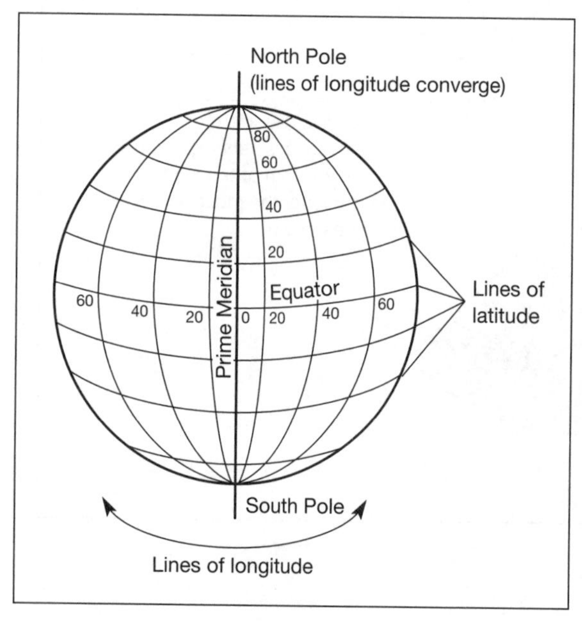

Figure 2.4 Latitude and longitude

Lines of latitude lie at right angles to lines of longitude and run parallel to one another. Each line of latitude represents a circle running round the globe. Each circle will have a different circumference and area depending on where it lies relative to the two poles. The circle with the greatest circumference is known as the equator (or central parallel) and lies equidistant from the two poles. At the two poles the lines of latitude are represented by a single point – the pole.

Using lines of latitude and longitude any point on the Earth's surface can be located by a reference given in degrees and minutes. For example, the city of Moscow represented as a point can be given a geographical co-ordinate reference using latitude and longitude of 55 degrees 45 minutes north and 36 degrees 0 minutes east (55° 45'N 36° 0'E). The first set of numbers, 55° 45'N, represents latitude. The N informs us that Moscow can be found north of the equator. The second set of numbers, 36° 0'E, tells us that Moscow lies to the east of the prime meridian. Therefore, the N and E together give the quarter of the globe in which Moscow is located (Figure 2.5a). The line of latitude on which Moscow lies is given by the degrees and minutes of this latitude away from the equator (Figure 2.5b). Finally, the line of longitude on which Moscow lies must be identified. Figure 2.5c shows how this angle is calculated based on relative distance from the prime meridian. Adopting this approach, all features on the surface of the Earth can be located relative to one another and the distance between them calculated. The shortest distance between two points on the Earth's surface is known as the great circle distance.

The latitude and longitude referencing system assumes that the Earth is a perfect sphere. Unfortunately this is not correct. The Earth is actually an oblate spheroid somewhat like an orange with flatter poles and outward bulges in equatorial regions. To complicate matters further the surface of the Earth is far from smooth and regular, as you will appreciate if you have visited any mountainous areas. At small scales these minor blemishes and imperfections in shape can be ignored. However, when dealing with large-scale maps of a small portion of the Earth's surface it is essential to make local corrections for these factors.

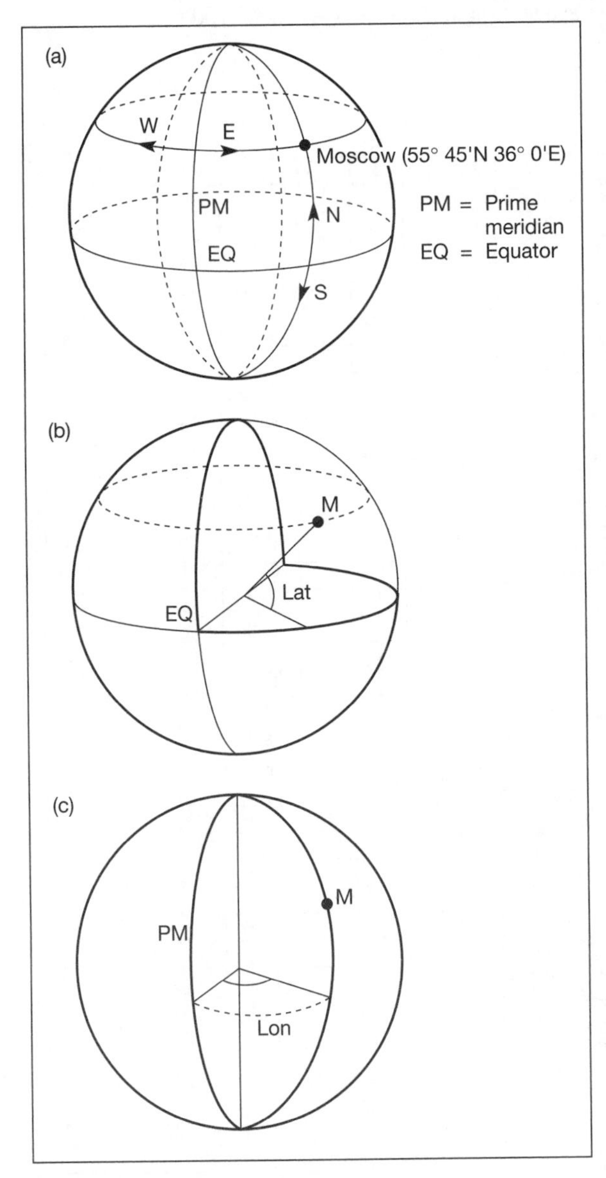

Figure 2.5 (a) Latitude and longitude of Moscow; (b) calculating the latitude; (c) calculating the longitude

The Quaternary Triangular Mesh referencing system (Goodchild and Yang, 1989) tries to deal with irregularities in the Earth's surface. It replaces lines of latitude and longitude with a mesh of regular-shaped triangles. The advantage of this referencing system is that each triangle occupies the same area on the Earth's surface. The individual triangles are also of the same size and shape. The flexible nature of a triangular mesh means that it can be moulded to fit the slight bumps and blemishes which form the true surface of the Earth. The use of triangles to model the surface of the Earth will be examined in more detail in Chapter 3 as it is an important concept in GIS.

At present, most of the spatial data available for use in GIS exist in two-dimensional form. In order to make use of these data a referencing system which uses *rectangular co-ordinates* is required. To obtain these a map graticule, or grid, is placed on top of the map. This graticule is obtained by projecting the lines of latitude or longitude from our representation of the world as a globe onto a flat surface using a map projection. The lines of latitude and longitude become the grid lines on a flat map. As already pointed out, the problem is that when you project from a sphere onto a flat surface the image becomes distorted. When small areas are being studied there will be only minor distortions in the layout of the grid. However, when large areas of the globe are projected onto a flat surface, the grid will tear and stretch. Therefore, all rectangular co-ordinate systems are designed to allow the mapping of specific geographical regions. A good example of a rectangular co-ordinate system is the UK Ordnance Survey's National Grid (Box 2.4). Another example is the Universal Transverse Mercator (UTM) plane grid system. This system uses the transverse Mercator projection and divides the Earth into 60 vertical zones that are 6 degrees of longitude wide, avoiding the poles. The system has been adopted by many organizations for remote sensing, topographic mapping and natural resource inventory (DeMers, 1997).

BOX 2.4

Ordnance Survey National Grid system

The Ordnance Survey National Grid is a rectangular grid system based on the transverse Mercator projection (Figure 2.6). The grid is 700 × 1300 km covering all of Great Britain from the Scilly Isles to Shetland. This is divided into 500 km squares, which are then divided into 25 100 km squares. Each 100 km square is identified by two letters. The first refers to the 500 km square and the second to the 100 km square. Each 100 km square is further divided into one hundred 10 km squares (10 km × 10 km), and the 10 km squares are divided into one hundred 1 km squares (1 km × 1 km). Grid references are commonly given as six figures prefixed by the letters denoting the 100 km square. An example could be SE 366 923. Here, the 'SE' denotes the 100 km square that has its origin 400 km east and 400 km north of the origin of the grid. The '366' and the '923' are the easting and northing recorded to the nearest 100 m.

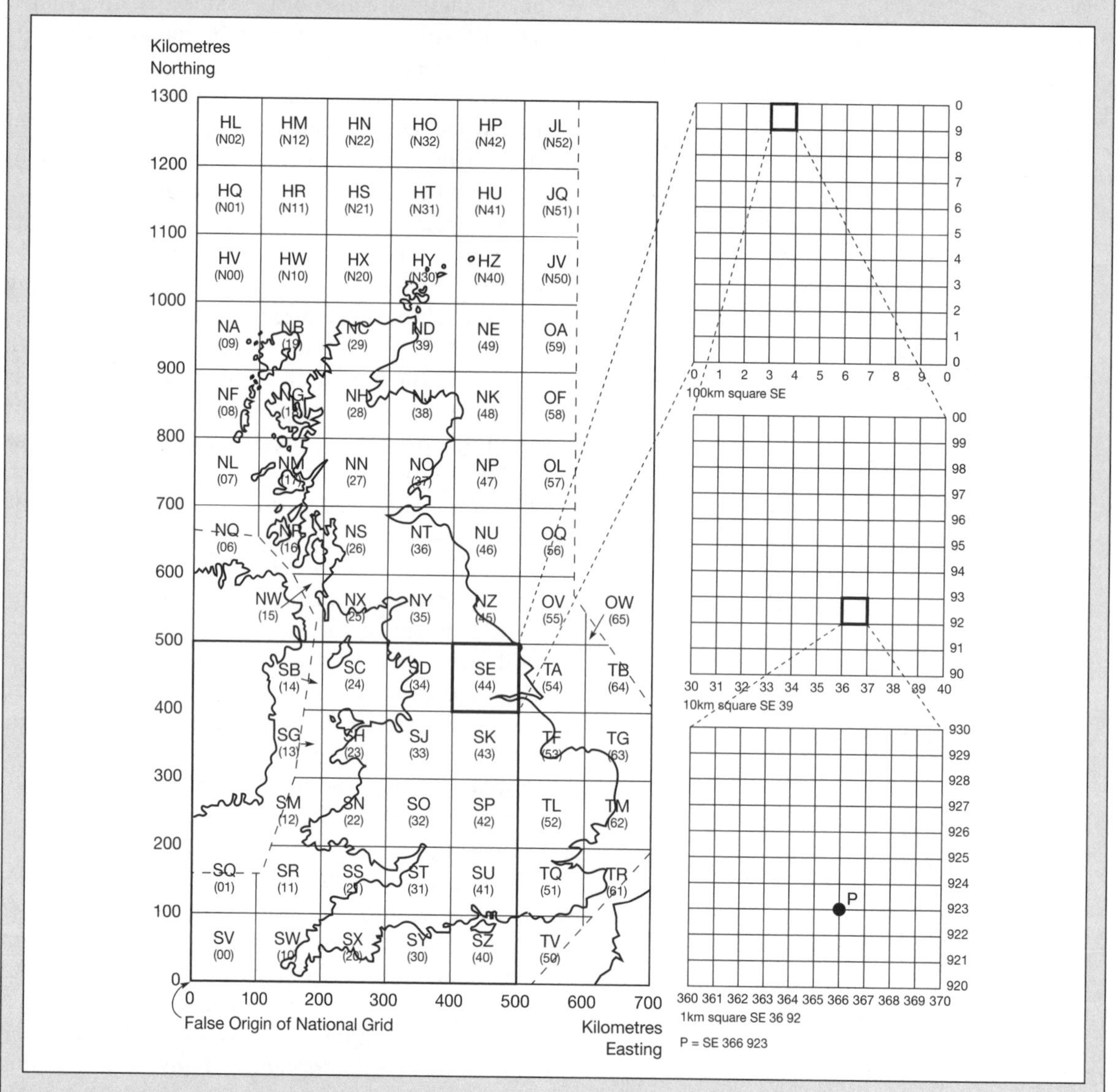

Figure 2.6 The Ordnance Survey National Grid system (adapted from Harley, 1975)

BOX 2.5

UK postcode system

In the UK the postcode system was developed about 25 years ago by the Royal Mail to help post sorting and delivery. Each code has two parts – the outward code and the inward code. The postcode system is hierarchical. The first one or two letters refer to a postcode area; these are followed by subsequent numbers and letters subdividing this into districts, sectors and unit postcodes (Department of the Environment, 1987) (Figure 2.7). The system is further complicated by the existence of single-user postcodes for business users and addresses which receive more than 15 items of mail per day (Raper *et al.* 1992). The system is very widely used in application areas such as health, marketing and education because of its ease of collection and widespread use for address-based data. However, as with any other postal code system there are problems:

- For entities without an address, a postcode system is useless. Entities without addresses include rivers, trees, fields and phone boxes.
- The spatial units – postal areas, districts and units – were designed to help mail delivery and bear no relationship to other spatial units commonly used by those handling spatial information. However, in the UK there is a link to the Ordnance Survey grid reference and census enumeration districts.
- Changes occur to postcodes. In the UK there is a three-month update cycle and approximately 18,000 changes are made each year. Changes may be corrections, due to the construction or demolition of properties, or to the movement of large users (who may be eligible to keep their postcode if they move within the same sector).
- Some buildings have more than one postcode. Office blocks containing different companies, or blocks of flats where there are separate entrances or letter boxes, may have several postcodes.
- When comparing and plotting population distribution maps, it should be remembered that unit postcodes, which cover approximately 15 houses, will represent very small areas in urban environments, but may be huge in rural areas.

(adapted from Raper *et al.*, 1992; Dale and McLaughlin, 1988)

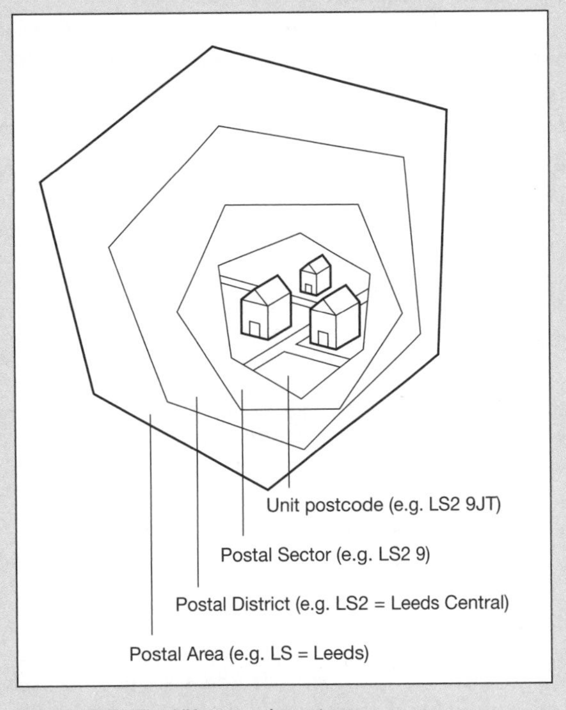

Figure 2.7 The UK postcode system

Non co-ordinate systems provide spatial references using a descriptive code rather than a co-ordinate. Postal codes, widely used throughout the world, are an example. Some postal codes are fully numeric, such as the American ZIP codes, while others are alphanumeric, as in the case of the UK postcode. All have the same basic purpose: to increase the efficiency of mail sorting and delivery rather than to be an effective spatial referencing system for GIS users. There are several advantages to such systems: they are important to the postal services and are therefore maintained and updated, and they offer coverage of all areas where people reside and work. Providing that individual codes do not refer to single addresses, they also provide a degree of confidentiality for data released using this as a referencing system. Box 2.5 provides more details on the UK postcode system.

In the western United States another non-co-ordinate referencing system is often used.

This is known as the Public Land Survey System (PLSS). Here, there has been a recursive subdivision of the land into quarter sections. By knowing which section you are in, you can reference yourself to the Earth's surface (DeMers, 1997). Other non-co-ordinate referencing systems in use are based on administrative areas, for example the units used for aggregation and presentation of population census data in different countries. For referencing within smaller areas, unique feature references may be used, for instance the property reference numbers used by a local authority, or the pipeline references used by a utility company.

All spatial referencing systems have problems associated with them. Some are specific to the referencing system, such as the updating problems with postcodes or the difficulties caused by geographical co-ordinates with respect to map projections. However, some of the problems stem from the nature of the spatial entities which require referencing:

- Spatial entities may be mobile. Animals, cars and people move, therefore any spatial reference they are tagged with will only represent their known location at a particular time.
- Spatial entities may change. Rivers meander, roads can be relocated and policy areas redefined.
- The same object may be referenced in different ways. A house may be represented and referenced as both a point and an area on maps of different scales.

An additional problem for the GIS user is the large number of different spatial referencing systems in use. Choosing an appropriate referencing system can be difficult, and it will frequently be necessary to integrate data collected using different referencing systems. This can be problematic. For some referencing systems, however, links have been developed to allow the integration and conversion of data. A good example of this is the linkage between the UK's postcode system and the Ordnance Survey National Grid. By extrapolating the start of a postcode to the corner of a 100 m grid square a grid reference has been allocated to each postcode.

Topology

In GIS, *topology* is the term used to describe the geometric characteristics of objects which do not change under transformations such as stretching or bending and are independent of any co-ordinate system (Bernhardsen, 1992). The topological characteristics of an object are also independent of scale of measurement (Chrisman, 1997). Topology, as it relates to spatial data, consists of three elements: adjacency, containment and connectivity (Burrough, 1986).

Adjacency and containment describe the geometric relationships which exist between area features. Areas can be described as being 'adjacent' when they share a common boundary. For example, the ski slopes and car parks in Happy Valley may be adjacent. Containment is an extension of the adjacency theme and describes area features which may be wholly contained within another area feature such as an island within a lake. Connectivity is a geometric property used to describe the linkages between line features. Roads are usually connected together to form a road network through which traffic can flow.

An understanding of the geometric relationships between spatial entities is important for analysis and integration in GIS. Without knowledge of how entities are geometrically related to each other, it is impossible to answer questions such as 'What is the shortest route from A to B?' or 'How many ski slopes lie within or are next to zones of high avalanche risk?'

Thematic characteristics of spatial data

Attributes are the non-spatial data associated with point, line and area entities. Attributes are the characteristics of an entity (Laurini and Thompson, 1992). For example, the attribute data associated with spatial entities used in the Happy Valley GIS might tell us that:

- a point represents a hotel;
- a line represents a ski lift; and
- an area represents a forest.

Each spatial entity may have more than one attribute associated with it. For example, a point representing the hotel may have a number of other attributes: the number of rooms; the standard of accommodation; the name and address of the owner. Attributes give additional information about the character of the entities. They also allow certain GIS operations to be performed where it is the characteristics of the entities which are under scrutiny. Questions such as 'Where are all the hotels with fewer than 20 bedrooms of luxury standard?' require the analysis of attribute data associated with the point entities used to represent the location of hotels (see Chapter 4).

The character of attribute data themselves can influence the utility of data sets in GIS analysis. One characteristic which is of considerable importance is the scale of measurement used to record and report the data. For example, every year the managers of Happy Valley must complete a table for a ski resort guide. For this table they need to provide the name of the ski area, its ranking (1st, 2nd, 3rd, 4th largest in the country), its average winter temperature and the size of the ski area. Each item of data uses a different scale of measurement. The names given to these scales are nominal, ordinal, interval and ratio. Table 2.1 shows each of these scales in relation to the data collated for the ski resort guide. Each scale of measurement dictates how the data can be used.

Table 2.1 Scales of measurement

Data	*Unit of measurement*	*Scale*
Resort name	text	nominal
Resort ranking	value	ordinal
Average winter temperature	°C	interval
Size of ski area	m^2	ratio

On a *nominal* scale numbers are used to establish identity. In Happy Valley, numbers on a nominal scale include telephone numbers or ski pass codes. These numbers cannot be processed in a mathematical sense, since they do not represent order or relative value. It is not possible to add, subtract or divide numbers on a nominal scale. Adding together two phone numbers is possible, but the answer is meaningless.

The numbers in an *ordinal* scale establish order. Location in a ski lift queue is an example. In Happy Valley, the ordinal scale is used to publish the top 10 cafes and ski runs based on the number of people using them each week. Using an ordinal scale you can obtain an impression of the order of numbers, but no information about relative sizes. The most popular ski run (ranked 1) is not necessarily twice as popular as the ski run which is ranked second. Arithmetic operations, while possible on ordinal data, will again give meaningless results.

On an *interval* scale the difference between numbers is meaningful but the scale does not have a real origin. Temperatures, in degrees Celsius, are a good example of data which are collected using an interval scale. On a temperature scale it is possible to say that there is a 10-degree difference between a thermometer which records a value of 10 degrees and one which records a value of 20 degrees. Thus, differences can be calculated. However, it would be incorrect to say that 20 degrees is twice as warm as 10 degrees, because zero degrees on the Celsius scale is not a true zero. There is still a temperature when the thermometer reads zero! Negative numbers are also possible on an interval scale.

On a *ratio* scale measurements can have an absolute or real zero, and the difference between the numbers is significant. Snow depth is an example. It is impossible to have a negative value for snow depth. However, something is known about relationships between data, for example a snow pack which is 3 m deep is twice as deep as one which is 1.5 m deep.

One of the problems with the scales of measurement used for the collection of attribute data is that the distinction between the various scales is not always obvious. Many data used in GIS are nominal or ordinal. It is important to take care when using these data in an analytical context. If the scale of measurement is not known, or the GIS user is unaware of what scale has been used or what operations can be carried out on data from that scale, the GIS is unlikely to indicate when impossible or meaningless operations have been carried out. To a computer numbers are all the same and will be treated in the same ways. So, ranked scores for city sizes may be added. Two different soil types might well have a numerical code to tag them to the appropriate area in the GIS. If clay soils have the value 2 and sandy soils 3 on a nominal scale, multiplying them together to give soil class 6 would be a meaningless operation. On the other hand, population and area (both on a ratio scale) can be divided to give population density, or elevation at one point may be subtracted from elevation at another point to give difference in elevation. In the second case the data have been collected using the interval scale.

Other sources of spatial data

So far in this chapter we have considered the characteristics of spatial data and their thematic dimension. To do this we have drawn heavily on the map metaphor. However, there are a number of other sources of spatial data, including census and survey data, aerial photographs, satellite images and global positioning systems, which have additional special characteristics. These are reviewed below.

Census and survey data

Census and survey data are collections of related information. They may be spatial in character if each item in the collection has a spatial reference which allows its location on the surface of the Earth to be identified. Examples are population census, employment data, agricultural census data or marketing data.

Population census data normally have some element of spatial referencing. Knowing how many people are in a given country may be useful in its own right, but details of where the population are will be of additional interest. The UK Census of Population is one such survey which provides a 'snapshot' of the distribution, size, structure and character of the people of Great Britain and Northern Ireland on a particular day once every 10 years (Openshaw, 1995). This is described in Box 2.6 together with examples from the USA. Most population censuses use a hierarchical series of spatial units to publish data. In the UK the smallest unit is the enumeration district (ED). This is the area over which one census enumerator delivers and collects census forms. Since normally this covers at least 150 households, this relatively small unit will guarantee privacy for individual respondents. Enumeration districts aggregate into wards and wards into local government districts (Martin, 1996). In the UK, census data are not usually released in map form, but in tables for the spatial areas you request. However, since a spatial reference is attached linking the data to the areal units of collection, the data are immediately useful for spatial studies. Census data are only one example of the myriad of spatial data which are collected through the use of survey techniques. Table 2.2 provides examples of the range of spatially referenced survey data held by the UK Department of the Environment in 1987.

Table 2.2 Examples of spatially referenced survey data

Home improvement grants	Derelict land in 1982
English house condition survey 1981	Land use change statistics
Land register database	Lead concentrations in drinking water
Urban development grants	Noise measurements 1975–79
Register of buildings of historical interest	Gypsy sites Lifestyle surveys

Source: adapted from Department of the Environment, 1987

Aerial photographs

Aerial photography was the first method of remote sensing. It is the capturing of images from a position above the Earth's surface, or without contact with the object of interest (Curran, 1989; Mather, 1991). Unlike a map, which is a model of the Earth's surface and contains only a selection of data, an aerial photograph is a 'snapshot' of the Earth at a particular instant in time. As such it contains a mass of data and it is necessary to carry out some form of interpretation to make effective use

BOX 2.6

Population Census

Census data describe the state of a whole nation, small area by area. No other data sets provide such comprehensive spatial coverage.

Census data are important for policy analysis. In the UK, for example, they are used by government in the allocation of billions of pounds of public expenditure. Census data are also very valuable commercially. They are essential ingredients in marketing analysis and retail modelling.

In the UK, the Census of Population is a simple questionnaire survey of the whole of the population that is held every ten years. The most recent census was held on April 29, 2001. The Census is administered separately in England and Wales, Scotland and Northern Ireland, but most of the statistics published are common to all countries.

The UK Census covers a wide range of subjects that describe the characteristics of the British population. These include demography, households, families, housing, ethnicity, birthplace, migration, illness, economic status, occupation, industry, workplace, mode of transport to work, cars and language. UK Census data are made available in computer format for a variety of geographical areas and spatial scales. These include:

- *Administrative areas*: small areas, wards, districts, counties, regions and countries.
- *Postal areas*: postcodes, output areas, postal sectors, postal districts and postal areas.
- *Electoral areas*: wards, parliamentary constituencies and European constituencies

In the USA, the Census has an additional role. The United States Constitution requires a census every 10 years to determine how many seats each state will have in the US House of Representatives. Census data are used to assist the apportionment (or distribution) of the 435 seats in the House of Representatives amongst the states. They are also used to assist the redistricting (or redrawing of political districts) within each state after apportionment.

Work began on developing the address list for Census 2000 in the USA in 1998. All states had received their redistricting counts by April 1, 2001.

There are two versions of the USA Census 2000 form. Eighty-three per cent of households received a short questionnaire that asked about name, sex, age, relationship, Hispanic origin, race and housing tenure. A longer questionnaire, received by 1 in 6 households, covered topics such as education, employment, ancestry, disability and heating fuel used.

Other countries that conducted censuses of population and housing in 2001 included Australia, New Zealand, Canada, Hong Kong, India and Sri Lanka.

(sources: http://www.geog.leeds.ac.uk/research/census/Census.html; http://www.census.gov)

of the information portrayed. Aerial photographs may be used in GIS as a background for other data, to give those data spatial context and to aid interpretation. Alternatively, the user may abstract information on land use, vegetation type, moisture or heat levels or other aspects of the landscape from the photograph. Aerial photographs are particularly useful for monitoring change, since repeated photographs of the same area are relatively inexpensive. For example, Gunn *et al.* (1994) have monitored changes in land use, particularly peat extraction, in County Fermanagh, Northern Ireland, from a time series of photographs. Interpretation of a sequence of photographs may allow the dating of events such as major floods which cause changes to the landscape.

Curran (1989) identifies six characteristics of aerial photographs which make them of immense value as a data source for GIS:

- wide availability;
- low cost (compared with other remotely sensed images);
- wide area views;
- time-freezing ability;
- high spectral and spatial resolution; and
- three-dimensional perspective.

Additionally, aerial photographs can be used to obtain data not available from other secondary sources, such as the location and extent of new housing estates, or the extent of forest fires. One characteristic of aerial photographs that constitutes a possible disadvantage is the fact that they do not provide spatially referenced data. Spatial referencing has to be added to features on the image by reference to other sources such as paper maps. Several different types of aerial photographs are available, from simple black and white, which may be used for a wide variety of purposes, to colour and thermal infrared for heat identification.

The angle at which the photograph was taken is important. A photograph is referred to as *vertical* if taken directly below the aeroplane, and *oblique* if taken at an angle. Oblique photographs generally

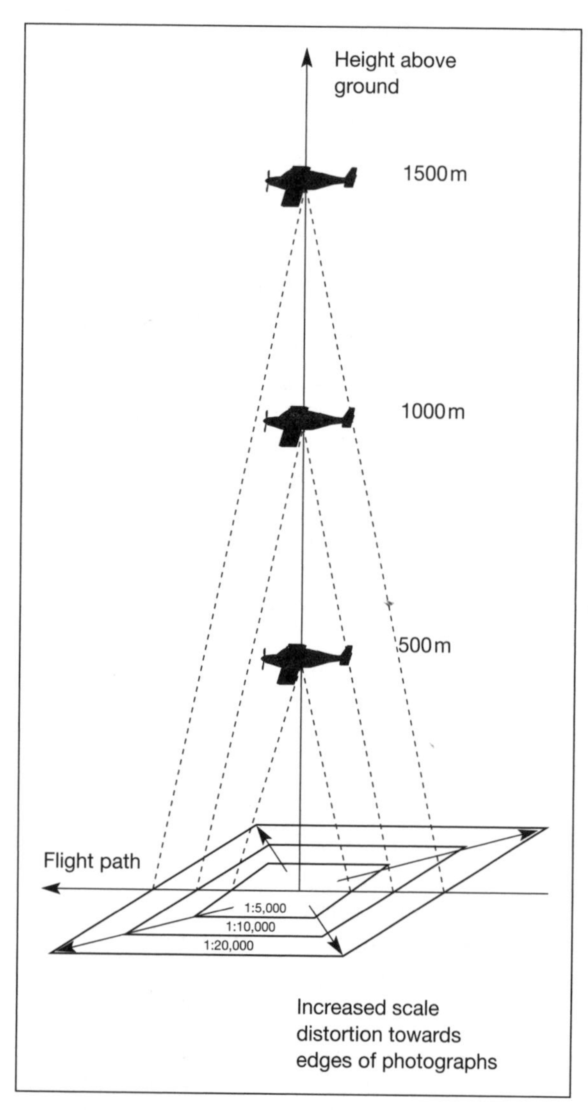

Figure 2.8 Varying scale on aerial photographs

cover larger areas and are cheaper than vertical photographs. Vertical photographs are, however, the most widely used for GIS applications.

Before any aerial photograph information can be used in a GIS a number of factors must be considered. The first of these is scale. Scale varies across an aerial photograph, owing to the distance of the camera from the ground (Figure 2.8). The scale will be constant only at the centre of the image, and the greater the flying height, the greater the scale difference between the centre and edges of the image. Second, factors which may influence interpretation need to be considered. These include time of day and time of year. On photographs taken in winter, long shadows may assist the identification of tall buildings and trees, but may obscure other features on the image. Conversely, in summer, when trees are in full leaf, features which may be visible from the air in winter will be obscured. The date of the photograph may be important to ensure the data taken from it are contemporaneous with the rest of the data in the GIS.

Aerial photographs represent a versatile, relatively inexpensive and detailed data source for many GIS applications. For example, local government bodies may organize aerial coverage of their districts to monitor changes in the extent of quarrying or building development. At a larger scale, photographs can be used to provide data on drainage or vegetation conditions within individual fields or parcels that could not be obtained from conventional topographic maps (Curran, 1989).

Satellite images

Satellite images are collected by sensors on board a satellite and then relayed to Earth as a series of electronic signals, which are processed by computer to produce an image. These data can be processed in a variety of ways, each giving a different digital version of the image.

There are large numbers of satellites orbiting the Earth continuously, collecting data and returning them to ground stations all over the world. Some satellites are stationary with respect to the Earth (geostationary), for example Meteosat, which produces images centred over Africa along the Greenwich meridian (Curran, 1989). Others orbit the Earth to provide full coverage over a period of a few days. Some of the well-known satellites, Landsat and SPOT, for example, operate in this way. Landsat offers repeat coverage of any area on a 16-day cycle (Mather, 1991).

Sensors on board these satellites detect radiation from the Earth for different parts of the electromagnetic spectrum, not only the portions visible to the human eye. The multispectral scanner (MSS) on board Landsat simultaneously detects radiation in four different wavebands: near infrared, red, green and blue (Curran, 1989). After processing, the images can be used to detect features not readily apparent to the naked eye, such as subtle changes in moisture content across a field, sediment dispersal in a lake or heat escaping from roofs in urban areas.

Scanned images are stored as a collection of pixels, which have a value representing the amount of radiation received by the sensor from that portion of the Earth's surface (Burrough, 1986). The size of the pixels gives a measure of the resolution of the image. The smaller the pixels the higher the resolution. The Landsat Thematic Mapper collects data for pixels of size 30 m by 30 m. Much greater resolution is possible, say 1 m by 1 m, but this has in the past been restricted to military use. Recent changes in US legislation and the availability of Russian military satellite data have made access to very high-resolution data easier. Resolution is an important spatial characteristic of remotely sensed data and determines its practical value. A Landsat Thematic Mapper image with a pixel size of 30 m by 30 m would be unsuitable for identifying individual houses but could be used to establish general patterns of urban and rural land use. Box 2.7 provides further discussion on the resolution issue.

A new form of remotely sensed data is LIDAR (Light Detection and Ranging) data. LIDAR is a remote sensing system that uses aircraft-mounted lasers to collect topographic data. The lasers are capable of recording elevation measurements with a vertical precision of 15 cm. Measurements are spatially referenced using high-precision GPS. The technology creates a highly detailed digital elevation model (DEM) that closely matches every undulation in the landscape. Even change in ground surface elevation detail caused by buildings and trees can be detected. This makes LIDAR extremely useful for large-scale mapping and engineering applications.

BOX 2.7

Resolution

Resolution is defined as the size of the smallest recording unit (Laurini and Thompson, 1992) or the smallest size of feature that can be mapped or measured (Burrough, 1986). In the Zdarske Vrchy case study introduced in Chapter 1, the data are stored as raster layers, the size of each individual cell being 30 m × 30 m. In this case, the resolution of each image is 30 m since this is the lowest level to which the data can be described. In the case of mapping census variables or other socio-economic data collected within administrative boundaries, the Department of the Environment (1987) refers to Basic Spatial Units (BSU) as the smallest spatial entity to which data are encoded. The BSUs should be constant across all the data used for a particular application so that the data collected are immediately comparable. In order to facilitate subsequent analysis it is necessary to choose the smallest unit possible when collecting the data. These BSU can then can be built up into any other unit by the process of aggregation. In the UK Census of Population the BSU for publicly accessible data is the enumeration district (ED), containing approximately 500 residents (150 in rural areas). EDs can easily be aggregated to form wards. Wards can then be aggregated into districts, and districts into counties. Disaggregation (the reverse process to aggregation) to areas smaller than the original BSU is fraught with difficulties and based on so many assumptions as to pose serious problems for data quality.

For GIS, remotely sensed data offers many advantages. First, images are always available in digital form, so transfer to a computer is not a problem. However, some processing is usually necessary to ensure integration with other data. Processing may be necessary to reduce data volumes, adjust resolution, change pixel shape or alter the projection of the data (Burrough, 1986). Second, there is the opportunity to process images or use different wavebands for the collection of data to highlight features of particular interest, for example water or vegetation. The repeated coverage of the Earth is a further advantage, allowing the monitoring of change at regular intervals, although for some types of imagery, cloud cover while the satellite passes overhead may prevent a useful image being obtained. Finally, the small scale of images provides data useful for regional studies, and applications have included mapping remote areas, geological surveys, land use monitoring and many others (see Curran, 1989; Mather, 1991; Maguire, 1989). Trotter (1991) considers the advantages of remotely sensed data for GIS applications in the area of natural resource management to be:

- low cost relative to other data sources;
- currency of images;
- accuracy;
- completeness of data; and
- uniform standards across an area of interest.

Field data sources: surveying and GPS

There are several methods of collecting raw data in the field for direct input into a GIS. These are most often used when the required data do not exist in any other readily available format such as a map or satellite image. Traditional manual surveying techniques using chains, plane tables, levels and theodolites are examples of direct field measurement, but the data collected need to be written down on paper first. Modern digital equivalents of these manual techniques have been adapted so that the data collected are stored in digital format ready for direct input into GIS. Examples include total stations (high-precision theodolites with electronic distance metering (EDM) and a data logger) and hand-held laser range finders.

A relatively new technique of field data collection which has found particular favour with GIS users is the use of satellite navigation systems or GPS (Global Positioning Systems). These are portable backpack or hand-held devices that use signals from GPS satellites to work out the exact location of the user on the Earth's surface in terms of (x,y,z) co-ordinates using trigonometry (Figure 2.9). Position fixes are obtained quickly and accu-

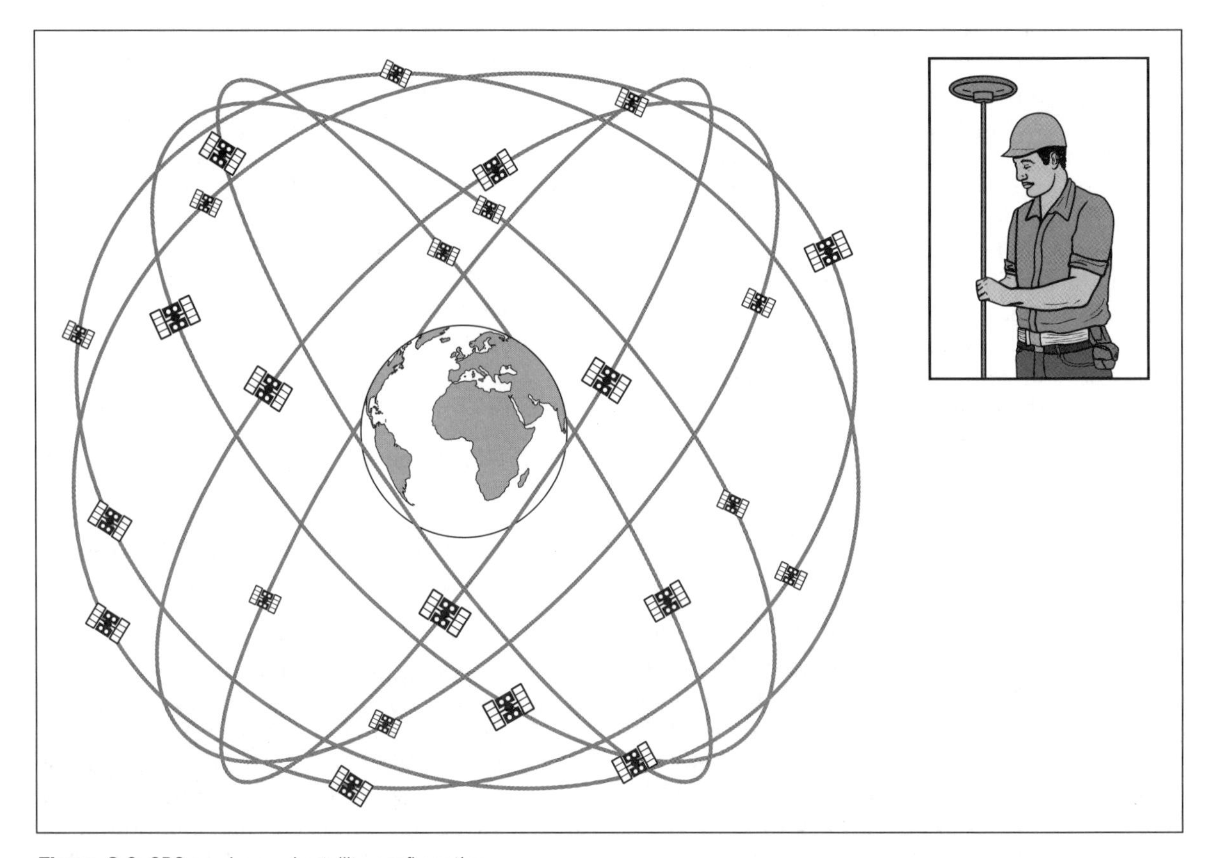

Figure 2.9 GPS receiver and satellite configuration

BOX 2.8

GPS basics

GPS is a set of satellites and control systems that allow a specially designed GPS receiver to determine its location anywhere on Earth 24 hours a day (Barnard, 1992). Two main systems exist, the American NAVSTAR system and the Russian GLONASS system. A European system is also planned for the near future. The American system consists of 24 satellites orbiting the Earth in high altitude orbits. These satellites have a 12-hour orbit time, and pass over control stations so that their orbits can be closely monitored and their positions precisely identified. Satellites and ground-based receivers transmit similarly coded radio signals, so that the time delay between transmission and receipt of the signals gives the distance between the satellite and the receiver. If a receiver can pick up signals from three or four satellites, trigonometry is used to calculate the location and height of the receiver.

A GPS user will see a position 'fix' displayed on their receiver. Until recently, all readings were affected by selective availability (SA), a deliberate error added to the signals by the US military. This has now been switched off and fixes of far greater accuracy can be obtained. One fix, obtained from a single receiver, will have an accuracy of about 25 m for 95 per cent of the time. When SA was in place accuracy was 100 m 95 per cent of the time. More advanced data collection methods, including the averaging of fixes and the use of two receivers in parallel (differential GPS), can be used to obtain even more accurate readings down to the sub-centimetre level.

Differential GPS techniques require two receivers, one fixed at a known location (the base station) and the other at an unknown location (the roving receiver). If both receivers are set up in exactly the same manner and use the same satellites to compute their location, then the positional error recorded will be the same for both receivers. A highly accurate positional fix can therefore be obtained for the roving receiver by subtracting the positional error calculated for the base station.

Despite 24-hour global coverage, GPS use can be hampered by certain factors. These include problems where the path between the satellite and the receiver is obstructed by buildings, dense tree cover or steep terrain, and in polar regions where favourable satellite configurations are not always available.

Despite practical difficulties, GPS is finding a wide range of applications, varying from navigation (air, sea and land), to mapping and surveying. Cornelius *et al.* (1994) and Carver *et al.* (1995) give some examples of the applications of GPS in geomorphology and fieldwork. Recent developments in portable computing and mobile communications have opened up a whole new area of application for GPS. Knowing exactly where you are in relation to a GIS database on your portable computer, or receiving location information via your WAP phone, can be of great value and is a highly saleable commodity. The use of this location technology as a form of GIS 'output' is discussed further in Chapter 5.

rately literally at the push of a button. The accuracy obtainable from civilian GPS receivers ranges from 100 m to as little as 0.5 m depending on how they are used, while there are military versions that are accurate to within a few centimetres. Originally designed for real-time navigation purposes, most GPS receivers will store collected co-ordinates and associated attribute information in their internal memory so they can be downloaded directly into a GIS database. The ability to walk or drive around collecting co-ordinate information at sample points in this manner has obvious appeal for those involved in field data collection for GIS projects. Box 2.8 gives further details on GPS.

GIS Data Standards

The number of formats available for GIS data is almost as large as the number of GIS packages on the market. This makes the sharing of data difficult and means that data created on one system is not always easily read by another system. This problem has been addressed in the past by including data conversion functions in GIS software. These conversion functions adopt commonly used exchange formats such as DXF and E00. As the range of data sources for GIS has increased, the need for widely applicable data standards to facilitate exchange of data has been recognized. Some of the standards in current use are listed in Table 2.3.

Table 2.3 Current geographical information standards

BS 7666	Spatial data sets for geographic referencing
CEN TC 287	European norms for geographic information
DIGEST	Digital Geographic Information Exchange Standards
DNF	Digital National Framework
GDF	Geographic Data File
GeoTIFF	Geographic Tagged Image Format File
GML	Geography Markup Language
ISO 6709	Standard representation of latitude, longitude and altitude
ISO 8211	Specification for a data descriptive file for information interchange
ISO 15046	Geographic information
NEN 1878	Netherlands transfer standard for geographic information
NTF	Neutral Transfer Format
OGIS	Open Geodata Interoperability Standard
RINEX	Receiver Independent Exchange Format
SDTS	Spatial Data Transfer Standard
UGDCS	Utilities Geospatial Data Content Standard

(Source: adapted from Harding and Wilkinson, 1997)

More details of one of these standards, BS 7666, are provided in Box 2.9.

There is still no universally accepted GIS data standard, although the Open GIS Consortium (OGC), formed in 1994 by a group of leading GIS software and data vendors, is working to deliver spatial interface specifications that are available for global use (OGC, 2001). The OGC has proposed the Geography Markup Language (GML) as a new GIS data standard. GML and early adopters of this new standard are described in Box 2.10.

Conclusions

In this chapter we have looked at the distinction between data and information, identified the three main dimensions of data (temporal, thematic and spatial) and looked in detail at how different spatial data sources portray the spatial dimension. The main characteristics of spatial data have been identified and a review of how the traditional map-making process has shaped these characteristics has been presented. In addition we have considered a range of other sources of spatial data. The discussion has shown that any source of spatial data may be influenced by some, or all, of the following factors:

- the purpose for which they have been collated;
- the scale at which they have been created;
- the resolution at which they have been captured;
- the projection which has been used to map them;
- the spatial referencing system used as a locational framework;
- the nature of the spatial entities used to represent real-world features;
- the generality with which these entities have been modelled; and
- the topological structure used to represent the relationship between entities.

In some data sources, one factor will dominate; in others it will be the interplay of factors which give the data their character. Appreciating the main characteristics of spatial data is important because these characteristics will determine how the data can be used in building a GIS model. For example, data collected at different resolutions should only be integrated and analyzed at the resolution of the coarsest data set. In the Zdarske Vrchy case study the 30 m by 30 m resolution of the land use map generated from TM satellite data dictated the resolution of the database for analysis.

Therefore, GIS models are only as good a representation of the real world as the spatial data used to construct them. Understanding the main characteristics of spatial data is an important first step in evaluating its usefulness for GIS. The next step is to understand how these data can be stored in a form suitable for use in the computer, as this will also influence the quality of the GIS model.

BOX 2.9

BS 7666

British Standard BS 7666 specifies a nationally accepted, standard referencing method for land and property in the UK. The standard was developed by a multi-disciplinary working party that included representatives from, amongst others, local government, the Ordnance Survey, Her Majesty's Land Registry, the Royal Mail, the Forestry Commission and academia. The standard provides a common specification for the key elements of data sets of land and property in Great Britain. It assures the quality of land and property information in terms of content, accuracy and format.

BS 7666 has four parts. It includes a specification for:

- A street gazetteer (BS 7666 Part 1).
- A land and property gazetteer (BS 7666 Part 2).
- Addresses (BS 7666 Part 3).
- A dataset for recording Public Rights of Way.

Part 3 provides the specification for addresses. The specification provides a nationally consistent means of structuring address-based information. Use of the standard should simplify the exchange and aggregation of address-based and related data.

The standard specifies that an address must contain sufficient information to ensure uniqueness within Great Britain. The combination of a primary addressable object name and a secondary addressable object name achieves this. An address must also contain the name of at least one or more of the street, locality, town and administrative area data, so that it is unique.

In addition, a postcode is mandatory for a mailing address, although a postcode may not exist for non-postal addresses. A postal address is a routing instruction for Royal Mail staff that must contain the minimum information necessary to ensure secure delivery. Its presentation and structure are specified in another international standard: ISO 11180.

The specification includes details of how locality name, town name, administrative area and postcode should be specified. There are also detailed text conventions, requiring for example the storage of text in upper case format; no abbreviations except for ST (Saint) and KM (Kilometre); and no underlining of text. Table 2.4 provides examples of land and property identifiers acceptable under BS 7666.

Table 2.4 Examples of Land and Property Identifiers

Unique Property Reference Number	*Secondary Addressable Object Name*	*Primary Addressable Object Name*	*Unique Street Reference Number*	*Street Name*
000100001		Palace Deluxe	00010001	Pine Avenue
000100002	Caretakers Flat	Palace Deluxe	00010001	Pine Avenue
000100003	Mountain View	23	00010006	High Street
000100004	Ski Lodge	10	00010002	Ski School Road

(Source: adapted from http://www.housing.dtlr.gov.uk/research)

BOX 2.10

GML

The Geography Markup Language (GML) is a non-proprietary computer language designed specifically for the transfer of spatial data over the Internet.

GML is based on XML (eXtensible Markup Language), the standard language of the Internet, and allows the exchange of spatial information and the construction of distributed spatial relationships. GML has been proposed by the Open GIS Consortium (OGC) as a universal spatial data standard. GML is likely to become very widely used because it is:

- Internet friendly;
- not tied to any proprietary GIS;
- specifically designed for feature-based spatial data;
- open to use by anyone;
- compatible with industry-wide IT standards.

It is also likely to set the standard for the delivery of spatial information content to PDA and WAP devices, and so form an important component of mobile and location-based (LBS) GIS technologies.

Among the organizations adopting GML is the Ordnance Survey (OS), the national mapping agency for the UK. The OS will deliver DNF (Digital National Framework) data in GML. DNF is a version of the OS's large-scale topographic database that will eventually encompass all types of spatial data and all data scales. In the DNF nearly 230,000 tiles of large-scale topographic data have been merged into a single, seamless topologically structured point, line and topographic database containing information on buildings, boundaries, roads, railways, water and other topographic features. Each feature in the DNF is assigned a unique 16-digit identifier that allows it to be unambiguously referenced and associated with other features.

By adopting GML, the OS is making the DNF accessible to more software systems and users than would be possible using any other single data standard.

(Sources: Holland, 2001; GISNews, 2001)

REVISION QUESTIONS

- Explain the difference between data and information.
- What are the three basic spatial entities and how are these used to portray geographical features on paper maps and in GIS?
- Explain the differences between geographic and rectangular co-ordinate systems. What are their relative advantages and disadvantages?
- Explain what is meant by adjacency, containment and connectivity.
- Why is a knowledge of the different scales of measurement important in GIS?
- Explain the importance of map projections for users of GIS.
- Describe the characteristics of three sources of spatial data.
- Using examples, outline the importance of standards for spatial data.

Further study

Gatrell (1991) provides a good, thought-provoking introduction to the concepts of space and geographical data. This discussion is a good starting point for anyone coming to GIS from a non-geographical background. Robinson *et al.* (1995) and Keates (1982) provide a comprehensive coverage of the subject of cartography from a more conventional viewpoint. Monmonier's book *How to Lie with Maps* (1991) offers a comprehensive and very readable introduction to the potential pitfalls of displaying data in map form. The discussion is just as applicable to maps on the computer screen as those on paper. Subjects such as scale, projections and generalization are covered in detail. A good discussion on UK spatial referencing can be found in Dale and McLaughlin (1988); DeMers (1997) provides a comparable review for the USA. The Chorley Report (Department of the Environment, 1987) provides brief details of postcodes and recommendations for use of spatial referencing in the UK. Raper *et al.* (1992) discuss the whole issue of UK postcodes in considerable depth and provide examples of address formats and postcode systems in a number of other countries including Austria, Germany, the Netherlands, Spain, Sweden and the USA. A similar discussion on the US ZIP code, though not in the same depth, can be found in DeMers (1997).

Comprehensive coverage of the principles and applications of remote sensing can be found in Curran (1989) or Clayton (1995). Curran (1989) contains a particularly useful chapter on aerial photography that discusses the characteristics and interpretation of aerial photographs. A good introduction to GPS and its importance for GIS can be found in Kennedy (1996). Seegar (1999) offers the basic principles of geodesy relevant to GPS.

Clayton K (1995) The land from space. In: O'Riordan T (ed.) *Environmental Science for Environmental Management*. Longman, London, pp. 198–222

Curran P (1989) *Principles of Remote Sensing*. Longman, London

Dale P F, McLaughlin J D (1988) *Land Information Management. An Introduction with Special Reference to Cadastral Problems in Third World Countries*. Clarendon Press, Oxford

DeMers M N (1997) *Fundamentals of Geographic Information Systems*. Wiley, New York

Department of the Environment (1987) *Handling Geographic Information*. Report of the Committee of Enquiry chaired by Lord Chorley. HMSO, London

Gatrell A C (1991) Concepts of space and geographical data. In: Maguire D J, Goodchild M F, Rhind D W (eds) *Geographical Information Systems: Principles and Applications*. Longman, London, Vol. 1, pp. 119–143

Keates J S (1982) *Understanding Maps*. Longman, London

Kennedy M (1996) *The Global Positioning System and GIS: An Introduction*. Ann Arbor Press, Ann Arbor, Michigan

Monmonier M (1991) *How to Lie with Maps*. University of Chicago Press, Chicago

Raper J F, Rhind D W, Shepherd J W (1992) *Postcodes: the New Geography*. Longman, Harlow, UK

Robinson A H, Morrison J L, Muehrecke P C, Kimerling A J, Guptill S C (1995) *Elements of Cartography*, 6th edn. Wiley, New York

Seegar H (1999) Spatial referencing and coordinate systems. In: Longley P A, Goodchild M F, Maguire D J, Rhind D W (eds) *Geographical Information Systems*. Wiley, New York, vol. 1, pp. 427–36

3 Spatial data modelling

KEY QUESTIONS AND ISSUES

- What is a spatial data model?
- How are spatial entities used to create a data model?
- What are rasters and vectors?
- What is a spatial data structure?
- What is topology and how is it stored in the computer?
- What are the advantages and disadvantages of different spatial data models?
- How are time and the third dimension handled in GIS?

Introduction

The previous chapter introduced the concept that a GIS, populated with data and ideas about how these data interact, is a model of reality. The aim of this model is to develop our understanding of geographical problems. Try to picture our Martian colleague contacting his friends back on his home planet while he tries to explain in conceptual form what a car is and what it can do.

In fact, the GIS model can be split into two parts: a model of spatial form and a model of spatial processes. The model of spatial form represents the structure and distribution of features in geographical space. In order to model spatial processes, the interaction between these features must be considered. More details of process models will be presented in Chapter 7. This chapter, and those which follow (particularly Chapter 6), focus on the representation and analysis of geographical features using models of form.

The construction of models of spatial form can be thought of as a series of stages of data abstraction (Peuquet, 1984). By applying this abstraction process the GIS designer moves from the position of observing the geographical complexities of the real world to one of simulating them in the computer. This process involves:

1. Identifying the spatial features from the real world that are of interest in the context of an application and choosing how to represent them in a conceptual model. In the Happy Valley GIS, hotel features have been selected for inclusion and will be represented by point entities.
2. Representing the conceptual model by an appropriate spatial data model. This involves choosing between one of two approaches: raster or vector. In many cases the GIS software used may dictate this choice.
3. Selecting an appropriate spatial data structure to store the model within the computer. The spatial data structure is the physical way in which entities are coded for the purpose of storage and manipulation.

Figure 3.1 provides an overview of the stages involved in creating a GIS data model. At each stage in the model-building process, we move further away from the physical representation of a feature in reality and closer to its abstract representation in the computer. This chapter examines the stages in the modelling process as shown in Figure 3.1. First, the definition of entities for representation in the computer is considered. Then, the different spatial data models and structures available are outlined. Finally, the modelling of more complex features and the difficulties of including the third and fourth dimensions in a GIS model are discussed.

Entity definition

In order to define entities for inclusion in a computer model features of interest need to be identified, then a method for their representation in the computer must be established. As suggested in Chapter 1, all geographical phenomena can be represented in two dimensions by three main entity types: points, lines and areas. Figure 3.2 shows how a spatial data model could be constructed for the Happy Valley ski area using points, lines and areas. Points are used to represent hotels, lines to represent ski lifts and areas to represent forests. Figure 3.2 also introduces two additional spatial entities: networks and surfaces. These are an extension of the area and line concepts and are introduced in Box 3.1.

There are a number of problems associated with simplifying the complexities of the real world into five basic, two-dimensional entity types. These include the dynamic nature of the real world, the identification of discrete and continuous features, and the scale at which a particular problem needs addressing.

BOX 3.1

Surfaces and networks

A surface entity is used to represent continuous features or phenomena. For these features there is a measurement or value at every location, as is the case with elevation, temperature and population density. In Happy Valley, elevation and snow depth are represented by surfaces. Both elevation and snow depth vary continuously over space – a value for each can be recorded at any location. This makes representation by a surface entity appropriate. The continuous nature of surface entities distinguishes them from other entity types (points, lines, areas and networks) which are discrete, that is either present or absent at a particular location.

A network is a series of interconnecting lines along which there is a flow of data, objects or materials. There are several networks in Happy Valley. One is the road network, along which there is a flow of traffic to and from the ski areas. Another is the river, along which there is a flow of water. Others, not visible on the land surface, include the sewerage and telephone systems.

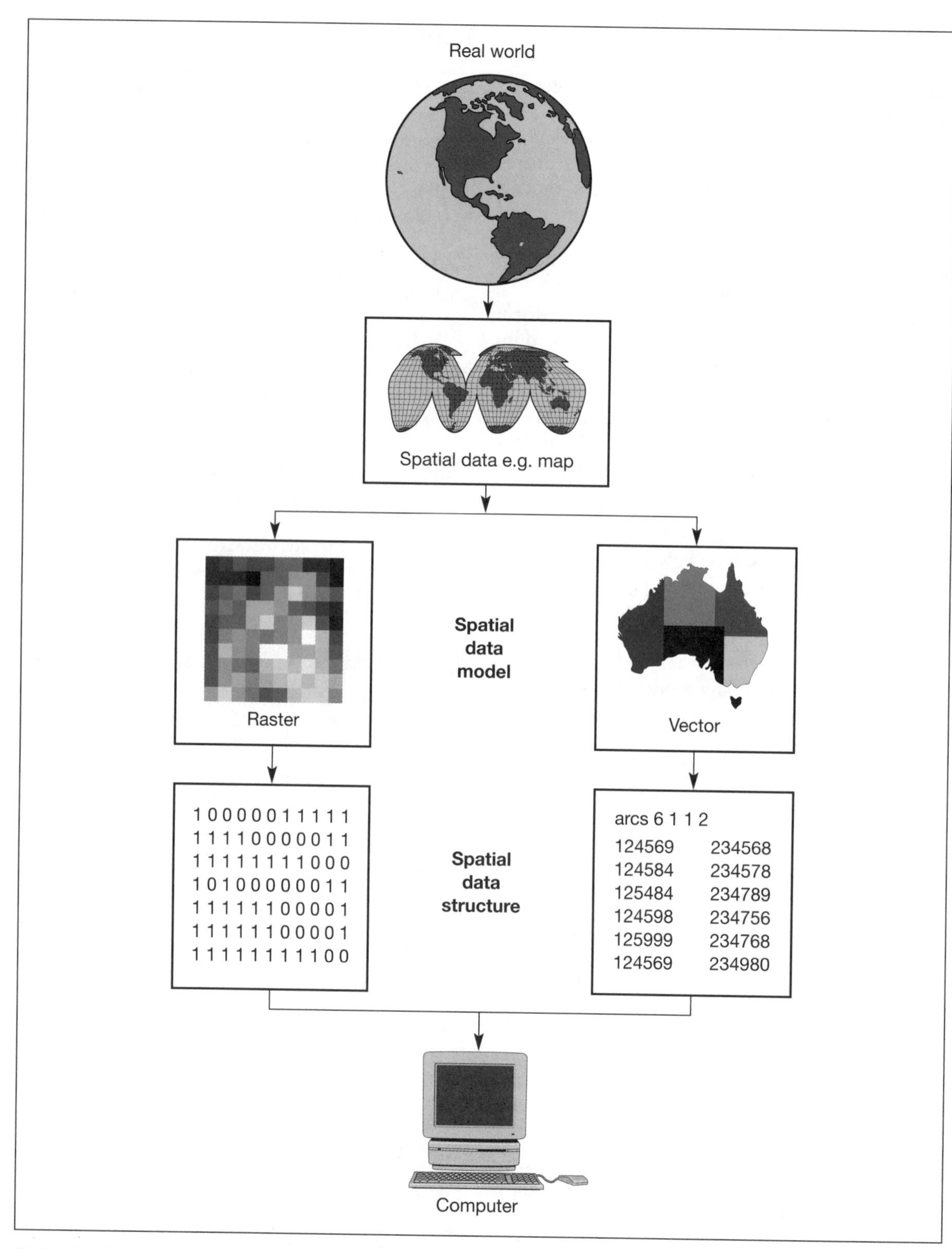

Figure 3.1 Stages involved in constructing a GIS data model

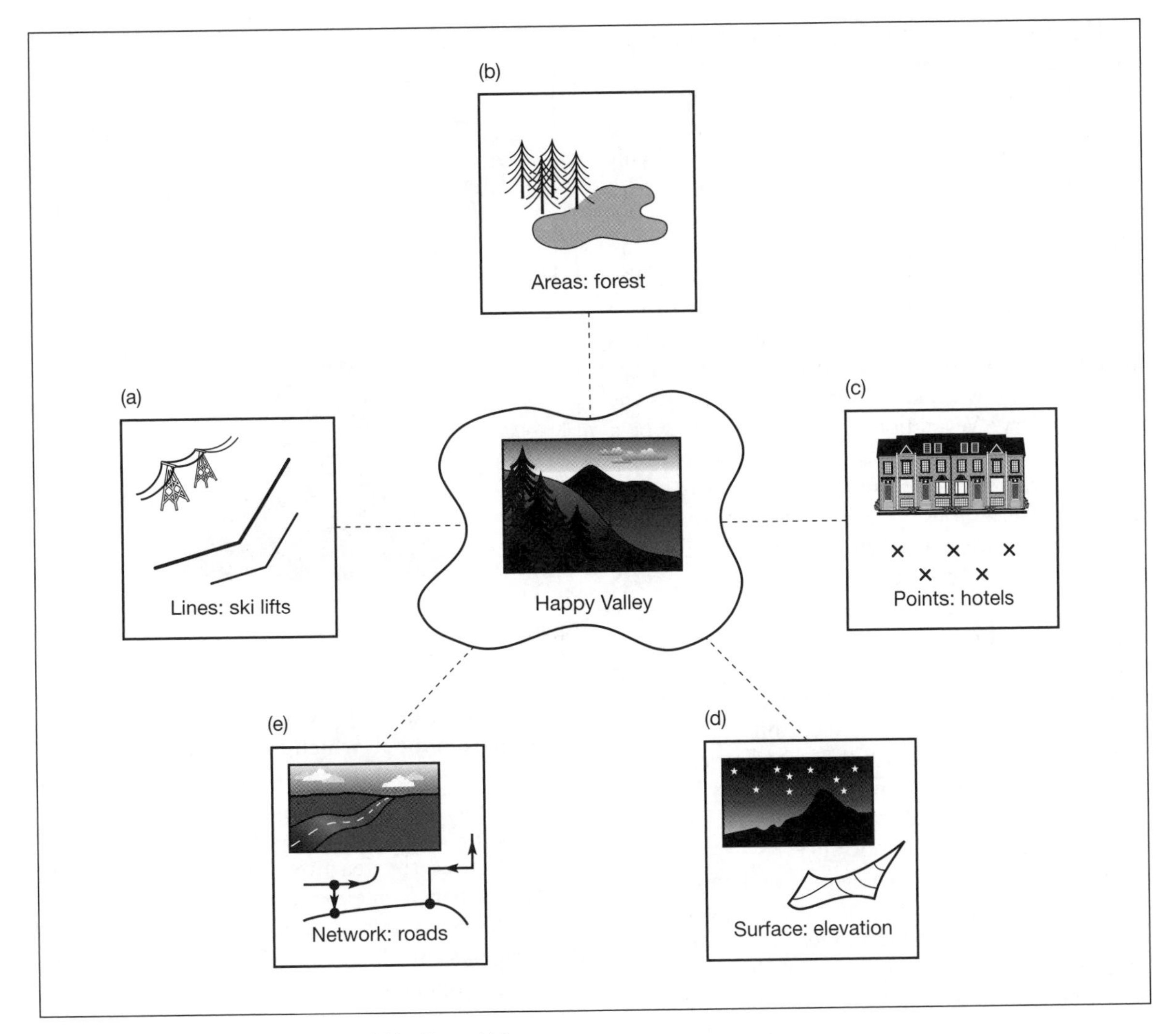

Figure 3.2 A simple spatial entity model for Happy Valley

The real world is not static. Forests grow and are felled; rivers flood; and cities expand. The dynamic nature of the world poses two problems for the entity-definition phase of a GIS project. The first is how to select the entity type that provides the most appropriate representation for the feature being modelled. Is it best to represent a forest as a collection of points (representing the location of individual trees), or as an area (the boundary of which defines the territory covered by the forest)? The second problem is how to represent changes over time. A forest, originally represented as an area, may decline until it is only a dispersed group of trees that are better represented using points.

In addition, the nature of the physical features themselves may remain static, but our perception of their geographic structure may differ according to our application area. For visitors to find their way around Happy Valley a map of the road network, with roads represented as a series of interconnected lines, would be sufficient. However, for road maintenance engineers, who must repair the surface after the winter season, it may be more appropriate to represent the roads as areas or even surfaces.

The definition of entity types for real-world features is also hampered by the fact that many real-world features simply do not fit into the categories of entities available. An area of natural woodland does not have a clear boundary as there is normally a transition zone where trees are interspersed with vegetation from a neighbouring habitat type. In this case, if we wish to represent the woodland by an area entity, where do we place the boundary? The question is avoided if the data are captured from a paper map where a boundary is clearly marked, as someone else will have made a decision about the location of the woodland boundary. But is this the true boundary? Vegetation to an ecologist may be a continuous feature (which could be represented by a surface), whereas vegetation to a forester is better represented as a series of discrete area entities.

Features with 'fuzzy' boundaries, such as the woodland, can create problems for the GIS designer and the definition of entities, and may have an impact on later analysis. Consider a grant that the Happy Valley ski company has received to pay for maintenance of the ski area. If the grant is awarded, *pro rata*, according to the area of maintained piste and GIS is used to calculate the area of piste, then how the boundary between maintained piste and off-piste is defined will clearly be of financial importance!

An example of the problem of scale and entity definition is illustrated by considering two ways in which the geographical location of the ski lifts might be modelled. First, imagine that the marketing department at Happy Valley has decided to create a GIS that will permit skiers to check the status of ski lifts (whether they are open or closed) over the Internet. For the production of this digital lift map a simple set of line entities, to which information can be attached about the status of the lift (open or closed), is appropriate. The simple line entity gives the user the appropriate geographical reference and information they require to assess whether it is worth visiting Happy Valley. Second, consider the geographical view of the ski lift network required by a lift engineer, who has to organize repairs to damaged lifts. This view might look quite different (Figure 3.3). Additional entities are required. These include a set of point entities to mark the locations of the supporting pylons; area entities to mark the embarkation and disembarkation zones; and the simple line segments used in the digital lift map.

Figure 3.3 An engineer's view of a ski lift

Therefore, deciding which entity type should be used to model which real-world feature is not always straightforward. The way in which individuals represent a spatial feature in two dimensions will have a lot to do with how they conceptualize the feature. In turn, this will be related to their own experience and how they wish to use the entity they produce. An appreciation of this issue is central to the design and development of all GIS applications.

Spatial data models

Burrough (1986) recognizes that the human eye is highly efficient at recognizing shapes and forms, but the computer needs to be instructed exactly how spatial patterns should be handled and displayed. Computers require unambiguous instructions on how to turn data about spatial entities into graphical representations. This process is the second stage in designing and implementing a data model. At present there are two main ways in which computers can handle and display spatial entities. These are the raster and vector approaches.

The raster spatial data model is one of a family of spatial data models described as tessellations. In the raster world individual cells are used as the building blocks for creating images of point, line, area, network and surface entities. Figure 3.4 shows how a range of different features from Happy Valley, represented by the five different entity types, can be modelled using the raster approach. Hotels are modelled by single, discrete cells; the ski lifts are modelled by linking cells into lines; the forest by grouping cells into blocks; and the road network by linking cells into networks. The relief of the area has been modelled by giving every cell in the raster image an altitude value. In Figure 3.4 the altitude values have been grouped and shaded to give the appearance of a contour map.

In the raster world the basic building block is the individual grid cell, and the shape and character of an entity is created by the grouping of cells. The size of the grid cell is very important as it influences how an entity appears. Figure 3.5 shows how the spatial character of the Happy Valley road network changes as the cell size of the raster is altered.

A vector spatial data model uses two-dimensional Cartesian (x,y) co-ordinates to store the shape of a spatial entity. In the vector world the point is the basic building block from which all spatial entities are constructed. The simplest spatial entity, the point, is represented by a single (x,y) co-ordinate pair. Line and area entities are constructed by connecting a series of points into chains and polygons. Figure 3.4 shows how the vector model has been used to represent various features for the Happy Valley ski area. The more complex the shape of a line or area feature the greater the number of points required to represent it. Selecting the appropriate number of points to construct an entity is one of the major dilemmas when using the vector approach. If too few points are chosen the character, shape and spatial properties of the entity (for example, area, length, perimeter) will be compromised. If too many points are used, unnecessary duplicate information will be stored and this will be costly in terms of data capture and computer storage. Figure 3.5 shows how part of the Happy Valley road network is affected by altering the number of points used in its construction. Methods have been developed to automate the procedure for selecting the optimum number of points to represent a line or area feature. The Douglas–Peucker algorithm is one example (see Chapter 5). This removes points that do not contribute to the overall shape of the line (Douglas and Peucker, 1973).

In the vector data model the representation of networks and surfaces is an extension of the approach used for storing line and area features. However, the method is more complex, and closely linked to the way the data are structured for computer encoding. Therefore, the approach used to store network and surface features is considered further in the following sections.

Spatial data structures

Data structures provide the information that the computer requires to reconstruct the spatial data model in digital form. There are many different data structures in use in GIS. This diversity is one of the reasons why exchanging spatial data between different GIS software can be problematic. However, despite this diversity data structures can be classified according to whether they are used to structure raster or vector data.

Raster data structures

In the raster world a range of different methods is used to encode a spatial entity for storage and representation in the computer. Figure 3.6 shows the most straightforward method of coding raster data. The cells in each line of the image (Figure 3.6a) are mirrored by an equivalent row of numbers in the file structure (Figure 3.6c). The first line of the file tells the computer that the image consists of 10 rows and 10 columns and that the maximum cell value is 1. In this example, a value of 0 has been used to record cells where the entity is not present, and a value of 1 for cells where the entity is present (Figure 3.6b).

In a simple raster data structure, such as that illustrated in Figure 3.6, different spatial features must be stored as separate data layers. Thus, to store all five of the raster entities shown in Figure 3.4, five separate data files would be required, each representing a different layer of spatial data.

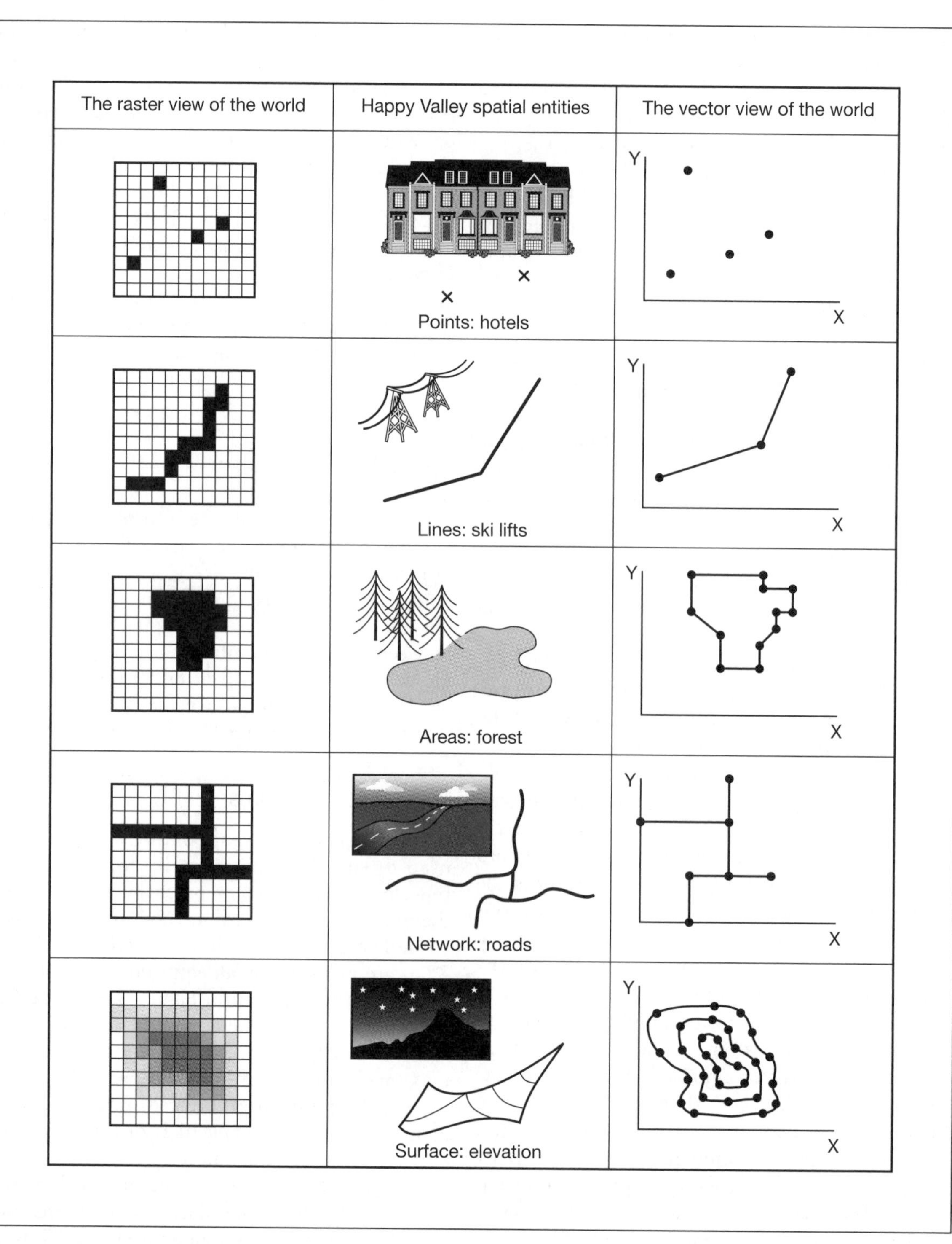

Figure 3.4 Raster and vector spatial data models

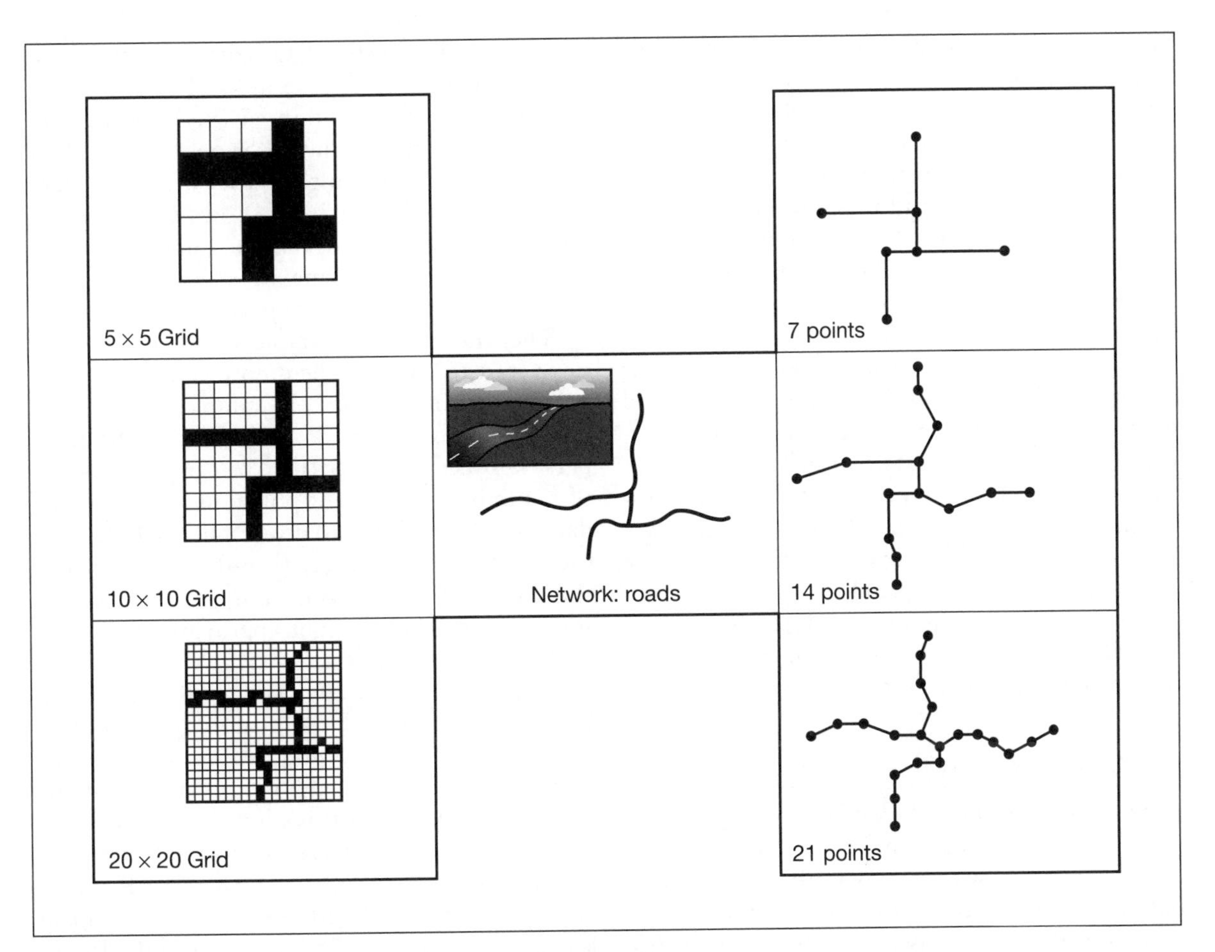

Figure 3.5 Effect of changing resolution in the vector and raster worlds

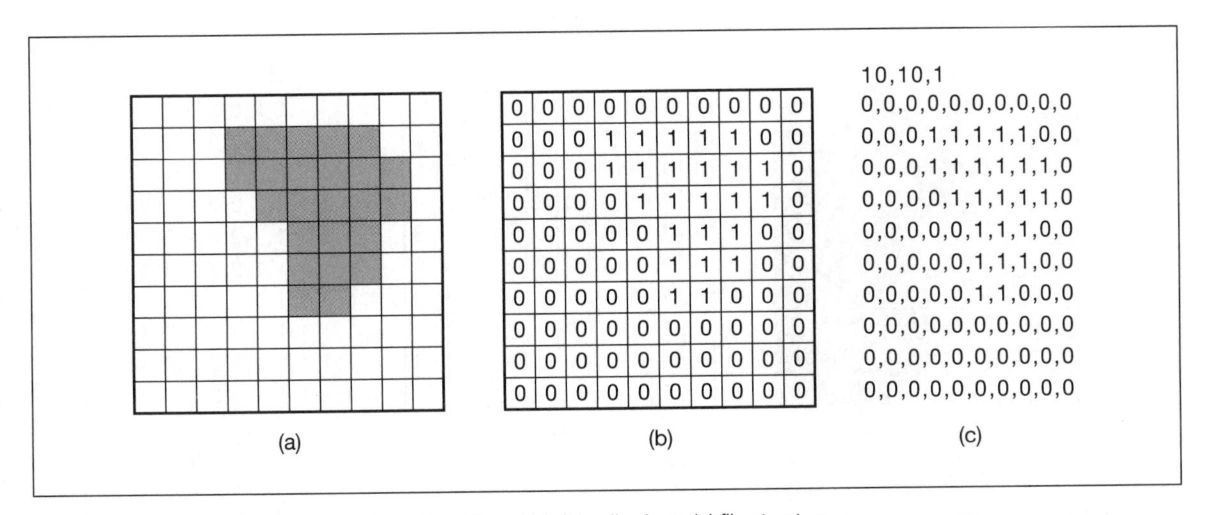

Figure 3.6 A simple raster data structure: (a) entity model; (b) cell values; (c) file structure

However, if the entities do not occupy the same geographic location (or cells in the raster model), then it is possible to store them all in a single layer, with an entity code given to each cell. This code informs the user which entity is present in which cell. Figure 3.7 shows how different land uses can be coded in a single raster layer. The values 1, 2 and 3 have been used to classify the raster cells according to the land use present at a given location. The value 1 represents residential area; 2, forest; and 3, farmland.

One of the major problems with raster data sets is their size, because a value must be recorded and stored for each cell in an image. Thus, a complex image made up of a mosaic of different features (such as a soil map with 20 distinct classes) requires the same amount of storage space as a similar raster map showing the location of a single forest. To address this problem a range of data compaction methods have been developed. These include run length encoding, block coding, chain coding and quadtree data structures (Box 3.2).

Vector data structure

Figure 3.4 shows how the different entity types – points, lines and areas – can be defined by co-ordinate geometry. However, as with the raster spatial data model, there are many potential vector data structures that can be used to store the geometric representation of entities in the computer.

The simplest vector data structure that can be used to reproduce a geographical image in the computer is a file containing (x,y) co-ordinate pairs that represent the location of individual point features (or the points used to construct lines or areas). Figure 3.10a shows such a vector data structure for the Happy Valley car park. Note how a closed ring of co-ordinate pairs defines the boundary of the polygon.

The limitations of simple vector data structures start to emerge when more complex spatial entities are considered. For example, consider the Happy Valley car park divided into different parking zones (Figure 3.10). The car park consists of a number of adjacent polygons. If the simple data structure, illustrated in Figure 3.10a, were used to capture this entity then the boundary line shared between adjacent polygons would be stored twice. This may not appear too much of a problem in the case of this example, but consider the implications for a map of the 50 states in the USA. The amount of duplicate data would be considerable. This method can be improved by adjacent polygons sharing common co-ordinate pairs (points). To do this all points in the data structure must be numbered sequentially and contain an explicit reference which records which points are associated with which polygon. This is known as a point dictionary (Burrough, 1986). The data structure in Figure 3.10b shows how

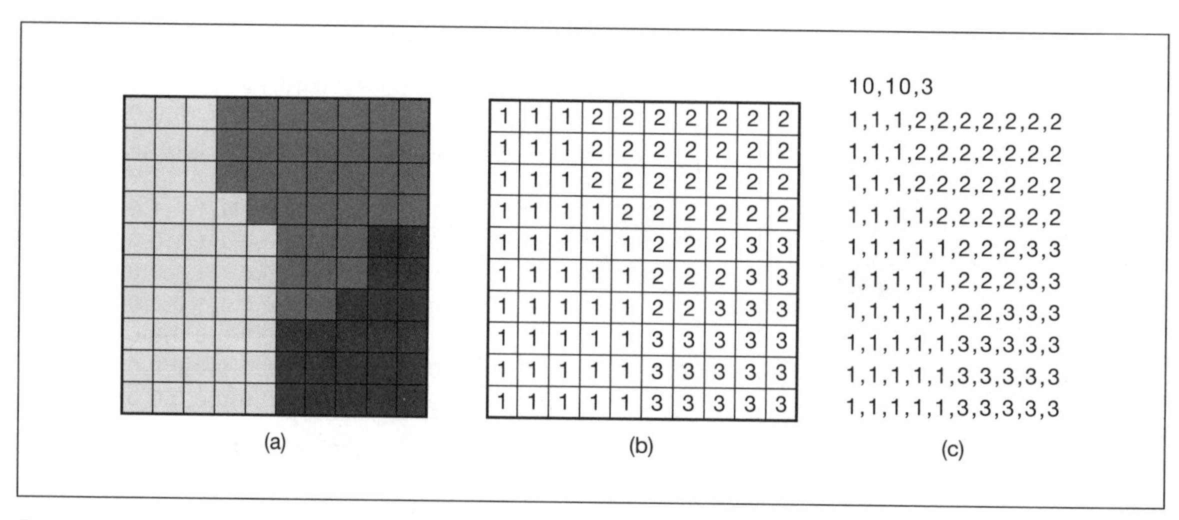

Figure 3.7 Feature coding of cells in the raster world: (a) entity model; (b) cell values; (c) file structure

BOX 3.2

Raster data compaction techniques

1 *Run length encoding*. This technique reduces data volume on a row by row basis. It stores a single value where there are a number of cells of a given type in a group, rather than storing a value for each individual cell. Figure 3.8a shows a run length encoded version of the forest in Happy Valley. The first line in the file represents the dimensions of the matrix (10 × 10) and number of entities present (1). In the second and subsequent lines of the

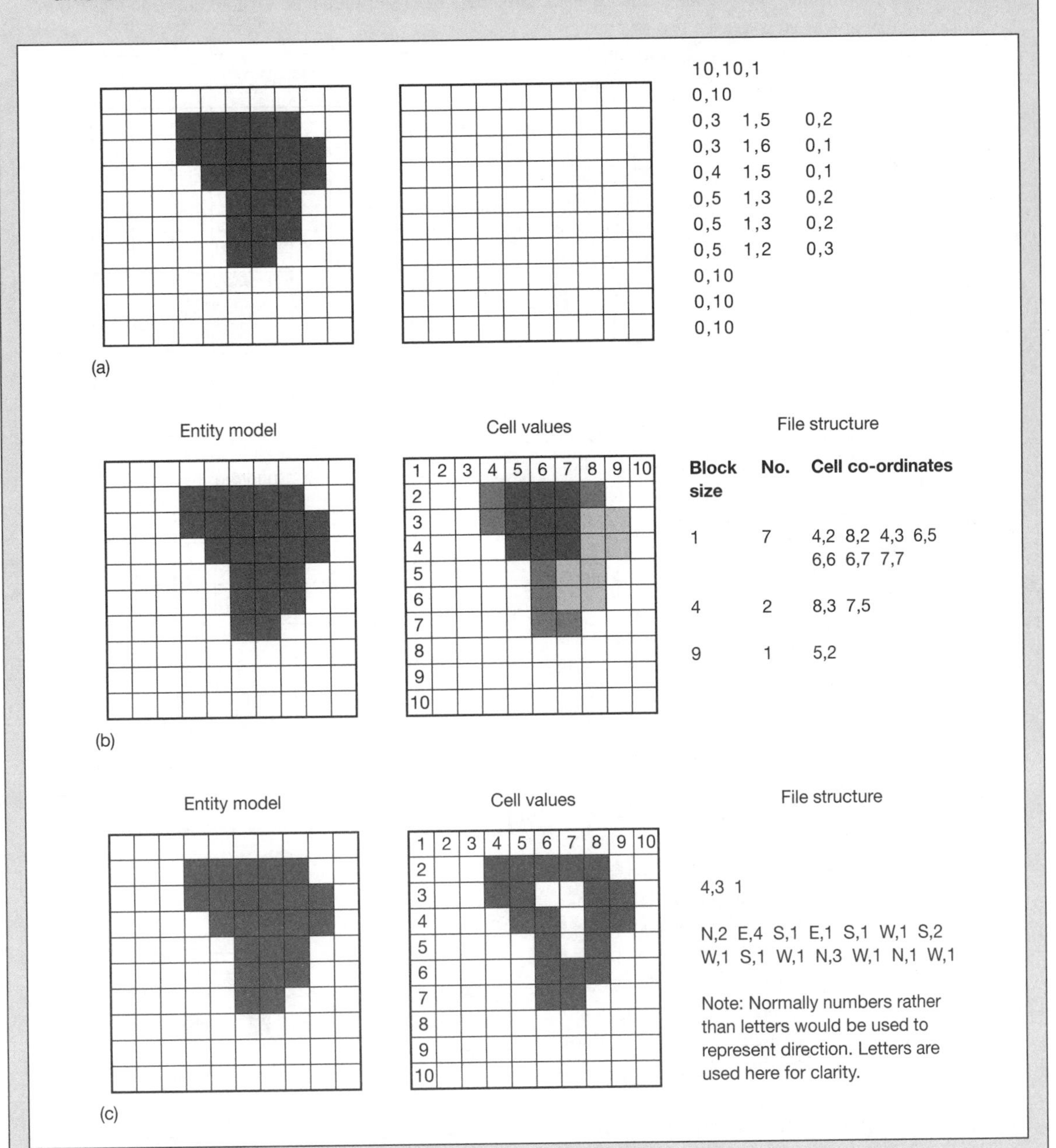

Figure 3.8 Raster data compaction techniques: (a) run length encoding; (b) block encoding; (c) chain encoding

BOX 3.2 CONTINUED

file, the first number in the pair (either 1 or 0 in this example) indicates the presence or absence of the forest. The second number indicates the number of cells referenced by the first. Therefore, the first pair of numbers at the start of the second line tell us that no entity is present in the first 10 cells of the first row of the image.

2 *Block coding*. This approach extends the run length encoding idea to two dimensions by using a series of square blocks to store data. Figure 3.8b shows how the simple raster map of the Happy Valley forest has been subdivided into a series of hierarchical square blocks. Ten data blocks are required to store data about the forest image. These are seven unit cells, two four-cell squares and one nine-cell square. Co-ordinates are required to locate the blocks in the raster matrix. In the example, the top left-hand cell in a block is used as the locational reference for the block.

3 *Chain coding*. The chain coding method of data reduction works by defining the boundary of the entity. The boundary is defined as a sequence of unit cells starting from and returning to a given origin. The direction of travel around the boundary is usually given using a numbering system (for example 0 = North, 1 = East, 2 = South and 3 = West). Figure 3.8c shows how the boundary cells for the Happy Valley forest would be coded using this method. Here, the directions are given letters (N, S, E and W) to avoid misunderstanding. The first line in the file structure tells us that the chain coding started at cell 4,3 and there is only one chain. On the second line the first letter in each sequence represents the direction and the number of cells lying in this direction.

4 *Quadtrees*. One of the advantages of the raster data model is that each cell can be subdivided into smaller cells of the same shape and orientation (Peuquet, 1990). This unique feature of the raster data model has produced a range of innovative data storage and data reduction methods that are based on regularly subdividing space. The most popular of these is the area or region quadtree. The area quadtree works on the principle of recursively subdividing the cells in a raster image into quads (or quarters). The subdivision process continues until each cell in the image can be classed as having the spatial entity either present or absent within the bounds of its geographical domain. The number of subdivisions required to represent an entity will be a trade-off between the complexity of the feature and the dimensions of the smallest grid cell. The quadtree principle is illustrated in Figure 3.9.

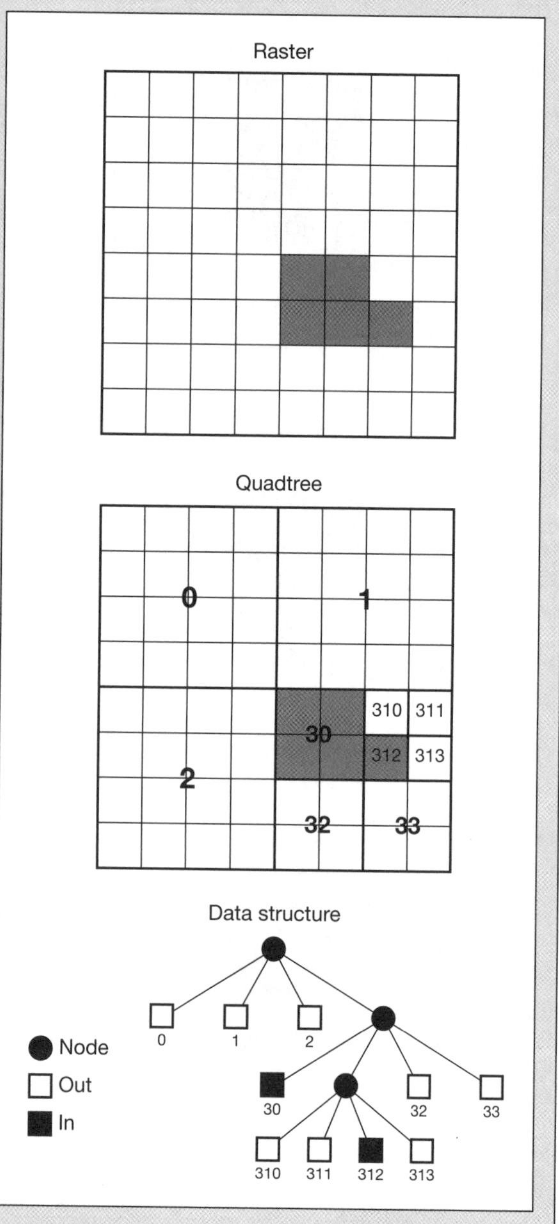

Figure 3.9 The quadtree

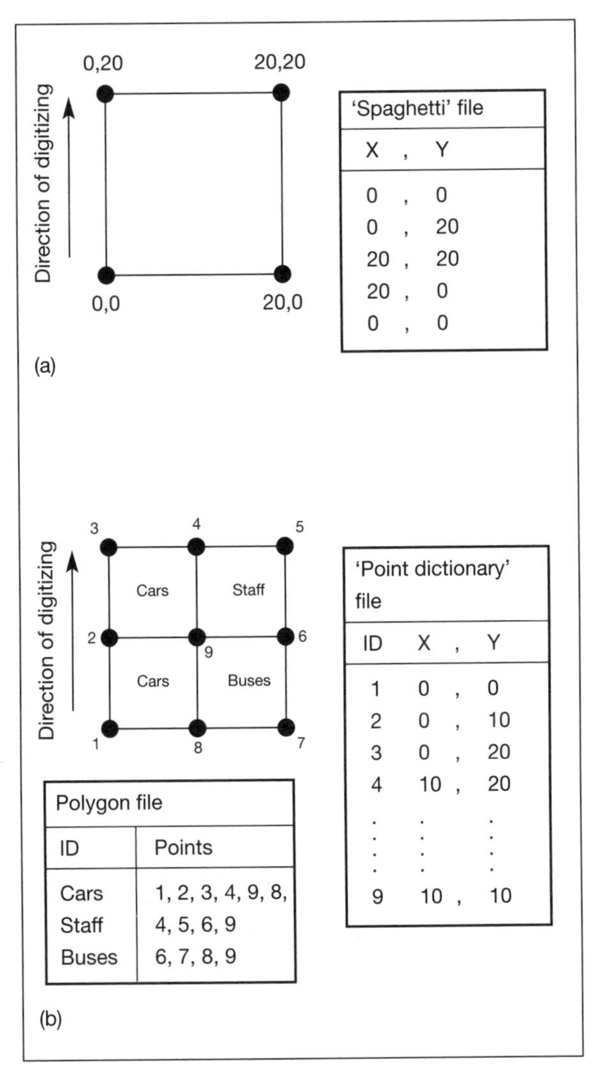

'Spaghetti' file	
X	Y
0	0
0	20
20	20
20	0
0	0

'Point dictionary' file		
ID	X	Y
1	0	0
2	0	10
3	0	20
4	10	20
:	:	:
9	10	10

Polygon file	
ID	Points
Cars	1, 2, 3, 4, 9, 8,
Staff	4, 5, 6, 9
Buses	6, 7, 8, 9

Figure 3.10 Data structures in the vector world: (a) simple data structure; (b) point dictionary

such an approach has been used to store data for the different zones in the Happy Valley car park.

The Happy Valley road network (Figure 3.5) illustrates a slightly different problem. The simple vector data structure illustrated in Figure 3.10a could be used to graphically reproduce the network without any duplication of data. However, it would not contain any information about the linkage between lines. Linkages would be implied only when the lines are displayed on the computer screen. In the same way, a series of polygons created using either the simple data structure described in Figure 3.10a or a point dictionary approach (Figure 3.10b) may appear connected on the screen when in fact the computer sees them as discrete entities, unaware of the presence of neighbouring polygons.

A further problem of area features is the island or hole situation. In Figure 3.11 a diamond-shaped area has been placed in the centre of the Happy Valley car park to represent the information kiosk. This feature is contained wholly within the polygons classified as the car park. While a simple vector file structure would recreate the image of the car park, it would not be able to inform the computer that the island polygon was 'contained' within the larger car park polygon.

For the representation of line networks, and adjacent and island polygons, a set of instructions is required which informs the computer where one polygon, or line, is with respect to its neighbours. Topological data structures contain this information. There are numerous ways of providing topological structure in a form that the computer can understand. The examples below have been selected to illustrate the basic principles underpinning topological data structures rather than describe the structures found in any one GIS environment.

Topology is concerned with connectivity between entities and not their physical shape. A useful way to help understand this idea is to visualize individual lines as pieces of spaghetti. Consider how you would create a model of the Happy Valley car park (Figure 3.11) using strands of spaghetti on a dinner plate. It is most likely that you would lay the various strands so that they overlapped. This would give the appearance of connectivity, at least while the spaghetti remained on the plate. If you dropped the plate it would be difficult to rebuild the image from the spaghetti that fell on the floor. Now imagine that the pieces of spaghetti are made of string rather than pasta and can be tied together to keep them in place. Now when the plate is dropped, the connectivity is maintained and it would be easier to reconstruct the model. No matter how you bend, stretch and twist the string, unless you break or cut one of the pieces, you cannot destroy the topology. It is this situation that computer

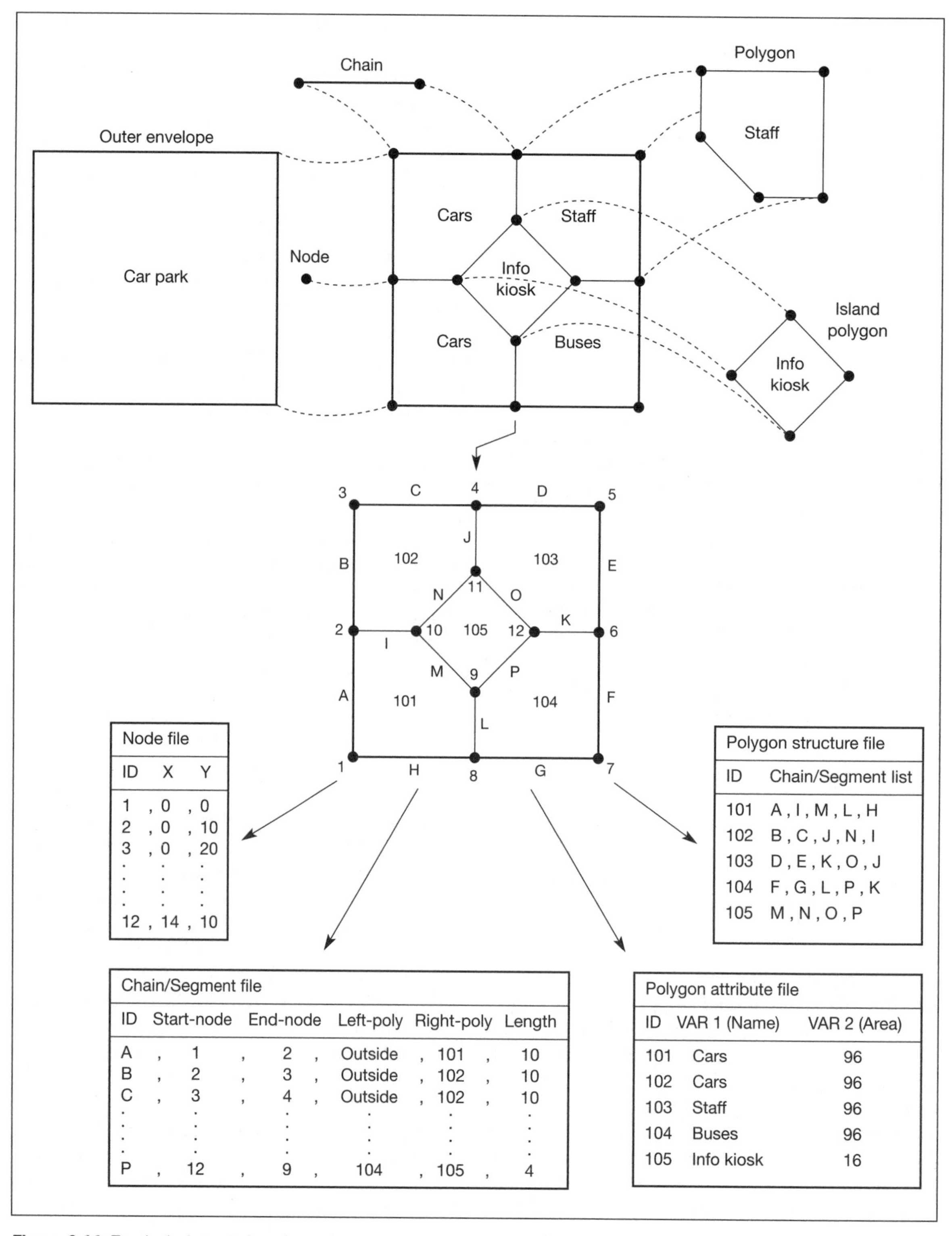

Figure 3.11 Topological structuring of complex areas

programmers are striving to mirror when they create topological data structures for the storage of vector data. The challenge is to maintain topology with the minimum of data to mimimize data volumes and processing requirements.

A point is the simplest spatial entity that can be represented in the vector world with topology. All a point requires to be topologically correct is a pointer, or geographical reference, which locates it with respect to other spatial entities. In order for a line entity to have topology it must consist of an ordered set of points (known as an arc, segment or chain) with defined start and end points (nodes). Knowledge of the start and end points gives line direction. For the creation of topologically correct area entities, data about the points and lines used in its construction, and how these connect to define the boundary, are required.

There is a considerable range of topological data structures in use by GIS. All the structures available try to ensure that:

- no node or line segment is duplicated;
- line segments and nodes can be referenced to more than one polygon;
- all polygons have unique identifiers; and
- island and hole polygons can be adequately represented.

Figure 3.11 shows one possible topological data structure for the vector representation of the Happy Valley car park. The creation of this structure for complex area features is carried out in a series of stages. Burrough (1986) identifies these stages as identifying a boundary network of arcs (the envelope polygon), checking polygons for closure and linking arcs into polygons. The area of polygons can then be calculated and unique identification numbers attached. This identifier would allow non-spatial information to be linked to a specific polygon.

This section has briefly considered a number of different approaches to the structuring of raster and vector data. It is important to remember that the examples presented here represent only a few of the data structures used by different GIS. There are many variations on these methods.

Modelling surfaces

The modelling of surface entities such as height, pollution and rainfall poses interesting problems in GIS. In this section we explore surface modelling by looking in detail at digital terrain models (DTM). These are used by way of an example because of their wide application in GIS.

The abbreviation DTM is used to describe a digital data set which is used to model a topographic surface (a surface representing height data). To model a surface accurately it would be necessary to store an infinite number of observations. Since this is impossible, a surface model approximates a continuous surface using a finite number of observations. Thus, an appropriate number of observations must be selected, along with their geographical location. The 'resolution' of a DTM is determined by the frequency of observations used. DTMs are created from a series of either regularly or irregularly spaced (x,y,z) data points (where x and y are the horizontal co-ordinates and z is the vertical or height co-ordinate). DTMs may be derived from a number of data sources. These include contour and spot height information found on topographic maps, stereoscopic aerial photography, satellite images and field surveys (see Box 3.3).

The raster approach to digital terrain modelling

There are differences between the way in which DTMs are constructed in raster-based and vector-based GIS. In raster GIS a DTM is simply a grid of height values in which each cell contains a single value representative of the height of the terrain that is covered by that cell. How accurately the terrain can be modelled using this method depends on the complexity of the terrain surface and the spacing (resolution) of the grid. If the terrain is complex, for example in a mountainous area like the European Alps or the North American Rocky Mountains, a fine grid will be required to represent it with any degree of accuracy. If the terrain is relatively even, such as in the Netherlands or the Great Plains of the USA and Canada, fewer points will be needed to model it to the same degree of accuracy. Thus, a coarser-resolution DTM may be used.

BOX 3.3

Data sources for DTMs

Contours and spot heights from topographic maps are the most likely source of raw data for a DTM. These may be supplied in digital form for the express purpose of building a DTM, or they may need to be digitized from a paper map (see Chapter 5). The digital set of contours and spot heights represents an irregular set of (*x*,*y*,*z*) observations. The digital contours are line features representing lines on the terrain surface of equal height. These can be built into a set of nested polygons. The attributes of these polygons are height ranges as described by the bounding line features of which the polygons are made. Digital contour data are often supplemented by spot heights, which give the height at specific points between the contour lines. These are used to give the location and height of significant terrain features such as peaks, depressions or isolated knolls.

Height data may also be derived from the stereoscopic interpretation of paired images of the Earth's surface. The paired images may be either aerial photographs or satellite images that have been taken of the same patch of ground but from slightly different angles. The technique relies on being able to calculate elevation based on the parallax displacement between the same point on both images. If done for many different points, a DTM may be built up for the area covered by the paired images. A recent example of this technique has been the introduction of detailed DTMs derived from high-resolution SPOT imagery.

GPS receivers give accurate (*x*,*y*) co-ordinate locations and provide height information. They are not as accurate in the vertical plane as they are in the horizontal, though height data may be used in conjunction with (*x*,*y*) co-ordinate data to produce a DTM if the user is prepared to take readings all over the area of interest.

Some examples of data source selection criteria for DTMs are presented in Table 3.1 (adapted from Stocks and Heywood, 1994).

Table 3.1 DTM data source selection criteria

Data source	*Capture method*	*DTM accuracy*	*Coverage*
Maps	1 Digitizing 2 Scanning	Low–moderate	All scales, but only where map coverage exists
Aerial photographs	Stereo plotters	High	Large areas
Satellite imagery	Stereo autocorrelation	Moderate	Large areas
GPS	Direct downloading from GPS receiver	High	Small project areas with moderate terrain
Ground survey	Direct downloading from field surveying equipment	Very high	Very small project areas, not feasible in rough terrain

Problems occur with raster DTMs where the terrain is varied. For example, if the area to be modelled includes an area of complex mountainous terrain to the north and a wide, gently sloping area to the south, choosing an appropriate resolution is difficult. A fine-resolution raster will accurately model the mountains in the north, but a high degree of data redundancy will be experienced in flat areas to the south. This could produce problems of large data storage requirements and long computing times. If you choose a low-resolution raster such as might be suitable for the flat southern part of the area, the detail of the mountains to the north will be lost. The final choice will inevitably be a compromise.

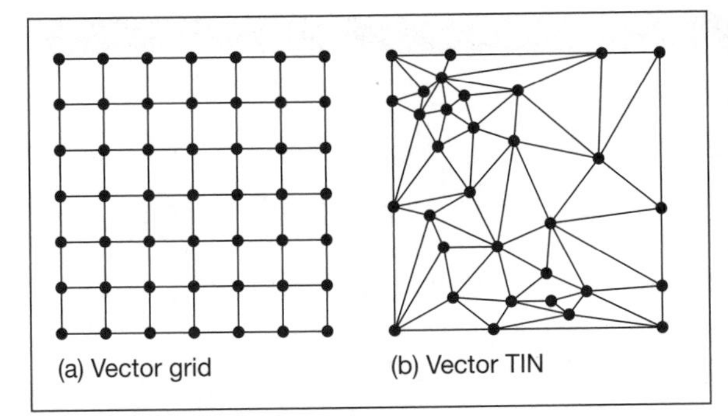

Figure 3.12 Digital terrain models: (a) vector grid; (b) vector TIN

The vector approach to digital terrain modelling

In its simplest form a vector DTM mimics the raster version by using a regularly spaced set of spot heights to represent the terrain surface (Figure 3.12a). A more advanced, more complex and more common form of vector DTM is the triangulated irregular network (TIN). In vector GIS a TIN is used to create a DTM from either regular or irregular height data. The TIN method joins the height observations together with straight lines to create a mosaic of irregular triangles. In the TIN model of a surface, the vertices of the triangles produced represent terrain features such as peaks, depressions and passes, and the edges represent ridges and valleys. The surfaces of individual triangles provide area, gradient (slope) and orientation (aspect). These values can be stored as TIN attributes or can be quickly calculated when the TIN is used in further analysis. An example of a TIN model is shown in Figure 3.12b and in Plate 13. A physical example is described in Box 3.4.

The main advantage of the TIN model is efficiency of data storage, since only a minimum number of significant points need to be stored to reproduce a surface. In the previous section we considered the difficulties of choosing an appropriate resolution for a raster DTM in an area of mixed terrain. Since TINs are created from irregularly spaced points, more points are used to represent the mountainous areas and fewer to represent the flatter areas.

To achieve this efficiency in storage, TIN models use only 'surface significant' points to reproduce a terrain surface. These points are selected by the TIN model from the input data on the basis of their spatial relationship with their neighbours (Lee, 1991). As in spatial interpolation (see Chapter 6), Tobler's law of geography (1976) regarding spatial autocorrelation comes into play. This states that points closer together are more likely to have similar characteristics (here in terms of their height) than those that are further apart. Consequently, if a height observation can be closely interpolated from its neighbours then it is not considered to be 'surface significant' and is dropped from the TIN model. Those points that cannot be closely interpolated from their neighbours are considered to be 'surface significant' and are used as TIN vertices. Box 3.5 looks at four of the most common methods used to identify which points are surface significant.

The methods used to identify a set of surface significant points (except for the hierarchy method)

BOX 3.4

Make your own TIN model

A simple experiment you can try yourself is to construct a TIN model using a sheet of paper. Take a sheet of paper and crumple it into a loose ball. Now carefully open it out again so that it resembles a sheet of paper again but still has all the creases you have just put into it. This is your TIN model. The creases are the ridges and valleys and the intersections of the creases are the peaks, depressions and passes. The areas of flat paper between the creases are the irregular triangles of the TIN model, which may be assigned area, slope and aspect values. Obviously this is not a perfect model as not all of the facets on your piece of paper will be triangles, but it does serve to illustrate the TIN principle.

Now hold the paper level with a light source, such as a window or desk lamp, to dramatize the effect. With a little imagination you can see a miniature terrain of peaks and valleys casting shadows in front of you. The tighter you crumple your paper the more complex the terrain you produce and the more you try to flatten it out the lower the relief. This experiment also demonstrates the 'two and a half' dimensional nature of terrain models in GIS. You can see clearly that the model you have just created is only a surface with no depth.

BOX 3.5

Methods for identifying surface significant points

The four most common methods used to identify surface significant points in the construction of a TIN are as follows.

1. *The skeleton method.* This method, developed by Fowler and Little (1979), is used to select surface significant points from a regular grid of height data using a 3 × 3 cell window. The window is passed over the data one cell at a time and the height value of the centre cell compared with its neighbours. A cell is labelled a peak if all the neighbouring cells in the window are lower; a pit (depression) if all the neighbouring cells are higher; and a pass if the heights of the neighbouring cells alternate lower and higher around the centre cell. Next, a 2 × 2 window is passed over the data. Cells in the 2 × 2 window that are highest are labelled as ridges and cells that are lowest are labelled as valleys. Peaks are then linked via ridges, and pits via valleys, to define the skeleton of the terrain. A line-thinning algorithm should then be used to identify the most crucial points from which to construct the TIN. This sounds simple enough but is a complex, computer-intensive, algorithm which is sensitive to noise in the data since it considers only the relative height differences between neighbouring cells.
2. *The filter or VIP.* The Very Important Point (VIP) method was developed by Chen and Guevara (1987) to identify points that define the character of a terrain (very important points). VIPs are determined by how well they can be approximated by their neighbours. If the predicted value of the central point is substantially different from its actual value then the point is considered to be a VIP and an essential component of the TIN. Points that are diametrically opposite the central point are connected by a straight line and the perpendicular distance of the central point in the third dimension calculated. This procedure is repeated for other diametrically opposed points and the average distance calculated. This average is then used as a measure of importance or significance of the point. The advantage of this method is that it is easy to compute and requires only one pass through the data. The number of points selected can also be specified by setting a 'significant threshold' at which points are considered to be VIPs.
3. *The hierarchy method.* This method, developed by DeFloriani *et al.* (1984), selects surface significant points by subdividing the total set of points into a series of successively smaller triangles in a manner similar to the quadtree data structure. The method evaluates all points located inside the first-level triangle by calculating the difference between their original elevations and the elevations interpolated from the enclosing triangle. This difference is used as a measure of the error associated with the fit of the TIN model to each point at this first level of triangulation. The point with the largest error is identified and if this is greater than a predefined level of precision it is used as the node by which to further subdivide the first-level triangle into three smaller triangles. This process continues until all triangles in the TIN meet the prespecified precision level. Advantages of this method include greater subdivision and smaller triangles in areas of more complex terrain and an efficient data structure allowing fast retrieval and access to information in the TIN. The main disadvantage is that the TIN inevitably consists of many long, thin triangles and the original first-level subdivision remains in the final TIN model. TIN models work better when the triangles are close to being equilateral (with angles of 60°) since this ensures that any point on the surface is as close as possible to a TIN vertex.
4. *The drop heuristic method.* This method works by starting with a full set of points and gradually discarding those of least importance. After each point is dropped a new TIN is constructed and the triangle containing the dropped point identified. The difference between the true elevation of the dropped point and the estimated elevation of the point based on the new TIN is calculated and a predefined tolerance level used to determine whether the point should remain in the TIN. The advantage of this technique is that it produces a TIN that best reflects the surface represented by the raw data, but it can be computationally very intensive.

do not help determine which points to connect to which. One of two methods is generally used to achieve the linking of TIN vertices. The first of these is known as distance ordering, where the distance between all pairs of points is calculated. These are sorted and the closest pair connected. The next closest pair are then connected if the resulting line does not cross any existing connections. This process is repeated until the TIN is complete. However, this method can produce too many thin triangles. The second method is Delaunay triangulation. Delaunay triangles are determined by a circle. Three points form the corners of a Delaunay triangle only when the circle that passes through them contains no other points.

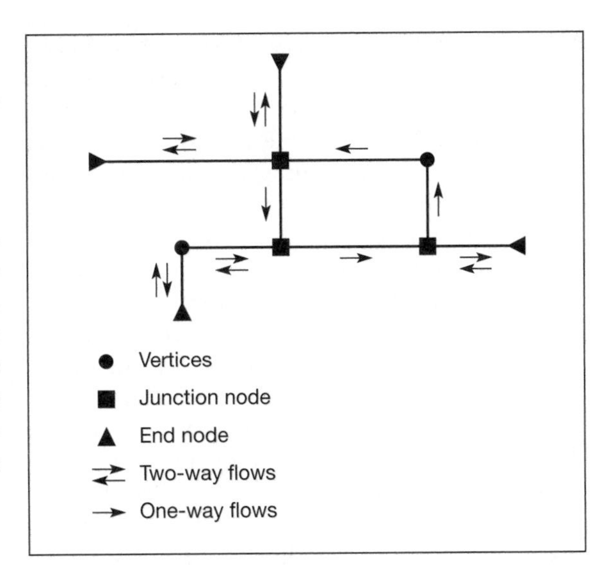

Figure 3.13 Network data model

Modelling networks

A network is a set of interconnected linear features through which materials, goods and people are transported or along which communication of information is achieved. Network models in GIS are abstract representations of the components and characteristics of their real-world counterparts. They are essentially adaptations of the vector data model and for this reason raster GIS are generally not very good at network analysis. The vector network model is made up of the same arc (line segments) and node elements as any other vector data model but with the addition of special attributes as illustrated in Figure 3.13.

In the network model the arcs become network links representing the roads, railways and air routes of transport networks; the power lines, cables and pipelines of the utilities networks; or the rivers and streams of hydrological systems. The nodes in turn become network nodes, stops and centres. Network nodes are simply the endpoints of network links and as such represent junctions in transport networks, confluences in stream networks, and switches and valves in utilities networks. Stops are locations on the network that may be visited during a journey. They may be stops on a bus route, pick-up and drop-off points on a delivery system, or sediment sources in a stream network. They are points where goods, people or resources are transferred to and from some form of transport system. Centres are discrete locations on a network at which there exists a resource supply or some form of attraction. Examples include shopping centres, airports, schools and hospitals. At a larger scale centres may be a whole city if the transport, resource or information networks for an entire country are being considered. Turns represent the transition from network link to network link at a network node. Turns therefore represent the relationships between network links and greatly affect movement through the network system. For example, turns across oncoming traffic on a road network take longer than turns down slipways, whereas turns that go against the flow of traffic on one-way streets are prohibited altogether.

All the data regarding the characteristics of network links, nodes, stops, centres and turns are stored as attribute information in the vector model database. Two key characteristics of network features are impedance and supply and demand.

Impedance is the cost associated with traversing a network link, stopping, turning or visiting a centre. For example, link impedance may be the time it takes to travel from one node to another along a network link. If we use the example of a delivery van travelling along a city street then the impedance value represents time, fuel used and the driver's pay. Factors influencing the impedance value will include traffic volume as determined by time of day and traffic control systems; direction (for instance, one-way streets); topography (more fuel is used going uphill); and weather (more fuel is used travelling into a strong headwind). Different links have different impedance values depending on local conditions. Turn impedance is also important and may be represented by the cost of making a particular turn. For example, when the delivery van reaches the end of the current link, the direction the driver chooses to turn may be strongly influenced by turn impedance. Again, stop impedance refers to the cost of making a stop and operates in a similar way to link and turn impedance. The delivery van driver may not be able to find a parking space and could get a parking ticket for stopping on double yellow lines. Impedance values are, therefore, very important in determining the outcome of route finding, allocation and spatial interaction operations. The various forms of impedance values that may be linked to a network are illustrated in Figure 3.14.

Supply and demand are equally important concepts in network analysis. Supply is the quantity of a resource available at a centre that is available to satisfy the demand associated with the links of a network. If we take the example of a hospital, then supply would be represented by the number of beds available. Demand, on the other hand, is the utilization of a resource by an entity that is associated with a network link or node. Using the hospital example again, demand is represented by the number of people requiring treatment living within the hospital's catchment area.

Correct topology and connectivity are extremely important for network analysis. Digital networks should be good topological representations of the real-world network they mimic. Correct geographical representation in network analysis is not so important, so long as key attributes such as impedance and distance are preserved. A classic example of this is the famous map of the London Underground system. This bears little resemblance to the real-world map of the underground system, which would be far too complex for underground users to follow. Instead the map was redrawn to simplify the network, making it easily understood whilst at the same time maintaining relative distance and connectivity between all the stations on the network.

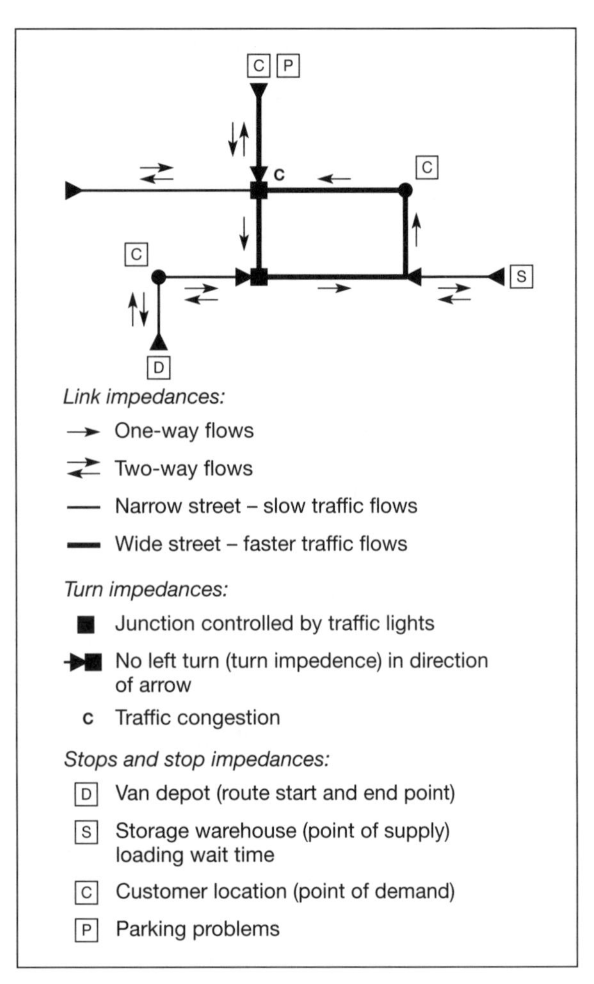

Figure 3.14 Link, turn and stop impedances affecting the journey of a delivery van

Building computer worlds

This chapter has explored the way in which different spatial data models (raster and vector) and their respective data structures are used to store entities. In this section we will consider the methods used to construct computer worlds by grouping these entities together. At the present time we have two options: layers and objects.

The most common method of structuring the geography of the real world in the computer is to use a layered approach (Laurini and Thompson, 1992). This layered approach is illustrated in Figure 3.15, in which Happy Valley has been dissected into a series of layers. Each layer is thematic and reflects either a particular use or a characteristic of the landscape. For example, layer one describes the general land use characteristics, layer two the soil type and layer three the hydrology. This layered approach to structuring spatial information has a long history characterized by the use of thematic maps to show different features for the same area. The map layer concept is helpful since it reminds us that many geographical features overlap with one another. For example, a village is built on the top of soil and geology. The concept of breaking the geography of the real world down into a series of thematic layers was the process that was used to develop the first map overlay analysis (McHarg, 1969).

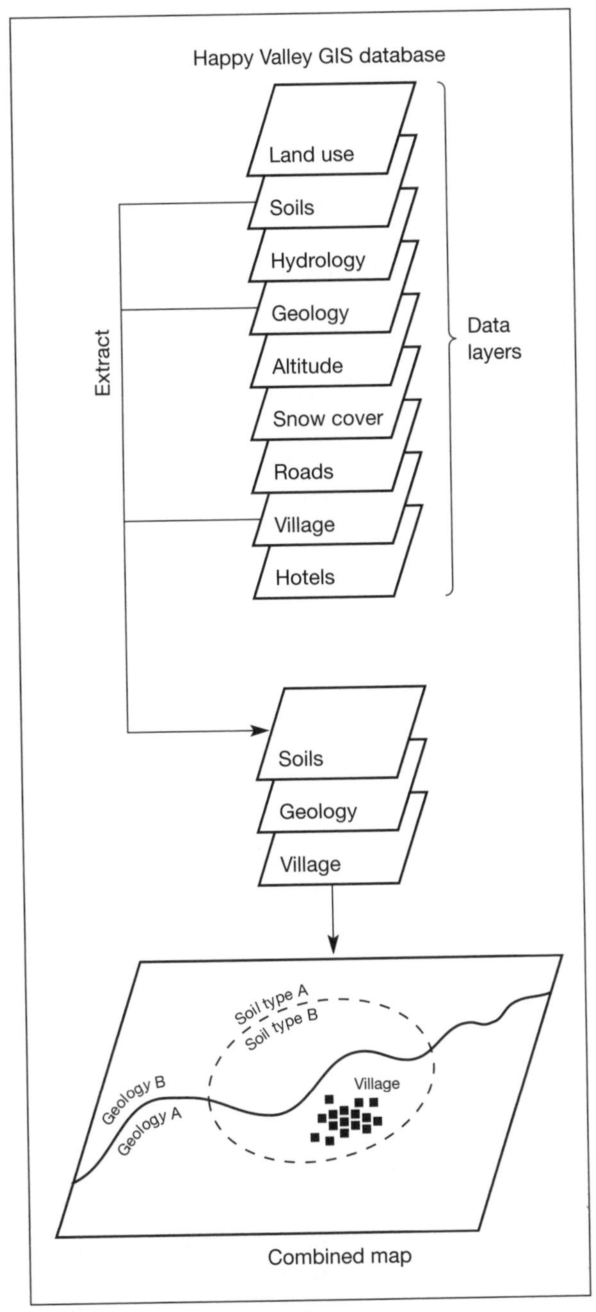

Figure 3.15 The layer-based approach

A logical extension to the layer concept is the use of tiles. This approach breaks down geographical space into a series of regular or irregular units that can be reassembled through the use of a coordinate system. The main purpose of the tiling concept is to assist with the storage of information in the computer. Spatial data sets are frequently large and many GIS systems require that data are broken down into a set of logical units to assist with physical display of data and retrieval and analysis. For example, a mosaic of individual map sheets may be used to provide geographical coverage of a country at a detailed scale.

Another, relatively new approach to structuring geographical space views the real world as a set of individual objects and groups of objects. This approach draws on the methods of object-oriented (OO) programming (Worboys, 1995). At the conceptual level, the object-oriented approach is straightforward. Figure 3.16 shows how such an approach might be used to structure some of the information for Happy Valley. Notice that features are not divided into separate layers but grouped into classes and hierarchies of objects. This has particular advantages for modelling the geography of the real world as it more accurately reflects the way we group spatial features using natural language (Frank and Mark, 1991). Adopting an object view of the world has many advantages over the more traditional layer-based approach. However, it does cause some problems associated with implementation into a workable GIS.

There are many different definitions of the object-oriented approach (Worboys *et al.*, 1991).

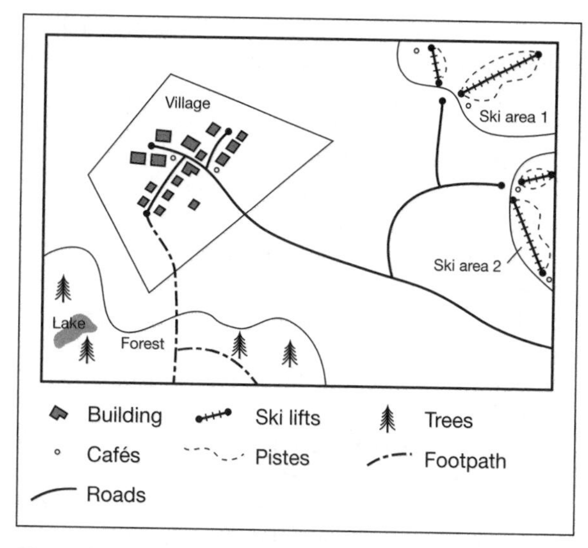

Figure 3.16 The object-oriented approach

However, all seem to have a similar aim: to bring the conceptual model of the modelling process closer to real-world things and events as opposed to the technicalities of computing (Reeve, 1997). Therefore, in theory at least, the OO approach to modelling should be easier to understand.

Consider the design of a GIS to assist in the management and maintenance of Happy Valley. Happy Valley consists of a collection of objects such as ski lifts, car parks and hotels. Each of these has unique characteristics: geographical (size, shape and location); topological (the ski slopes are adjacent to the car parks); related to the activities which go on within objects (hotels may be used for holidaymakers, private parties or conferences); and related to the behaviour of objects in relation to one another (the lifts are used to get the skiers to the pistes). The OO approach tries to retain as much as possible of the character of these objects and their relationships to each other.

Let us assume that a traditional GIS model of Happy Valley has been produced using the concepts discussed earlier in this chapter. The GIS comprises several different data layers made up of point, line and area entities. These spatial entities have specific attributes that inform the user what they are. In addition, the spatial entities have some geometric and topological information stored about them so that, for instance, the car park knows where it is and which entities are next to it. This is all the data that are recorded and can be derived and used through the application of GIS. The OO approach suggests that we should stop thinking about these real-world features in an abstract manner, but instead think of them as objects, about which much more data can be stored. For example, because real-world objects are usually classified into groups the OO model requests information about the classes to which an object belongs. This class membership can take one of two forms: that which reflects the real-world behaviour of the object; or that which provides information about how the object is to be treated in the computer world. Therefore, if we take the car park object as an example, we can indicate that our polygon is a car park and that it belongs to the class of objects known as ski resort.

In addition, entities may be grouped into classes according to their computer representation. Thus, car parks will also belong to the class of entity known to the computer as *areas*. Areas, in turn, are composed of another set of entities of the class *lines*. The beauty of defining things as objects and providing them with membership of a specific class is that it becomes much easier to perform a whole range of spatial and aspatial operations. Therefore, use of a command such as 'show all car parks' generates a picture of all the car parks in Happy Valley. A command that requested that a map of Happy Valley should be produced would automatically draw the car parks as part of Happy Valley.

In the discussion above we have only introduced the basics of an OO approach to GIS. The idea is revisited in Chapter 4 and if you wish to know more you should take a closer look at Worboys (1995).

Modelling the third dimension

So far in our discussion of computer-based models of the real world we have largely avoided any discussion of the importance of the third and fourth (time) dimensions. There is historical justification for this as most of the GIS modelling concepts explored in the preceding sections of this chapter have their origins in the techniques used for the design and construction of paper maps. Chapter 2 demonstrated how the map metaphor has shaped the way GIS views the world. Thus, most GIS tend to take a two-dimensional perspective on the world at a particular time. This two-dimensional (2D) snapshot view of the world has, to some extent, constrained our thinking about how spatial information should be structured in the computer.

In terms of its display capabilities the computer screen is a two-dimensional display device even though, through the use of clever graphics, it is possible to simulate the appearance of the third dimension. However, technically this is nothing more than good perspective drawing. This approach has become known by the somewhat tongue-in-cheek term 'two and a half' dimensions (2.5D). Classic uses of the 2.5D approach are wire frame diagrams to show relief, an example of which is shown in Plate 1.

All the features we are trying to model have a third dimension. To produce systems capable of representing the complexities of the real world we need to portray the third dimension in more than a visual way. The representation of the third dimension of an entity can also help us model the form of an entity and associated spatial processes. For example, a 3D scale model could be used for visualization of a proposed extension to the Happy Valley ski area. The use of the third dimension in this model is crucial to allow people to evaluate the potential impact the development will have on the landscape. Unfortunately, current GIS software products do not permit us to model the third dimension. Partly this is due to the technical difficulties of constructing full three-dimensional models of geographical space. Those including full three-dimensional topology are particularly difficult. Imagine, for example, constructing a three-dimensional quadtree – an octree, as referred to by Raper and Kelk (1991) and Worboys (1995) – or providing full topology for a set of interconnecting three-dimensional polygons. It would appear that until recently computer power has been insufficient to convince the major GIS software developers that investing in such data models was worthwhile. However, as Raper and Kelk (1991) identify, work has been taking place in several sectors, notably the geoscientific, civil engineering and landscape architecture disciplines. Worboys (1995) describes how the simple raster grid cell can be extended into a 3D representation, known as a voxel. The problem with the voxel is that it is unable to record topological information. Worboys suggests that the solution to this problem lies in extending the vector approach by using constructive solid geometry to create three-dimensional objects such as cubes, spheres and cylinders.

As Raper and Kelk (1991) pointed out, it is only a matter of time before we will see the emergence of 3D modelling as an integral part of the GIS toolbox. To some extent this has already begun to happen. For example, Jacobsen *et al.* (1993) described a new three-dimensional data structure employed to improve modelling and understanding of the behaviour of glaciers.

Modelling the fourth dimension

A closer look at virtually any GIS database will reveal that the data layers represent no more than a collage of the state of various entities at a given time. This, in many ways, has much in common with a photograph as a GIS database is often a record of the state of entities or groups of entities at a particular time. If we are lucky, our GIS collage will be made up of entities captured at the same time. However, as the production of a GIS is often a long-term process it is more than likely that this collage of data will include entities at different periods in time. Modelling time is made more complex since there are several different sorts of time that GIS developers need to consider (Langran, 1992): work practice time, database time and future time.

Work practice time is the temporal state of a GIS database used by many people. In this situation, there may be several different versions of the GIS in existence at any one moment in time with several people (perhaps in different offices) working to update information. In a planning context the situation could arise where the same land parcel (area entity) is being edited by two people at the same time.

Database time is the period for which the database is thought to be correct. Inevitably, this lags behind real time owing to the data capture process. Problems with database time occur if there is more than one version of the database in existence and no procedures to govern who can update it and when.

Many applications of GIS seek to model future time. Applications such as the forecasting of avalanches in Happy Valley require predictions to be made. This can be a complex problem since there may be several possible alternative scenarios. For example, a GIS used to model avalanches under different climatic conditions will yield several different future images that the GIS needs to handle.

The problems with developing a data structure that is flexible enough to handle the time elements of entities include *what* to include, *how frequently* to update the information and *when* to stop storing old information. In addition, the problem is complicated as an organization may need its GIS to be accessible by individuals at different geographic locations, and these individuals may make changes to the time status of an entity. To date, most of these problems are still within the research domain of GIS. Most users of GIS represent the different temporal states of a spatial entity either by creating a separate data layer for each time period or by recording the state of the entity at a given time as a separate field in the database (see Chapter 4 and Box 3.6).

BOX 3.6

Time and GIS

Integrating the dimension of time into GIS presents challenges. There are two main reasons for this. First, it is still rare for the GIS analyst to be in a position where data about a spatial object are available for a continuous period (though this position is changing – see Chapter 13). Second, data models and structures that allow us to record, store and visualize information about an object in different temporal states are still in their infancy. This problem is bad enough when the geographic entity under investigation is fixed with respect to location, but is considerably more complex when the object is either mobile or changes its entity type through time.

Peuquet (1999) suggests that there are four types of temporal event. This provides an indication of the types of changes that may affect an entity:

- Continuous – these events go on throughout some interval of time.
- Majorative – these events go on most of the time.
- Sporadic – these events occur some of the time.
- Unique – events that occur only once.

For a skier visiting Happy Valley, skiing would probably represent the majorative event during a single day, with visits to cafes and bars being sporadic events and falling being (hopefully) a unique event.

Handling time in GIS

In a raster or vector layer-based GIS, one option for handling time is to store multiple layers for the theme in which you are interested. For example, in the case of the skier a layer could be stored that shows the location of the skier at a particular time. As a number of layers build up for different times, each layer becomes a little like the individual frame in a movie. The problem with this approach is the need for a lot of duplicate data. One solution is to store only information that changes to reduce the data storage requirements.

In a GIS using an object-oriented data model, you can use a slightly different approach. The various elements and attributes which make up an object can each be given a time tag (Worboys, 1995). When the GIS analyst explores the database, the position of the skier changes as the 'time of view' is changed. The presence or absence of the skier is switched on or off at a given location in relation to the time tag. For example, if you choose to view a map of Happy Valley at 1630 hours, when the skier is in the bar, they are linked with the bar entity. When a map of 'building occupancy' is generated for this time the skier is aggregated into the appropriate count.

Much remains to be done before true temporal GIS can be realised (Peuquet, 1999), but a large part of the solution may lie in exploring how the temporal dimension of information has been dealt with by other disciplines.

Conclusions

This chapter has taken a detailed look at how to model spatial form in the computer. We have reviewed how the basic spatial entities (points, lines and areas) can be extended to include two other entity types (networks and surfaces) to permit the modelling of more complex spatial features. In addition, the use of the raster and vector approaches in spatial modelling have been examined. Finally, modelling of three- and four-dimensional data has been introduced.

Although few current GIS can handle three- and four-dimensional data, or even objects, most can now handle raster and vector data in two dimensions. It is useful at this point to remember that raster and vector data structures are two alternative methods for storing and representing geographical phenomena. As models they both have strengths and weaknesses in terms of:

- their data storage requirements;
- the faithfulness with which they reflect the character and location of the real-world features;
- the ability to change the scale of visual representation; and
- the analytical opportunities they present.

As a rule, raster data structures generally demand more data storage than their vector counterparts. However, the more complex a spatial entity becomes, the closer the data volume requirements of the different data structuring techniques. During the 1970s and first half of the 1980s the data storage issue was at the forefront of the raster versus vector debate. However, with the increase in the processing power of the desktop PC and the falling price of data storage, the data volume debate is becoming less of an issue.

The world is not made up of cells, or co-ordinates joined by straight line segments. Therefore, any data model (raster or vector) in GIS will only be an approximation of reality. However, in certain circumstances one model may give a better representation than another. For example, the road and river network in Happy Valley is better represented using a vector data model. On the other hand, where you have high spatial variability in the feature being modelled a raster model may provide a better representation. Vegetation maps, where boundaries are difficult to identify and there is rapid and frequent change between vegetation types, may be better represented by a raster model.

The vector spatial data model handles changes in the scale of visual representation much more easily than its raster counterpart. This is because of the precise way in which data are recorded as a set of (x,y) co-ordinates. Changes of scale pose a problem in the raster world if a resolution is requested below the size of the cell specified at the outset of the project. Changes in scale in the raster world are often typified by the appearance of a blockier image.

There is a large distinction between the analytical capabilities of raster and vector GIS. Traditionally, vector spatial data models have been considered more appropriate for answering topological questions about containment, adjacency and connectivity. However, with the advent of more intelligent raster data structures such as the quadtree that contains information about the relationships between cells, this distinction is closing. Chapter 6 looks in detail at analysis functions in the raster and vector world and compares and contrasts their suitability in the context of different applications.

Future developments are needed in spatial data models to ensure that our computer worlds accurately reflect the reality of geographical space. This is an area of GIS where it is likely that we will see much more complex systems capable of extending our computer worlds into the third and fourth dimensions (Chapter 12). One area for development is in the linkage between spatial and attribute data in GIS. The next chapter considers the structuring and storage of attribute data using a database approach.

REVISION QUESTIONS

- Describe how the raster and vector approaches are used to construct point, line and area entities for representation in the computer.
- What are data structures? Outline their importance in GIS.
- What is a TIN? How are TINs constructed?
- What are surface significant points and how can they be defined?
- How can networks be modelled in GIS?
- Why is it difficult to model the third and fourth dimensions in GIS?

Further study

Peuquet (1990) provides an excellent description of the stages involved in moving from a conceptual spatial model of the real world through to its representation in the computer. Burrough (1986) and Aronoff (1989) provide good introductions to data modelling, particularly with regard to the distinction between the raster and vector data models. Burrough (1986) also provides a good introduction to the TIN data model and extends the discussion to look in more detail at digital elevation models.

Laurini and Thompson (1992) take a look at the spatial data modelling process from the viewpoint of a computer scientist rather than a geographer. They provide a good introduction to the concept of object-oriented database design for handling spatial information. However, material about data modelling is dispersed throughout their book, so it may take you some time to find what you are looking for. If you want to dig deeper into spatial data structures then Worboys (1995) gives a clear and readable account of a range of different approaches. Samet (1989) provides more advanced material, particularly on quadtree structures.

To explore the history and issues associated with the evolution of 3D GIS the paper by Raper and Kelk (1991) is a good introduction. Langran (1992) provides a comparable introduction to the issues surrounding the representation of time in the GIS environment.

Aronoff S (1989) *Geographic Information Systems: A Management Perspective*. WDL Publications, Ottawa

Burrough P A (1986) *Principles of Geographical Information Systems for Land Resources Assessment*. Clarendon Press, Oxford

Langran G (1992) *Time in Geographical Information Systems*. Taylor and Francis, London

Laurini R, Thompson D (1992) *Fundamentals of Spatial Information Systems*. Academic Press, London

Peuquet D J (1990) A conceptual framework and comparison of spatial data models. In: Peuquet D J, Marble D F (eds) *Introductory Readings in Geographical Information Systems*. Taylor and Francis, London, pp. 209–14

Raper J F, Kelk B (1991) Three-dimensional GIS. In: Maguire D J, Goodchild M F, Rhind D W (eds) *Geographical Information Systems: Principles and Applications*. Longman, London, Vol. 1, pp. 299–317

Samet H (1989) *Applications of Spatial Data Structures: Computer Graphics Image Processing and Other Areas*. Addison-Wesley, London

Worboys M F (1995) *GIS: A Computing Perspective*. Taylor and Francis, London

4 Attribute data management

KEY QUESTIONS AND ISSUES

- What are databases and database management systems?
- Why are databases important in GIS?
- What types of databases are used with GIS?
- How does the relational database model work?
- How do you set up a relational database?
- How are databases linked with GIS?
- What are the important considerations for GIS databases?

Introduction

Data about our world are being produced continuously. They are being collected by remote sensing satellites; from automatic environmental monitoring equipment; during automated business transactions; and by individuals engaged in research and survey work. A large proportion of these data are managed in databases. In addition, many of our everyday activities produce data that automatically find their way into databases. Imagine, for example, purchasing some new ski equipment from a major sports retailer. If you paid by credit card, debit card or cheque, the data in your bank's database would be updated. If the bar code on the item you purchased was scanned, information in the store's database would be updated and might trigger reordering from a supplier's database. Information from the guarantee card you complete and send back by post may trigger a marketing database to start sending targeted mail to you. If you used a store card to make your purchase, extra data about your buying habits are now available, and the store can

also target mail to you. Four, or even more, databases may be updated as a result of one purchase. Since each database holds your address, your postal code may be used as a spatial reference to link these new data to other data in a GIS.

The ski buying example illustrates the amount of data which may be generated by one sales transaction. Scaling this up to consider the data generated by all the sales transactions in a day begins to give an idea of the amount of data that our society is generating. These data, if transformed into information, are a valuable resource, which can be traded in the same way that commodities were traded in the past. Market research companies now sell information about the characteristics of the population; environmental agencies sell information relating to nature conservation; and mapping agencies sell topographic information. We are in an age where the information resource will not run out, rather the problem is that we have too much of it. We need to be able to manage data and information efficiently to realize their value. One method of management is to use a database. Linking databases to GIS to provide additional spatial capabilities can enhance the value of data and provide useful information.

Large organizations such as utilities, multinational companies, retailers and government agencies have adopted GIS for a whole range of applications, from automated mapping and facilities management type applications to decision support applications. An important issue for successful implementation of many of these new systems is their ability to integrate GIS with existing internal databases. Many of these existing databases (managing customer databases, inventories of parts or financial information) are relational, as the relational database approach to data management has permeated all sectors of the business world. These business databases may be data sources for GIS, or be required to integrate with GIS for decision support within the framework of a larger organizational 'information management strategy'. Issues of database management for large-scale, corporate applications of GIS are addressed in this chapter. The demands of large-scale applications are examined, along with the strengths and weaknesses of the current database approaches.

First, this chapter considers the conceptual and theoretical ideas behind databases in order to explain why they are used and how they work. Whichever type of GIS is being used, an understanding of databases is important. This chapter will introduce the main types of database in current use and explain their characteristics, advantages and disadvantages. As the relational database is the most commonly used at present, the steps involved in establishing a relational database will be discussed in some detail.

Terminology recap

In GIS, there are two types of data to be managed: spatial data and attribute data. An entity (point, line or area) has both spatial and attribute data to describe it (Chapter 2). Spatial data are the 'where things are' data and attribute data the 'what things are'. For example, a latitude and longitude reference gives the location of a point entity and to accompany this there would be attribute data about the nature of the real-world feature the point represents.

Conventionally, databases have stored only non-spatial entities; however, all entities in GIS are spatial. As a result the conventional database has been adapted to GIS in various ways. Some GIS are good at handling attribute data; some rely on links with conventional databases; and others have very limited database capabilities but good analysis facilities.

Databases offer more than just a method of handling the attributes of spatial entities; they can help to convert data into information with value. Chapter 2 introduced the concept that data are raw facts that have little value without structure and context. A single number, for example '10', is data. With some explanation, this number becomes information. 'Ten degrees Celsius' is information, since now there is some meaning that can be interpreted by a user. Information results from the analysis or organization of data, and, in a database, data can be ordered, re-ordered, summarized and combined to provide information. A database could be used to sort a range of temperature values into order; to calculate maximum and minimum values or average temperature; or to convert degrees Celsius to degrees Fahrenheit. With the additional mapping capabilities of a GIS, the locations of the points at which the temperatures have been monitored could be mapped, thus adding further value to the original data. Decision makers using GIS need information, not data, so a database offers one method of providing that information.

There are nearly as many definitions of a database as there are of GIS. Perhaps the simplest definition is that a database is a set of structured data. An organized filing cabinet is a database, as is a dictionary, telephone directory or address book. Thus, databases can be computer-based or manual.

Why choose a database approach?

Elsewhere in this book are examples of the difficulties faced when handling spatial data manually (see especially Chapter 9). Problems can also be encountered when processing attribute data manually. Imagine Happy Valley, the ski resort, as it might have been in the days before computerized databases. Different organizations and companies within the resort could be producing and using similar data for different purposes. There are ski schools, hotels and travel companies, all handling data such as clients' names and addresses and details of where they are staying. All these data are stored in different formats and separately by each organization. A box file may be used by one organization, a set of index cards by another and a third may have the data stored in the head of one of the employees! Each organization uses the data for different purposes: the ski school for booking lessons; the hotels for booking rooms; and the travel companies for arranging flight details and allocating hotels. The situation is illustrated schematically in Figure 4.1.

There is considerable duplication of data using traditional data management approaches. For example, a single visitor's address may be held three times, once by the travel company, then again

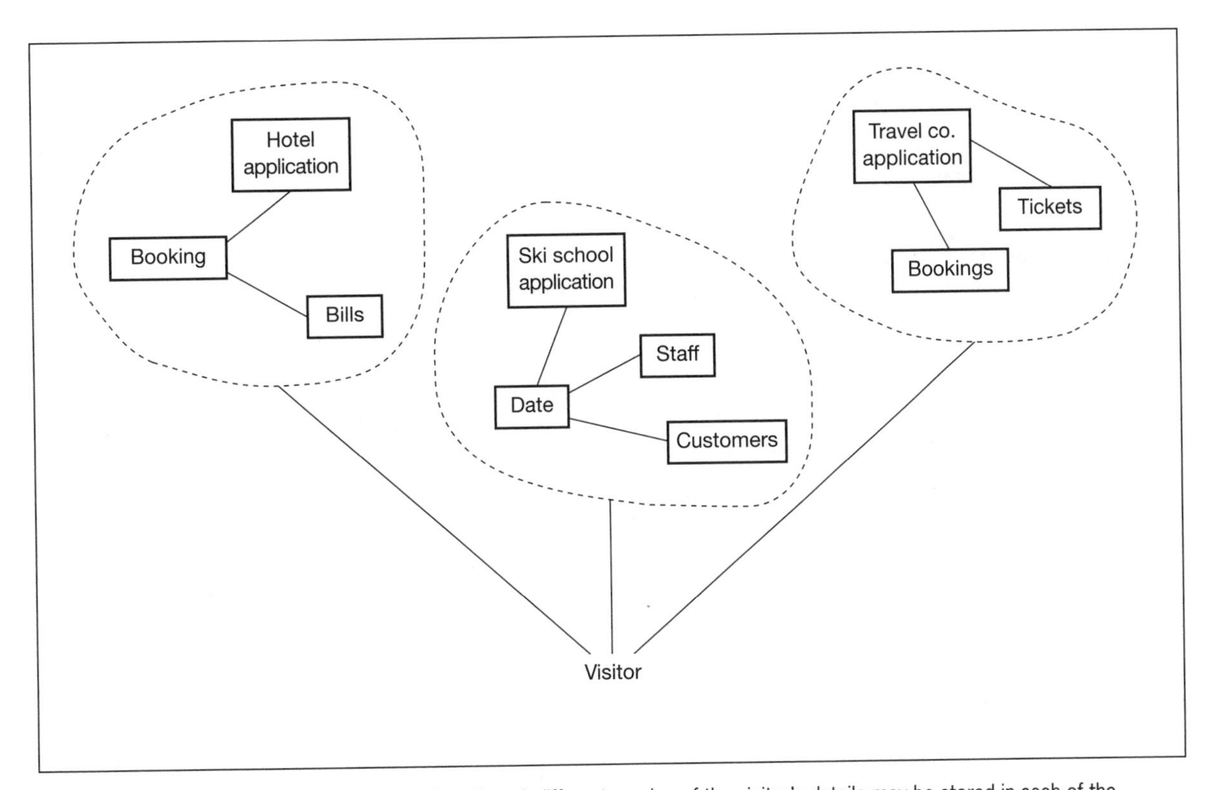

Figure 4.1 The traditional approach to data handling. A different version of the visitor's details may be stored in each of the separate databases

by the hotel and ski school. It is likely that there will be errors made during the transcription and copying of an address, or even that different parts of the address will be held or omitted in each case. If a visitor changes their home address, they may remember to tell the travel company, but forget to tell the hotel or the ski school. Thus, different versions of the data could exist in the three companies. Errors are inevitable, and there is considerable effort involved in handling the data. Difficulties would be encountered if the organizations attempted to share their data. Even if the data were in computer files of one sort or another there might be incompatibilities in data formats which would prevent efficient data exchange.

Date (1986) summarizes the problems with the traditional approach to data management as:

- redundancy (the unnecessary repetition or duplication of data);
- high maintenance costs;
- long learning times and difficulties in moving from one system to another;
- the possibility that enhancements and improvements to individual files of data will be made in an *ad hoc* manner;
- data-sharing difficulties;
- a lack of security and standards; and
- the lack of coherent corporate views of data management.

In addition, the data storage mechanisms may be inflexible, creating difficulties in dealing with *ad hoc* 'one-off' queries. Oxborrow (1989) also identifies the problem of modelling the real world with traditional data management methods – many are simply not suitable. Data should be structured in such a way that they represent features from the real world and the relationships between them. Traditional methods of storing data cannot represent these relationships, so a database approach is necessary.

The database approach

A database is a simply a collection of related data. This can include non-computerized data such as those found in a telephone directory or address book. A generally accepted feature of data in a database is that they can be shared by different users (Stern and Stern, 1993). Data within a database are also organized to promote ease of access and prevent unnecessary duplication. In a paper-based telephone directory the entries are organized, usually alphabetically by surname, to allow ease of access to different users. However, there are problems with this paper-based database – it is not possible to search by first name, or using only part of an address. It is not possible to extract a list of all those people who live in a particular district. There is no security, as anyone who can read can access the information (the only security mechanism is to be ex-directory, then the data set is incomplete). It is not possible to adapt the telephone book for other uses, such as direct mailing. Updating requires expensive and time-consuming reprinting. A computer approach can overcome these problems. Some of the most often cited advantages of computer-based databases are summarized in Box 4.1.

As Box 4.1 indicates, a database is a shared collection of data often with secure controlled access. Since all the data are (normally) stored in one place, standards are possible and data exchange is facilitated. Most importantly, the data in a database are stored independently of the application for which they will be used. The electronic telephone directory enquiry service available in most countries provides an example of how the problems of a non-computerized database have been overcome.

Database management systems

The data in a computer database are managed and accessed through a database management system (DBMS). Individual application programs will access the data in the database through the DBMS. For example, to book a new client's ski lessons, the booking clerk will use an application produced using capabilities offered by the DBMS. This will instruct them to fill in a data entry form, which will then automatically update data in the database. The clerk will not need to interact with the database directly or understand how data are structured within the database. A conceptual view of the Happy Valley data as they might be organized for such a computer database, and the role of the DBMS, is shown in Figure 4.3 (adapted from Reeve, 1996).

There are many definitions of a DBMS. Dale and McLaughlin (1988) define a DBMS as a com-

BOX 4.1

Advantages of computer-based databases

Imagine that there is a box containing index cards with the names, addresses and telephone numbers of all the ski school clients in Happy Valley, together with details of the ski lessons they have undertaken. A single index card is held for each individual, and cards are stored alphabetically by surname in the box. A typical card is shown in Figure 4.2.

If the data from the cards are transferred to a computer database, the following benefits could be achieved:

Surname	Smith
First name	Jane
Home Address	12 High Street, Northton, Northtonshire, NN1 1NN
Telephone No.	~~0181 234 443~~ 020 8222 442
Lessons taken	Beginners 1 week course (Jan 2001) Private lesson × 2 (March 2002)

Figure 4.2 Card index record from ski school manual database

1 *Different data access methods will be possible.* Data can be accessed by country of residence or lessons undertaken, not just surname of client as in the index box.
2 *Data are stored independently of the application for which they will be used.* The database may have been established to assist the sending of bills to clients. However, the database could also be used to produce mailing lists for the promotion of special events, or for summarizing client data (where clients came from, number of beginners courses undertaken, number of one-off lessons) for annual reports.
3 *Redundancy (the unnecessary duplication of data) will be minimized.* In a paper-based records system it may have been necessary to keep two sets of records for one client. For example, when one index card is full, another one may be started and the old one left behind. Alternatively, should a client change address, a new card may be added, and the old one not removed.
4 *Access to data will be controlled and centralized.* A card index box kept on a desk is not very secure, and, unless it can be locked, can be viewed by anyone in the office. A computer database can have security built in – passwords can prevent access to all or part of the information, or to functions that will allow the updating or deletion of data. Allowing only one individual to access the raw data for updates and changes will also improve the reliability of the data.
5 *A computer database is relatively easy to maintain and updating is possible.* The cards in the index box will soon become unreadable if clients' details change repeatedly. The computer version will prevent problems caused by unreadable text, and allow efficient updating.
6 *Simple query systems and standardized query languages are available.* Queries, such as 'which clients have completed advanced training courses?', or 'how many Americans have taken training courses?' would be time-consuming with the card index box. A computerized system will offer simple query methods, or a standardized query language.

(*Sources*: Oxborrow, 1989; Healey, 1991)

puter program to control the storage, retrieval and modification of data (in a database). Stern and Stern (1993) consider that a DBMS will allow users to join, manipulate or otherwise access the data in any number of database files. A DBMS must allow the definition of data and their attributes and relationships, as well as providing security and an interface between the end users and their applications and the data itself. From such definitions the functions of a DBMS can be summarized as:

- file handling and file management (for creating, modifying or deleting the database structure);

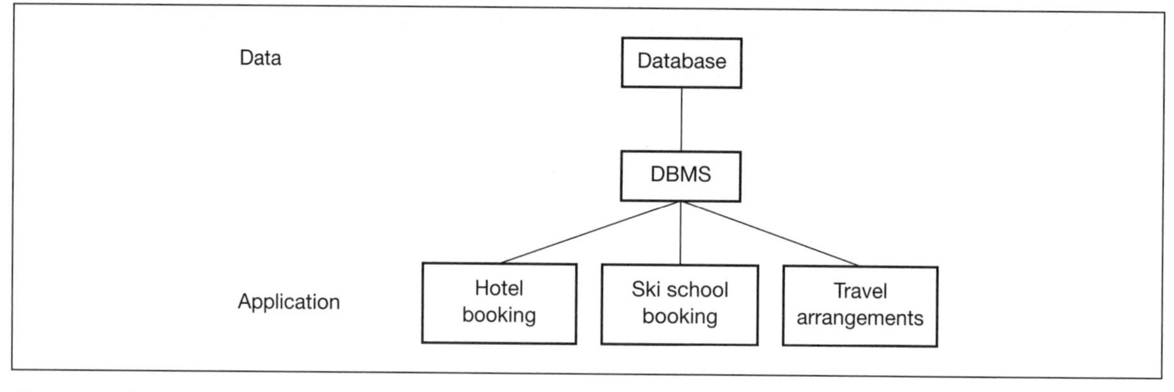

Figure 4.3 The database approach to data handling

- adding, updating and deleting records;
- the extraction of information from data (sorting, summarizing and querying data);
- maintenance of data security and integrity (housekeeping, logs, backup); and
- application building.

The overall objective of a DBMS is to allow users to deal with data without needing to know how the data are physically stored and structured in the computer. To achieve this, DBMS usually comprise software tools for structuring, relating and querying data; tools for the design of data entry and report forms; and application generators for the creation of customized applications.

A DBMS manages data that are organized using a database data model. This is analogous to the way in which spatial data are organized in a GIS according to a spatial data model (for example, raster or vector). Database data models for GIS are similar to those used for databases elsewhere.

Database data models

There are a number of different database data models. Amongst those that have been used for attribute data in GIS are the hierarchical, network, relational and object-oriented data models. Of these the relational data model has become the most widely used and will be considered in detail here, while the object-oriented model is an emerging trend in GIS and a topic of current research. Further details of the hierarchical and network models can be found in other GIS texts (Aronoff, 1989; Bernhardsen, 1992; DeMers, 1997).

The Relational Database Model

At present the relational database model dominates GIS. Many GIS software packages link directly to commercial relational database packages, and others include their own custom-designed relational database software. Some GIS use a relational database to handle spatial as well as attribute data.

The relational data model is based on concepts proposed by Codd (1970). Data are organized in a series of two-dimensional tables, each of which contains records for one entity. These tables are linked by common data known as keys. Queries are possible on individual tables or on groups of tables. For the Happy Valley data, Figure 4.4 illustrates an example of one such table.

Each table in a relational database contains data for one entity. In the example in Figure 4.4 this entity is 'hotel'. The data are organized into rows and columns, with the columns containing the attributes of the entity. Each of the columns has a distinctive name, and each of the entries in a single column must be drawn from the same domain (where a domain may be all integer values, or dates or text). Within a table, the order of the columns has no special significance. Other characteristics are listed by Reeve (1996). There can be only one entry per cell; each row must be distinctive (so that keys that use unique row entries are possible – in GIS

Hotel ID	Name	Address	Number of rooms	Standard
001	Mountain View	23 High Street	15	budget
002	Palace Deluxe	Pine Avenue	12	luxury
003	Ski Lodge	10 Ski School Road	40	standard

Figure 4.4 Relational database table data for Happy Valley

location is often the key); and null values are possible where data values are not known.

The terminology of relational databases can be confusing, since different software vendors have adopted different terms for the same thing. Table 4.1 illustrates the relationship between relational database terminology and the traditional table, or simple computer file. Figure 4.5 applies this terminology to the table suggested for the Happy Valley database. A useful shorthand way of describing a table is using its 'intension'. For the table in Figure 4.4 this would be:

> HOTEL (Hotel ID, Name, Address, No.rooms, Standard).

The data in a relational database are stored as a set of base tables with the characteristics described above. Other tables are created as the database is queried and these represent virtual views. The table structure is extremely flexible and allows a wide variety of queries on the data. Queries are possible on one table at a time (for example, you might ask 'which hotels have more than 14 rooms?' or 'which hotels are luxury standard?'), or on more than one table by linking through key fields (for instance, 'which passengers originating from the UK are staying in luxury hotels?' or 'which ski lessons have pupils who are over 50 years of age?'). Queries generate further tables, but these new tables are not usually stored. There are few restrictions on the types of query possible.

Table 4.1 Relational database terminology

Paper version	*File version*	*RDBMS*
Table	File	Relation
Row	Record/case	Tuple
Column	Field	Attribute
Number of columns	Number of fields	Degree
Number of rows	Number of cases	Cardinality
Unique ID	Primary key	Index
	Possible values	Domain

Source: adapted from Date, 1986

With many relational databases querying is facilitated by menu systems and icons, or 'query by example' systems. Frequently, queries are built up of expressions based on relational algebra, using commands such as SELECT (to select a subset of rows), PROJECT (to select a subset of columns) or JOIN (to join tables based on key fields). SQL (standard query language) has been developed to facilitate the querying of relational databases. The advantages of SQL for database users are its completeness, simplicity, pseudo English-language style and wide application. However, SQL has not really developed to handle geographical concepts such as 'near to', 'far from', or 'connected to'.

The availability of SQL is one of the advantages of the relational database model. Additionally, the model has a sound theoretical base in mathematics, and a simple logical data model that is easy to understand. The relational database model is more flexible than either of the previously used hierarchical or network models. However, the model will always produce some data redundancy and can be slow and difficult to implement. There are also problems with the handling of complex objects such as those found in GIS (and CAD and knowledge-based applications) as there is a limited range of data types, and difficulties with the handling of time. Seaborn (1995) considers that many of the limitations of relational databases in GIS stem from the fact that they were developed to handle

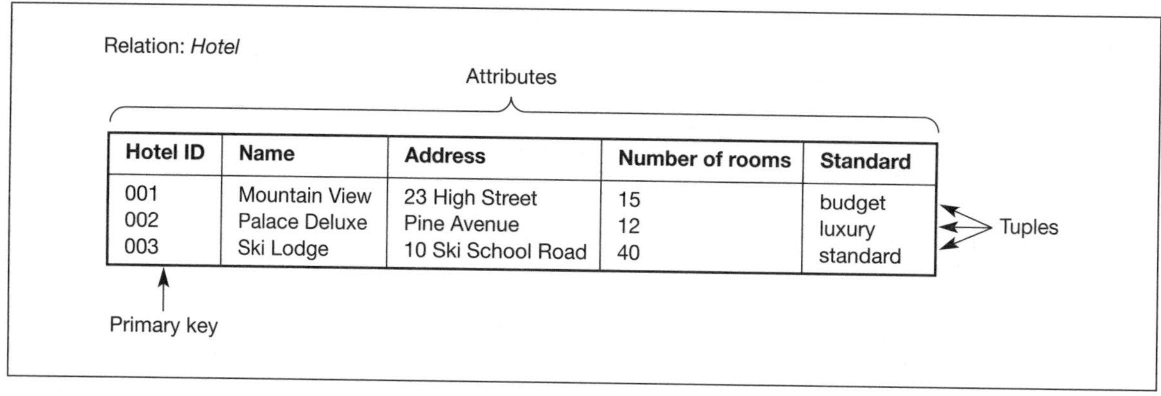

Hotel ID	Name	Address	Number of rooms	Standard
001	Mountain View	23 High Street	15	budget
002	Palace Deluxe	Pine Avenue	12	luxury
003	Ski Lodge	10 Ski School Road	40	standard

Figure 4.5 Database terminology applied to Happy Valley table

simple business data, and not complex multi-dimensional spatial data. However, they have been widely adopted and successfully used.

Relational databases are predominantly used for the handling of attribute data in GIS. For example, ARC/INFO maintains an attribute table in relational database software, using a unique ID to link this to spatial data.

Creating a database

Database design and implementation are guided by the relationships between the data to be stored in the database. The database design process is concerned with expressing these relationships, then implementation with setting up a new structure for these relationships within the chosen database software. The stages of database creation are summarized in Box 4.2.

The first stage in designing a database, data investigation, can result in a mass of unstructured information on information flows, relationships and possible entities. Thus, a key part of database development is data modelling. There is a wide range of techniques available to assist this data modelling, including entity relationship modelling (Chen, 1976) and normalization.

There are four stages to entity relationship modelling (or entity attribute modelling, EAM): the identification of entities; the identification of relationships between entities; the identification of attributes of entities; and the derivation of tables from this. In a database for Happy Valley we may wish 'hotels', 'tour companies', 'ski schools' and 'visitors' to be regarded as distinct entities. Each entity has distinctive characteristics and can usually be described by a noun. Its characteristics are the attributes (for example a hotel will have a name, address, number of rooms and standard), and its domain is the set of possible values (for example the standard may be budget, standard, business or luxury). The relationships between the entities can be described using verbs. Thus, a hotel *is located in* a resort; a visitor *stays* at a hotel, and a ski school *teaches* visitors. Three types of relationship are possible: one to one (one visitor stays at one hotel); one to many (one ski school teaches many visitors); or many to many (many tour companies use many hotels). These relationship can be expressed by the symbols: '1:1', '1:M' and 'M:N', and by using simple diagrams. Figure 4.6 illustrates a possible entity relationship model for the Happy Valley example, showing the following relationships:

- many visitors stay at one hotel (M:1);
- one travel company organizes holidays for many visitors (1:M);
- one ski school teaches many visitors (1:M); and
- several different travel companies may use more than one ski school (M:N).

The entity relationship model diagram is helpful in deciding what will be appropriate tables for a relational database. Where the relationship is 1:1, tables for each entity can be joined together or kept separate. Where the relationship is 1:M, two tables are needed with a key field to allow a relational

BOX 4.2

The steps involved in database creation

The steps involved in database creation, suggested by Oxborrow (1989) and Reeve (1996), are summarized in Table 4.2 and described below.

Table 4.2 Steps in database creation

Oxborrow (1989)	*Reeve (1997)*
Data investigation	Needs analysis
Data modelling	Logical design
Database design	Physical design Testing
Database implementation	Implementation
Database monitoring	Maintenance

1 *Data investigation* is the 'fact finding' stage of database creation. Here the task is to consider the type, quantity and qualities of data to be included in the database. The nature of entities and attributes is decided.
2 *Data modelling* is the process of forming a conceptual model of data by examining the relationships between entities and the characteristics of entities and attributes. This stage, like the data investigation stage, can be carried out independently of the software to be used.
3 *Database design* is the creation of a practical design for the database. This will depend on the database software being used, and its data model. This is the process of translating the logical design for the database (produced during the data modelling stage) into a design for the chosen DBMS. Field names, types and structure are decided. In practice, the design will be a compromise to fit the database design model with the chosen DBMS.
4 *Database implementation* is the procedure of populating the database with attribute data, and this is always followed by monitoring and upkeep, including fine tuning, modification and updating.

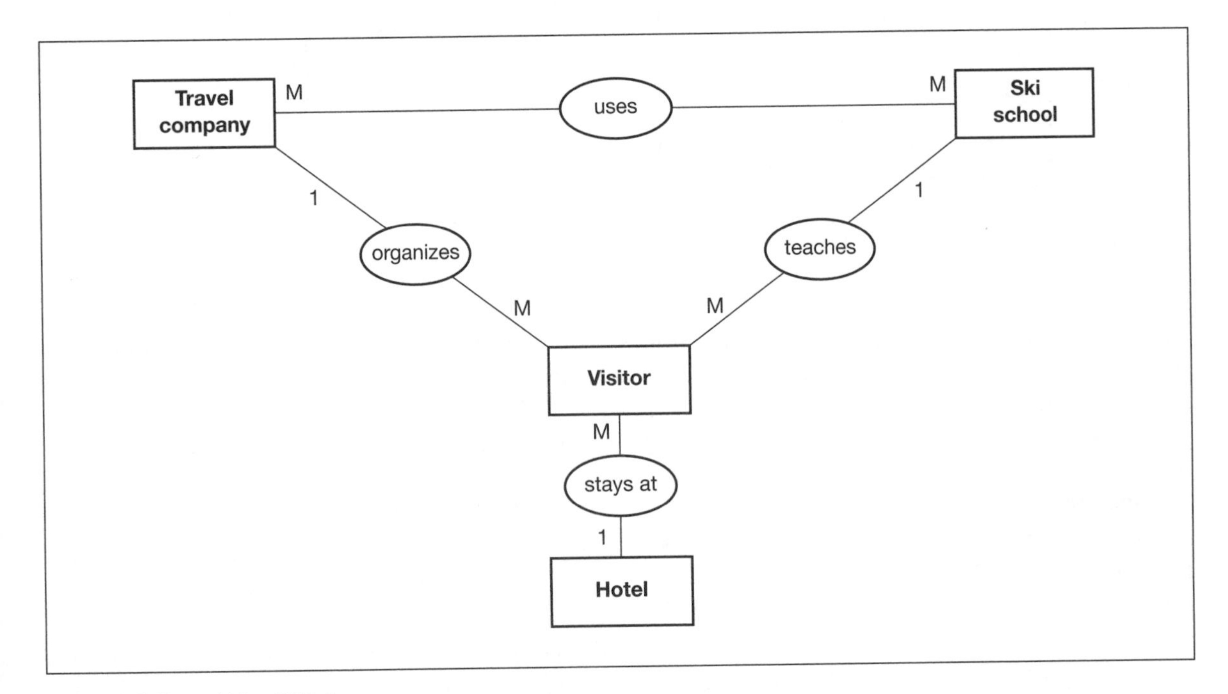

Figure 4.6 Happy Valley EAM diagram

Table 1: Hotel

Hotel ID	Name	Other attributes...
001	Mountain View	.....
002	Palace Deluxe	.....
003	Ski Lodge	.....
.....	.....	.....

Table 2: Travel company

Travel co. ID	Travel Co. Name	Other attributes...
T01	Ski Tours	.....
T02	Snow Breaks	.....
.....	.....	.....

Table 3: Visitors

Visitor ID	Visitor Name	Hotel ID	Travel Co. ID	Ski School ID	Other attributes...
V001	Smith J	002	T01	S02	.....
V002	Schmidt S	001	T02	S02	.....
.....	.....	.....	.....	.....	.....

Table 4: Ski schools

Ski School ID	Ski School name	Other attributes...
S01	Snow Fun	.....
S02	Bert's Ski School	.....
S03	Ski School Professional	.....
.....	.....	.....

Table 5: Link table

Travel Co. ID	Ski School ID
T01	S01
T01	S02
T02	S02
.....	.....

Figure 4.7 Relational tables for Happy Valley database

join. Where the relationship is M:N, the tables should be separated. Where repeating fields occur, tables need to be broken down further to avoid redundancy. Figure 4.7 shows possible tables developed from the EAM model for Happy Valley.

Once the tables have been decided upon, the attributes needed should be identified. This is the process of developing intensions that was outlined earlier. The intensions for the Happy Valley example are:

HOTEL (<u>Hotel ID</u>, Name, other attributes ...)
TRAVELCO (<u>TravelCo ID</u>, TravelCo Name, other attributes ...)
SKISCHOOL (<u>SkiSchool ID</u>, SkiSchool Name, other attributes ...)
VISITOR (<u>Visitor ID</u>, Visitor Name, Hotel ID, TravelCo ID, SkiSchool ID, other attributes ...)
LINK (<u>TravelCo ID</u>, SkiSchool ID)

The final result of entity relationship modelling is an EAM model diagram, a set of table definitions and details of attributes (names, size and domain). The database can then be implemented. Another example of the use of entity relationship modelling to develop a GIS database is given by Healey (1991).

Linking spatial and attribute data

The relationship between GIS and databases varies. For a simple raster GIS, where one cell in a layer of data contains a single value that represents the attributes of that cell, a database is not necessary. Here the attribute values are likely to be held in the same file as the data layer itself. However, there are few 'real' GIS like this, and those which exist are designed for analysis, rather than attribute data handling. An improvement on this approach is the ability to handle attribute values in a file separate from the raster image. Although this method also lacks the flexibility of a true relational DBMS, it is possible to link the GIS software with proprietary relational DBMS to upgrade the capabilities. Most GIS, particularly vector-based systems, offer a hybrid approach (Batty, 1990; Maguire *et al.*, 1990; Cassettari, 1993). In this case spatial data are stored as part of the GIS data structure and attribute data are stored in a relational DBMS. This approach allows integration of existing databases with graphics by the allocation of a unique identifier to each feature in the GIS (Figure 4.8).

Finally, an alternative approach is an extended GIS, where all aspects of the spatial and attribute data are in a single DBMS. Seaborn (1995) considers these 'all-relational' GIS to have considerable potential, and cites examples of major organizations such as British Telecom, Electricité de France and New Zealand Lands who have adopted this approach.

GIS database applications

GIS databases cover a wide spectrum of applications, from those involving a single user with a PC and low-cost software working on small-scale research projects to huge corporate databases with data distributed over several sites, each with different computer systems and different users. Worboys (1995) offers a classification of potential database applications. This includes single-user small databases, corporate databases, office information systems, engineering databases, bibliographic databases, scientific databases, and image and multimedia databases, and has a separate category

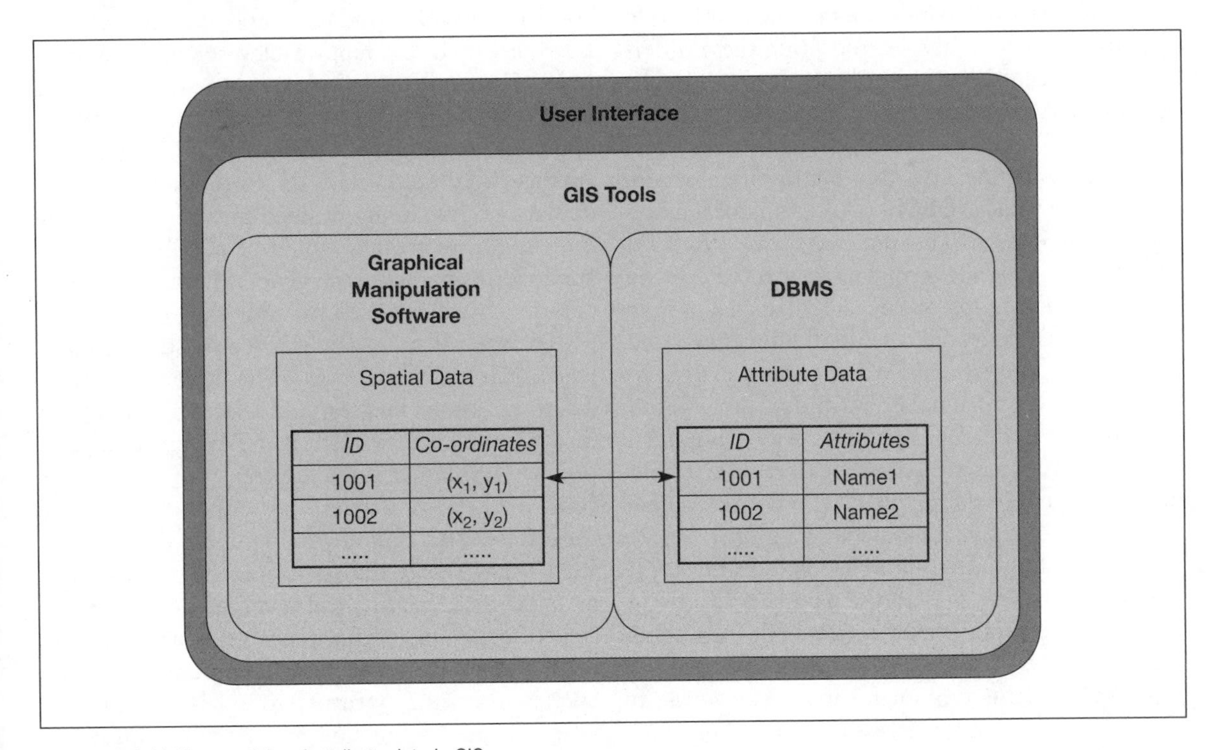

Figure 4.8 Linking spatial and attribute data in GIS

for geographic databases. Despite this separation of geographic databases, geographic data could be found in most of the other types of databases – single-user, corporate, office, image and multimedia. The differing nature of potential geographical applications leads to a range of issues for the GIS database user to consider.

For a single user, working on a PC database, issues and design considerations will be very different from those for a large, multi-user corporate database. For the single user, flexibility and ease of use may be important, whereas security, reliability (the probability of the system running at a given time), integrity, performance and concurrent access by different users may be required by a large-scale multi-user application. Many GIS applications are now large-scale corporate systems, for instance in the gas, electricity, water and telecommunications industries. Large corporate GIS projects have special database demands. In some organizations GIS may be operated by a specialist department which controls access and manages the data. Increasingly, however, GIS are being integrated into overall information strategies, requiring the integration of GIS and general business data. There are two options for such databases, reflecting these differences: the centralized database system or the distributed database system.

In a centralized database system all the system components reside at a single computer or site (this includes data, DBMS software, storage and backup facilities). The database can often be accessed remotely via terminals connected directly to the site. In Happy Valley, a centralized system may be available at the town hall, controlled and updated by council employees, and accessible to other users via terminals distributed throughout the town. The current trend is away from such central servers to databases located in different places but connected together by networks. These are distributed database systems.

Using a distributed database system the local user should believe that all the data they are using are located on their machine, even though some may be on the other side of the world. Using the Internet gives this type of feeling – you work at your own computer, but access data stored on computers all around the world. It is not important where these other computers are, as long as you can obtain the data you need. A distributed database system uses communications networks that connect computers over several sites. Elmasri and Navathe (1994) describe such a system as one in which a collection of data that belongs logically to the same system is physically spread over the sites of a computer network. If the multinational company, SkiResorts Inc., were to buy the whole of Happy Valley (ski schools, hotels, tourist offices, etc.) it may decide to implement a distributed database for visitor monitoring and management. In this case, individual databases held by the ski schools and hotels could still be maintained by the originators but made accessible to other authorized users, including the multinational company's headquarters, via a communications network. Worboys (1995) considers that the distributed database approach is naturally suited to many GIS applications. One of the major advantages of a distributed database is that the reliability of the system is good, since if one of the database sites goes down, others may still be running. The approach also allows data sharing, whilst maintaining some measure of local control, and improved database performance may be experienced. In Happy Valley, some of the problems of different users holding their own databases could be overcome by a distributed database approach. For example, data duplication and replications could be tracked, the consistency of copies of data could be maintained and recovery from individual site crashes would be possible. Box 4.3 summarizes some of the general issues that large corporate databases need to address.

The specialist needs identified in Box 4.3 make large-scale GIS databases difficult to implement. There are other factors, not related to databases directly, which may restrict the evolution of corporate GIS databases. McLaren (1990) outlined some of these as different basic spatial units used for different data, data copyright and data ownership issues, large data volumes, formulation of projects at departmental level and institutional structures. Many of these problems are not technical but human and organizational in nature. Such issues will be returned to in Chapter 11.

A variety of considerations will combine to ensure a successful database application. These

BOX 4.3

Issues for large corporate databases

Large corporate database applications are characterized by the following needs:

1. *The need for concurrent access and multi-user update.* This will ensure that when several users are active in the same part of the database, and could inadvertently update the same data simultaneously, they do so in an orderly manner, with the result they collectively intended.
2. *The need to manage long transactions.* This is a database update composed of many smaller updates. The update takes place over a period of hours or days (not seconds, as in short transactions), most often by one user, but often by several users working in parallel, but independently. This situation might arise if major revisions to base map data were in progress.
3. *The need for multiple views or different windows into the same database.* For example, a bank employee may be able to access and update customer information for any customer, but an individual customer can only access, but not update, their own records.

(adapted from Seaborn, 1995)

are generally considered to be contemporaneous data, data as detailed as necessary for potential applications, data that are positionally accurate and internally accurate, up-to-date and maintained data, and data that are accessible to users. Databases should be usable by non-technical experts who do not need to know the principles of databases to add, edit, query or output data from the database.

The general problems with a database approach are complexity, cost, inefficiencies in processing and rigidity (Oxborrow, 1989). The complexity of DBMS means that training is required to design and maintain the database and applications. The costs include those of the software, development and design phases and maintenance and data storage. Processing inefficiencies may be caused by changes in user requirements. If changes require restructuring of the database, this may be difficult to implement. Finally, the data formats offered by DBMS are fixed. There are limited capabilities, in most DBMS, for the handling of long text strings, graphics and other types of data. All of these general problems apply to GIS databases, and there is an additional set of problems associated with data backup, recovery, auditing, security, data integrity and concurrent update (Batty, 1990). Batty also considers that GIS are no different from other data management systems, and must face the challenges of sharing data between applications and handling data distributed across several computers.

Geographical data have three elements: space, theme and time (Chapter 2). Whilst we have discussed the handling of spatial data in Chapter 3, and thematic or attribute data in this chapter, temporal data have not been addressed in any detail. One of the limitations of a relational database, in fact of most database systems available, is the inability to handle temporal data effectively. This is an issue that is being addressed by current research activities.

Another area of current interest is the implementation of databases and GIS using the Internet. Box 4.4 provides an insight into how this might work.

Developments in databases

Worboys (1995) suggests that there are problems with the relational approach to the management of spatial data. Spatial data do not naturally fit into tabular structures, and, as discussed earlier, the SQL query language does not have capabilities for spatial ideas and concepts. Some researchers have suggested extensions to SQL to help with the handling of spatial data (Raper and Rhind, 1990) whilst Worboys (1995) suggests that current research on 'extensible relational DBMS' may provide the way forward. However, more attention has been focused on the development of object-oriented (OO) approaches to database design and there are some examples of OO database principles in current GIS.

The object-oriented database approach offers the opportunity to move away from the 'geometry-centric' data models that present the world

BOX 4.4

Web GIS

Most web-based GIS are client server systems (Figure 4.9). Servers hold data (possibly in a relational database management system) and desktop clients use standard browser software to view those data. The Internet Map Server typically provides functions to allow, for instance, zooming or panning around map images.

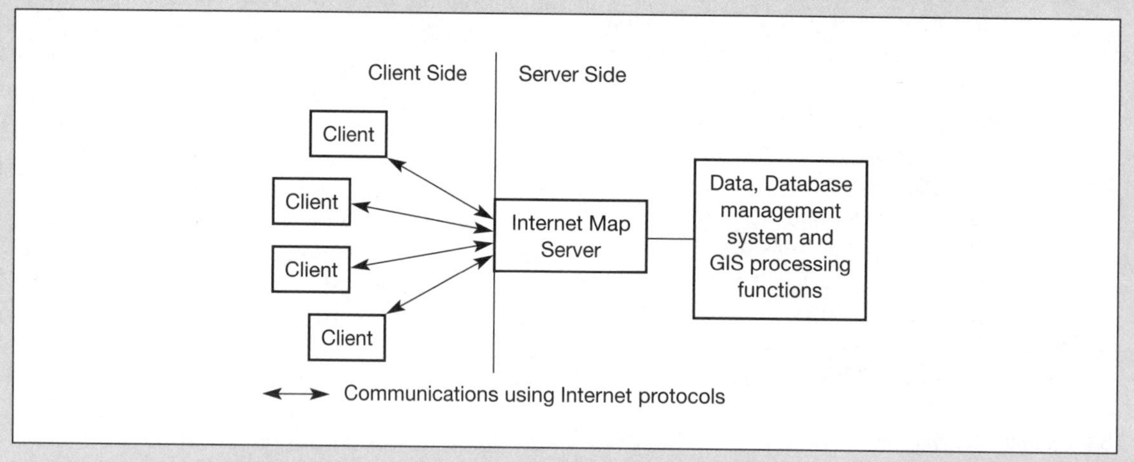

Figure 4.9 Client–Server Web GIS

Using a client server system a client requests a map image from a server. The Internet Map Server creates a map based on the request from the client as a graphics file in, for example, GIF or JPEG format.

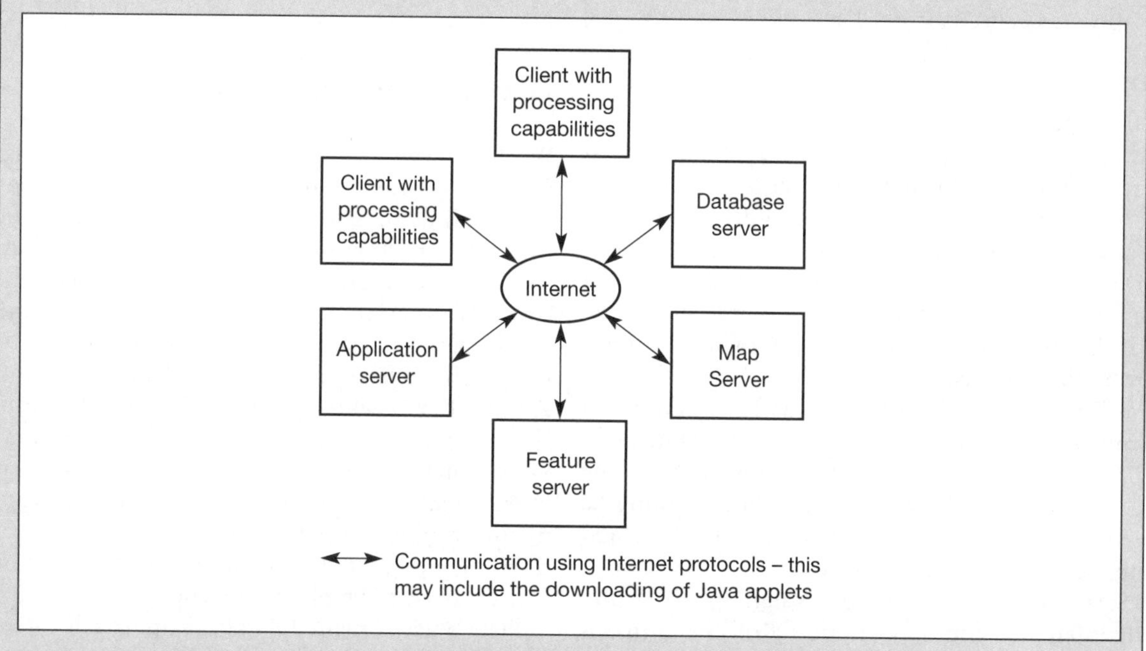

Figure 4.10 A networked Web GIS

BOX 4.4 CONTINUED

The server program transmits the image over the Internet back to the client's browser. The client and server systems communicate using a protocol – an established communication language such as HTTP (Hypertext Transfer Protocol).

If the user obtains a map and then decides to zoom in, another request would be sent to the server. The server would send a new version of the map image back to the client.

In some cases more of the processing can be done on the client side. In this case the Internet Map Server sends more complex data or structured data, for example in vector format, to the client. Users may have to obtain applications, plug-ins or Java applets for their client systems to allow them to process these data.

More complex networked systems include more than one server. Separate servers may be used as in the example in Figure 4.10. Here the clients have processing capabilities: data, features, maps and applications are all located on separate servers. The feature server may provide the results of a database query.

The applications of Web GIS are growing rapidly. They include:

- displaying static maps which users can pan or zoom whilst online;
- creating user-defined maps online which are in turn used to generate reports and new maps from data on a server;
- integrating users' local data with data from the Internet;
- providing data that are kept secure at the server site;
- providing maps through high-speed intranets within organizations;
- providing maps and data across the Internet to a global audience.

(*Sources*: Harder, 1998; Limp, 1999; Longley *et al.*, 2001)

as collections of points, lines and polygons. The fundamental aim of the OO model is to allow data modelling that is closer to real-world things and events (Longley *et al.*, 2001). Thus, the way events and entities are represented in a database should be closer to the way we think about them. The method for achieving this is to group together all the data describing a real-world entity, together with any operations that are appropriate to the entity, into an object. The key feature of an OO database is the power given to the user to specify both the structure of these objects and the operations that can be applied to them (Elmasri and Navathe, 1994). So, each entity is modelled as an object and can be represented by the simple formula:

Object = state + behaviour

The 'state' of an object is the set of values of its attributes and the 'behaviour' represents the methods of operating on it. The identity of an object is unique and does not change during its lifetime. Composite objects, which are made up of more than one other object, can be created. In addition, an object can belong to a superclass of objects, and then will inherit all the properties of this superclass. What does this mean in practice? Consider again the Happy Valley ski resort. One possible method for representing the entity 'hotel' was considered earlier during the discussion of the relational database model. An intension was proposed to describe a table for the entity 'hotel' in a relational database:

HOTEL (Hotel ID, Name, Address, No.rooms, Standard).

This tells us that the entity 'hotel' has attributes ID-number, name, address, number of rooms and standard. In the OO model, these attributes become the states and 'hotel' becomes a class of objects. In addition to the states of an object, information about its behaviour is stored. These are the operations that can be performed on the object. For example, it is possible for a hotel to open additional rooms, have its standard revised or be closed to visitors. So, for the object 'hotel':

Table 4.3 Definition of the object 'hotel' for the Happy Valley OO database

Object	*State*	*Behaviour*
Hotel	Name	Can be plotted on map
	Address	Can be added to database
	Number of bedrooms	Can be deleted from database
	Standard of accommodation	Standard can be upgraded or downgraded
		Number of bedrooms can be increased or decreased

Hotel = (Hotel ID, Name, Address, No.rooms, Standard) +
(Open additional rooms, Change standard, Close to visitors)

In addition, the object 'hotel' may belong to a superclass of objects known as 'buildings'. All objects in this superclass will share similar properties – states and behaviours (for example buildings all have an identity, size, shape and location, and they can be constructed, demolished or drawn on a map). Any object in the class 'hotel' will inherit the properties of the superclass 'building', thus the states and operations applicable to 'buildings' will also apply to 'hotels'. Therefore, a hotel has a size, shape, identity and location, and can be constructed, demolished or drawn on a map. The class of object 'hotel' has *inherited* the properties of the superclass 'buildings'. Thus a *hierarchy* of objects is emerging. This hierarchy can work in the other direction too, and the object in the class 'hotel' may be *composite* – made up, for example, of subclasses 'restaurant', 'swimming pool' and 'garden'. This situation is illustrated in Table 4.3 and Figure 4.11.

In a GIS each class of object is stored in the form of a database table: each row represents an object and each column is a state. The methods that can be applied are attached to the objects when they are created in memory for use in an application (Longley *et al.*, 2001).

The OO approach is possibly more appropriate for geographical data than the relational model, since it allows the modelling of complex, real-world objects, does not distinguish between spatial and attribute data and is appropriate for graphics operations. Longley *et al.* (2001) list the three particular features of object data models that make them particularly good for modelling geographic systems:

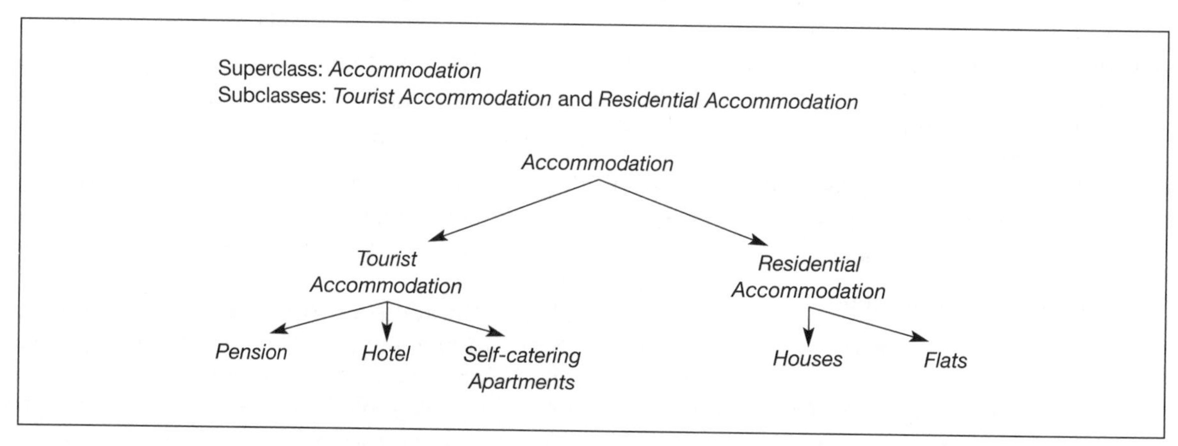

Figure 4.11 Object hierarchy for Happy Valley OO database

- Encapsulation – the packaging together of the description of state and behaviour in each object.
- Inheritance – the ability to re-use some or all of the characteristics of one object in another object.
- Polymorphism – the specific implementation of operations like draw, create or delete for each object.

However, Goodchild (1990) points out that there are problems with the approach since many geographical ideas have implicit uncertainty and the spatial objects that we require our databases to model are often the products of interpretation or generalizations. Therefore, it can be difficult to represent the world as rigidly bounded objects. Other disadvantages are that the methods are still under research and development. For the user there are the additional problems that, to date, there is no standard data model, no clear theoretical base for the OO model and no standard query language.

Conclusions

This chapter has introduced the methods available for the handling of attribute data in GIS. The need for formal methods for database management has been discussed, then the principles and implementation of a relational database model considered in detail, since this is the most frequently used in current GIS. Options for large-scale users have been presented, including the use of centralized and distributed database systems. Finally, a brief introduction to one of the trends in database management in GIS, the object-oriented approach, has been given.

An understanding of the database models for the management and handling of attribute data in GIS is essential for effective use of the systems available. It is important to understand how both spatial data and attribute data are structured within GIS to enable you to specify appropriate and achievable questions for your GIS to answer, and to implement these effectively. The way in which the spatial and attribute data are linked in an individual system is also important and an area where considerable differences are seen between systems. Returning to the story of the Martian, let us imagine that he has decided to buy a car and needs to know what he can do with it – can he ask it to go backwards, or sideways? How does the engine work, and how is the car made to move? How is the car structured, how effective is it and how complex is it to maintain? To keep the car running it is essential to know some of this technical information. Similarly, for a GIS user, a GIS will not be able to function unless the data are appropriately modelled and structured. But first we need to get some data into the GIS, or some fuel into the car!

REVISION QUESTIONS

- What is the difference between a database and a database management system?
- Describe the main characteristics of the relational database model. Why have relational databases dominated in GIS?
- Draw up the tables you would expect to see in a simple relational database containing data for the following entities: student, course, teacher, department. Note which fields would be used as keys, and give examples of the types of queries that could be asked of the database.
- What are the main issues to be considered by users of large corporate GIS databases?
- Consider possible future directions and trends for GIS databases.
- List the main features of the object-oriented approach to databases.

Further study

A chapter that covers similar ground to this one, offering the development of a National Parks database as a useful example, is Healey (1991). Other texts that contain chapters introducing databases in GIS, without too much technical detail, are Aronoff (1989), Burrough (1986) and Dale and McLaughlin (1988).

For a more technical introduction, from a computer science perspective, there is a huge range of texts on databases available. Some readable and useful examples are Oxborrow (1989) and Elmasri and Navathe (1994). Worboys (1995) is an interesting book, considering all aspects of GIS from a computer science perspective, and this has a comprehensive section on databases.

Batty (1990) offers an article on exploiting relational database technology for GIS and Worboys (1999) reviews relational and object-oriented approaches. Further details of the object-oriented approach, including an example of an object data model, can be found in Longley *et al.* (2001). Corporate database issues have been considered by McLaren (1990).

Aronoff S (1989) *Geographic Information Systems: a Management Perspective.* WDL Publications, Ottawa

Batty P (1990) Exploiting relational database technology in GIS. *Mapping Awareness* 4 (6): 25–32

Burrough P A (1986) *Principles of Geographical Information Systems for Land Resources Assessment.* Clarendon Press, Oxford

Dale P F, McLaughlin J D (1988) *Land Information Management: An Introduction with Special Reference to Cadastral Problems in Third World Countries.* Clarendon Press, Oxford

Elmasri R, Navathe S B (1994) *Fundamentals of Database Systems*, 2nd edn. Benjamin/Cummings, California

Healey R G (1991) Database management systems. In: Maguire D J, Goodchild M F, Rhind D W (eds) *Geographical Information Systems: Principles and Applications.* Longman, London, Vol. 1, pp. 251–67

Longley P A, Goodchild M F, Maguire D J, Rhind D W (2001) *Geographical Information Systems and Science.* Wiley, Chichester

McLaren R A (1990) Establishing a corporate GIS from component datasets – the database issues. *Mapping Awareness* 4 (2): 52–8

Oxborrow E P (1989) *Databases and Database Systems: Concepts and Issues*, 2nd edn. Chartwell Bratt, Sweden

Worboys M F (1995) *GIS: a Computing Perspective.* Taylor and Francis, London

Worboys M F (1999) Relational Databases and Beyond. In: Longley P A, Goodchild M F, Maguire D J, Rhind D W (eds) *Geographical Information Systems.* New York, Wiley, Vol. 1, pp. 373–84

5 Data input and editing

KEY QUESTIONS AND ISSUES

- What is the data stream?
- What is data encoding?
- What methods of data encoding are available?
- How are paper maps digitized?
- What problems are faced when encoding digital and analogue data?
- What methods of data editing and conversion are used in GIS?
- How is an integrated GIS created?

Introduction

So far this book has presented the characteristics of spatial and attribute data and considered how to represent and structure them in the computer. This chapter introduces the methods for getting data into the computer – a process known as data encoding. Normally, of course, data would be entered into a GIS before being structured. However, since the characteristics of data and the way they are modelled influence data encoding methods, these have been covered first in this book.

A GIS without data can be likened to a car without fuel – without fuel you cannot go anywhere; without data a GIS will not produce output. However, this is perhaps where the similarity ends, as there is only one place to obtain fuel (a petrol station) and only one method of putting fuel into a car (using a petrol pump). Spatial data, on the other hand, can be obtained from many different sources (see Chapter 2), in different formats, and can be input to GIS using a number of different methods. Maps, which may come as paper sheets or digital files, may be input by digitizing, scanning or direct file transfer; aerial photographs may be scanned into a GIS; and satellite images may be downloaded from digital media. In addition, data can be directly

input to GIS from field equipment such as GPS, or from sources of ready-prepared data from data 'retailers' or across the Internet. Once in a GIS, data almost always need to be corrected and manipulated to ensure that they can be structured according to the required data model. Problems that may have to be addressed at this stage of a GIS project include:

- the re-projection of data from different map sources to a common projection;
- the generalization of complex data to provide a simpler data set; or
- the matching and joining of adjacent map sheets once the data are in digital form.

This chapter looks in detail at the range of methods available to get data into a GIS. These include keyboard entry, digitizing, scanning and electronic data transfer. Then, methods of data editing and manipulation are reviewed, including re-projection, transformation and edge matching. The whole process of data encoding and editing is often called the 'data stream'. This is outlined in Figure 5.1 and used as a framework for the chapter.

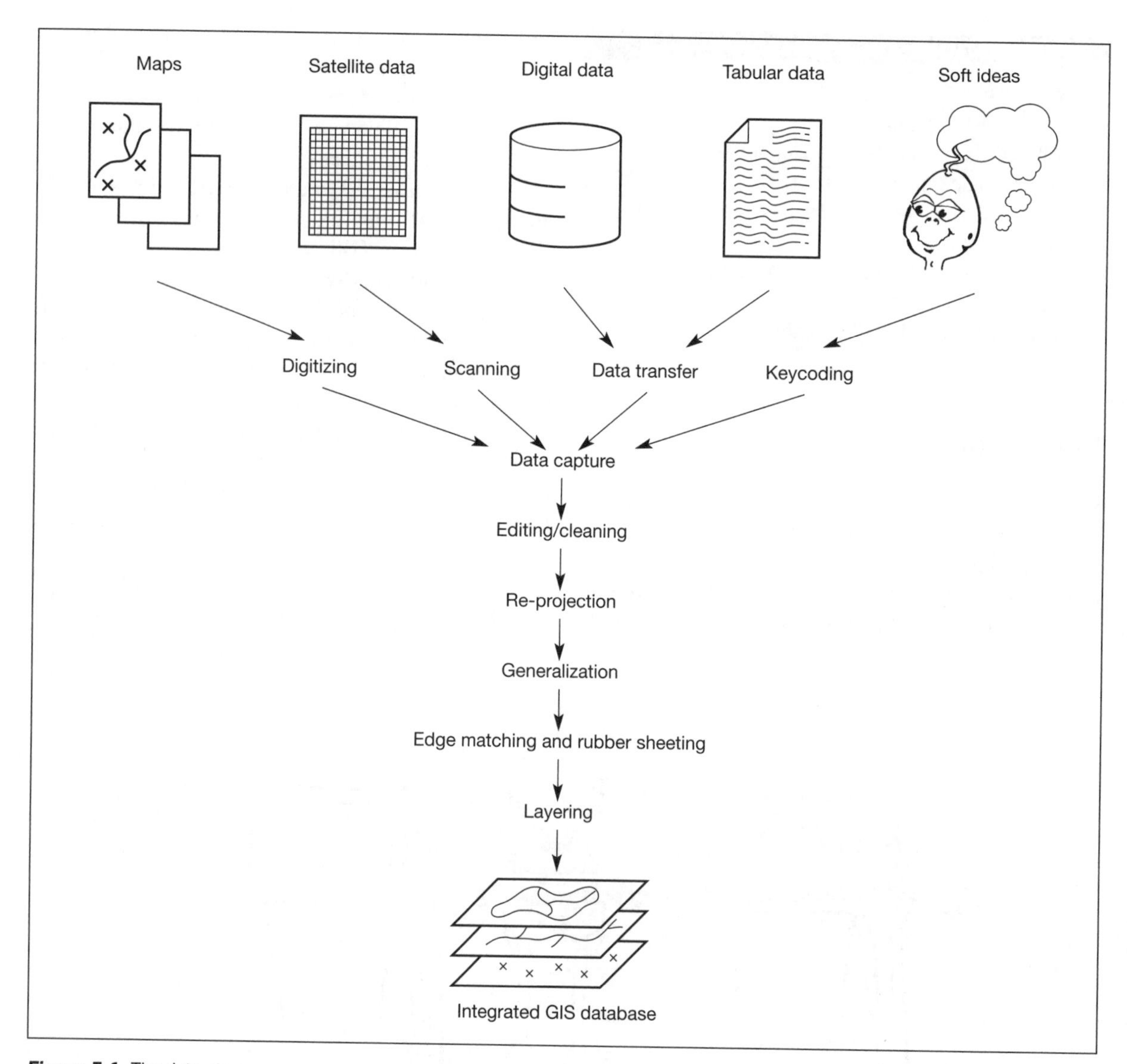

Figure 5.1 The data stream

Before further explanation of the stages in the data stream, it is necessary to make a distinction between analogue (non-digital) and digital sources of spatial data. Analogue data are normally in paper form, and include paper maps, tables of statistics and hard-copy (printed) aerial photographs. These data all need to be converted to digital form before use in a GIS, thus the data encoding and correction procedures are longer than those for digital data. Digital data are already in computer-readable formats and are supplied on diskette, magnetic tape or CD-ROM or across a computer network. Map data, aerial photographs, satellite imagery, data from databases and automatic data collection devices (such as data loggers and GPS) are all available in digital form. If data were all of the same type, format, scale and resolution, then data encoding and integration would be simple. However, since the characteristics of spatial data are as varied as their sources, the task is complex. This variety has implications for the way data are encoded and manipulated to develop an integrated GIS database. Much effort has been made in recent years to develop universal GIS data standards as described in Chapter 2.

Methods of data input

Data in analogue or digital form need to be encoded to be compatible with the GIS being used. This would be a relatively straightforward exercise if all GIS packages used the same spatial and attribute data models. However, there are many different GIS packages and many different approaches to the handling of spatial and attribute data. Chapter 3 reviewed the two main data models in use in GIS (raster and vector), and Chapter 4 considered the relational database model

Table 5.1 Possible encoding methods for different data sources

Data source	*Analogue or digital source*	*Possible encoding methods*	*Examples*
Tabular data	Analogue	• Keyboard entry • Text scanning	• Address lists of hotel guests • Tables of regional tourism statistics from official publications
Map data	Analogue	• Manual digitizing • Automatic digitizing • Scanning	• Historical maps of settlement and agriculture • Infrastructure and administrative maps
Aerial photographs	Analogue	• Manual digitizing • Automatic digitizing • Scanning	• Ski piste locations • Extent of spring floods
Tabular data	Digital	• Digital file transfer (with reformatting if necessary)	• National population census data • Data from meteorological station recording equipment
Map data	Digital	• Digital file transfer (with reformatting)	• Digital topographic data from national mapping agency • Digital height (DTM) data
Aerial photographs	Digital	• Digital file transfer (with reformatting)	• Background contextual data • Ski piste locations
Satellite imagery	Digital	• Digital file transfer • Image processing and reformatting	• Land use data • Forest condition data

widely used for attribute data. While the general principles behind these models remain the same in different GIS packages, the implementation may be very different. Different too are the problems faced when transferring data from one package to another, or when preparing data for data analysis. The situation is being eased by the development of standards for spatial data encoding, a topic that will be returned to later in this chapter and in Chapter 10. Table 5.1 summarizes possible data encoding methods and gives an indication of their appropriateness for different data sources.

All data in analogue form need to be converted to digital form before they can be input into GIS. Four methods are widely used: keyboard entry, manual digitizing, automatic digitizing and scanning. Keyboard entry may be appropriate for tabular data, or for small numbers of co-ordinate pairs read from a paper map source or pocket GPS. Digitizing is widely used for the encoding of paper maps and data from interpreted air photographs. Scanning represents a faster encoding method for these data sources, although the resulting digital data may require considerable processing before analysis is possible.

Digital data must be downloaded from their source media (diskette, CD-ROM or the Internet) and may require reformatting to convert them to an appropriate format for the GIS being used. Reformatting or conversion may also be required after analogue data have been converted to digital form. For example, after scanning a paper map, the file produced by the scanning equipment may not be compatible with the GIS, so reformatting may be necessary. Keyboard entry, manual digitizing, automatic digitizing (including scanning) and digital conversion are covered in greater detail below.

Keyboard entry

Keyboard entry, often referred to as keycoding, is the entry of data into a file at a computer terminal. This technique is used for attribute data that are only available on paper. If details of the hotels in Happy Valley were obtained from a tourist guide, both spatial data (the locations of the hotels – probably given as postal codes) and attributes of the hotels (number of rooms, standard and full address) would be entered at a keyboard. For a small number of hotels keyboard entry is a manageable task, although typographical errors are very likely. If there were hundreds of hotels to be coded an alternative method would probably be sought. Text scanners and optical character recognition (OCR) software can be used to read in data automatically. Attribute data, once in a digital format, are linked to the relevant map features in the spatial database using identification codes. These are unique codes that are allocated to each point, line and area feature in the data set.

The co-ordinates of spatial entities can be encoded by keyboard entry, although this method is used only when co-ordinates are known and there are not too many of them. If the locations of the Happy Valley hotels were to be entered as co-ordinates then these could be read from a paper map and input at the keyboard. Where there are large numbers of co-ordinates and features to be encoded it is more common to use manual or automatic digitizing.

Manual digitizing

The most common method of encoding spatial features from paper maps is manual digitizing. It is an appropriate technique when a selection of features are required from a paper map. For example, the Happy Valley road network might be required from a topographical map of the area. Manual digitizing is also used for map encoding where it is important to reflect the topology of features, since information about the direction of line features can be included. It is also used for digitizing features of interest from hard-copy aerial photographs (for example, snow cover on a particular date could be encoded from aerial photographs).

Manual digitizing requires a table digitizer that is linked to a computer workstation (Figure 5.2). The table digitizer is essentially a large flat tablet, the surface of which is underlain by a very fine mesh of wires. Attached to the digitizer via a cable is a cursor that can be moved freely over the surface of the table. Buttons on the cursor allow the user to send instructions to the computer. The position of the cursor on the table is registered by reference to its position above the wire mesh.

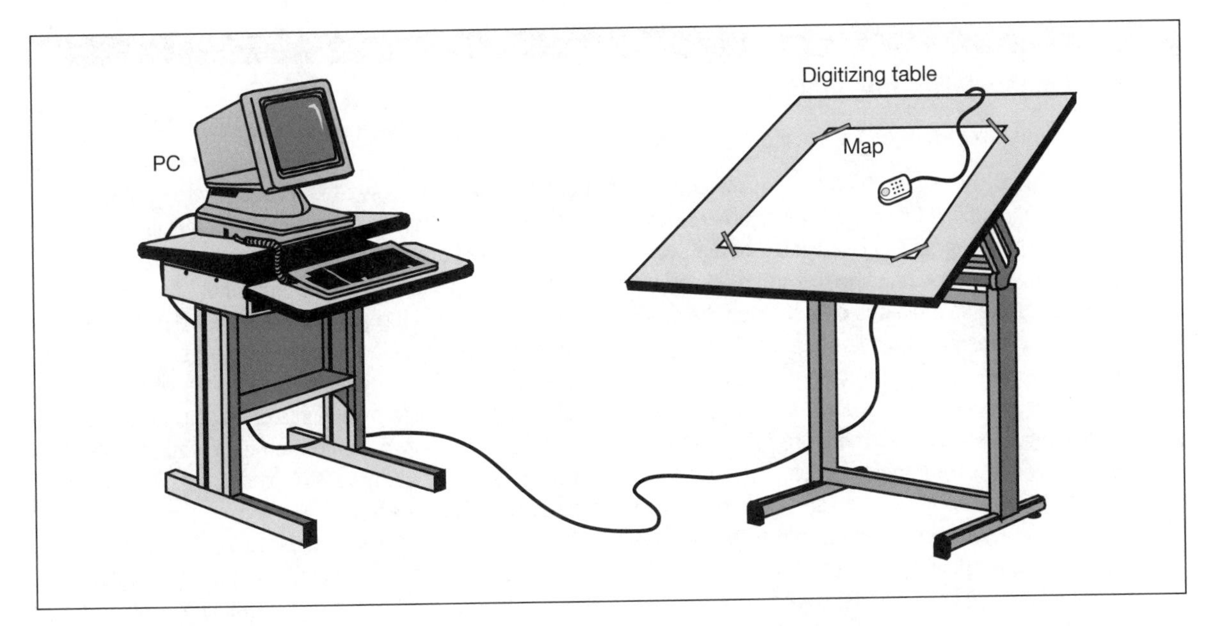

Figure 5.2 Digitizing table and PC workstation

Using a digitizer is quite straightforward. However, to achieve good results experience and awareness of some of the problems and limitations of digitizing are required. The general procedure for digitizing is summarized in Box 5.1.

Most manual digitizers may be used in one of two modes: point mode or stream mode. In point mode the user begins digitizing each line segment with a start node, records each change in direction of the line with a digitized point and finishes the segment with an end node. Thus, a straight line can be digitized with just two points, the start and end nodes. For more complex lines, a greater number of points are required between the start and end nodes. Smooth curves are problematic since they require an infinite number of points to record their true shape. In practice, the user must choose a sensible number of points to represent the curve (a form of user generalization). In addition, the digitizing equipment will have a minimum resolution governed by the distance between the wires in the digitizing table (typically about 0.1 mm). However, some digitizing packages allow the user to record smooth curves as mathematically defined splines or Bézier curves.

In stream mode the digitizer is set up to record points according to a stated time interval or on a distance basis (Jackson and Woodsford, 1991). Once the user has recorded the start of a line the digitizer might be set to record a point automatically every 0.5 seconds and the user must move the cursor along the line to record its shape. An end node is required to stop the digitizer recording further points. The speed at which the cursor is moved along the line determines the number of points recorded. Thus, where the line is more complex and the cursor needs to be moved more slowly and with more care, a greater number of points will be recorded. Conversely, where the line is straight, the cursor can be moved more quickly and fewer points are recorded. The choice between point mode and stream mode digitizing is largely a matter of personal preference. Stream mode digitizing requires more skill than point mode digitizing, and for an experienced user may be a faster method. Stream mode will usually generate more points, and hence larger files, than point mode.

The manual digitizing of paper maps is one of the main sources of positional error in GIS. The accuracy of data generated by this method of encoding is dependent on many factors, including the scale and resolution of the source map, and the quality of the equipment and software

BOX 5.1

Using a manual digitizing table

The procedure followed when digitizing a paper map using a manual digitizer has five stages:

1. *Registration.* The map to be digitized is fixed firmly to the table top with sticky tape. Five or more control points are identified (usually the four corners of the map sheet and one or more grid intersections in the middle). The geographic co-ordinates of the control points are noted and their locations digitized by positioning the cross-hairs on the cursor exactly over them and pressing the 'digitize' button on the cursor. This sends the co-ordinates of a point on the table to the computer and stores them in a file as 'digitizer co-ordinates'. These co-ordinates are the positions of the cursor cross-hairs relative to the wire mesh in the table. They are usually stored as decimal inches or centimetres from the bottom left-hand corner of the table. Later, the geographic co-ordinates of the control points are used to transform all digitizer co-ordinates into geographic co-ordinates. Therefore, it is essential that the map is carefully registered on the digitizer table to ensure an accurate transformation of digitized features from digitizer to geographic co-ordinates. Once the map has been registered the user may begin digitizing the desired features from the map.
2. *Digitizing point features.* Point features, for example spot heights, hotel locations or meteorological stations, are recorded as a single digitized point. A unique code number or identifier is added so that attribute information may be attached later. For instance, the hotel with ID number '1' would later be identified as 'Mountain View'.
3. *Digitizing line features.* Line features (such as roads or rivers) are digitized as a series of points that the software will join with straight line segments. In some GIS packages lines are referred to as arcs, and their start and end points as nodes. This gives rise to the term arc–node topology, used to describe a method of structuring line features (see Chapter 3). As with point features, a unique code number or identifier is added to each line during the digitizing process and attribute data attached using this code. For a road, data describing road category, number of carriageways, surface type, date of construction and last resurfacing might be added.
4. *Digitizing area (polygon) features.* Area features or polygons, for example forested areas or administrative boundaries, are digitized as a series of points linked together by line segments in the same way as line features. Here it is important that the start and end points join to form a complete area. Polygons can be digitized as a series of individual lines, which are later joined to form areas. In this case it is important that each line segment is digitized only once.
5. *Adding attribute information.* Attribute data may be added to digitized polygon features by linking them to a centroid (or seed point) in each polygon. These are either digitized manually (after digitizing the polygon boundaries) or created automatically once the polygons have been encoded. Using a unique identifier or code number, attribute data can then be linked to the polygon centroids of appropriate polygons. In this way, the forest stand may have data relating to tree species, tree ages, tree numbers and timber volume attached to a point within the polygon.

being used. Errors can be introduced during the digitizing process by incorrect registration of the map document on the digitizer table or 'hand-wobble'.

A shaky hand will produce differences between the line on the map and its digitized facsimile. Published estimates of the accuracy of manual digitizing range from as high as ± 0.8 mm (Dunn *et al.*, 1990) to as low as ± 0.054 mm (Bolstad *et al.*, 1990). As a rule an experienced digitizing technician should be able to encode data with an accuracy equal to the width of the line they are digitizing. Walsby (1995) considers that line characteristics have a significant influence on error creation, with the greatest number of errors being produced during the digitizing of small thin polygons and sinuous lines. In addition, she found that operators with limited cartographic experience who were also infrequent digitizers were most likely to create errors. Errors produced during manual digitizing are covered in more detail in Chapter 10.

Automatic digitizing

Manual digitizing is a time-consuming and tedious process. If large numbers of complex maps need to be digitized it is worth considering alternative, although often more expensive, methods. Two automatic digitizing methods are considered here: scanning and automatic line following.

Scanning is the most commonly used method of automatic digitizing. Scanning is an appropriate method of data encoding when raster data are required, since this is the automatic output format from most scanning software. Thus, scanning may be used to input a complete topographic map that will be used as a background raster data set for the over-plotting of vector infrastructure data such as pipelines or cables. In this case a raster background map is extremely useful as a contextual basis for the data of real interest.

A scanner is a piece of hardware for converting an analogue source document into digital raster format (Jackson and Woodsford, 1991). All scanners work by sampling the source document using transmitted or reflected light. Jackson and Woodsford (1991) provide a comprehensive review of scanning technology. The cheapest scanners are small flat-bed scanners – a common PC peripheral. High-quality and large-format scanners require the source document to be placed on a rotating drum, and a sensor moves along the axis of rotation.

Practical problems faced when scanning source documents include:

- the possibility of optical distortion when using flat-bed scanners;
- the automatic scanning of unwanted information (for example, hand-drawn annotations, folds in maps or coffee stains);
- the selection of appropriate scanning tolerances to ensure important data are encoded, and background data ignored;
- the format of files produced and the input of data to GIS software; and
- the amount of editing required to produce data suitable for analysis.

The general procedure for using a scanner is described in Box 5.2.

BOX 5.2

Using a Scanner

There are three different types of scanner in widespread use:

- flat-bed scanners;
- rotating drum scanners;
- large-format feed scanners.

Large-format feed scanners are most suitable for capturing data for input to GIS as they are relatively cheap, quick and accurate. Flat-bed scanners, while being a common PC peripheral, are too small and inaccurate; and drum scanners, despite being very accurate, tend to be too slow and expensive.

All scanners work on the same principles (Hohl, 1998). A scanner has a light source, a background (or source document) and a lens. During scanning the absence or presence of light is detected as one of the three components moves past the other two.

The user should be aware of a number of issues when scanning maps for use in GIS:

Output quality

If the output is to be used as a visual backdrop for other data in a GIS then quality may not be a major issue. If, on the other hand, scanned output is to be used to produce vector data then the output must be as sharp and clear as possible. Quality can be influenced by the setting of a threshold above which all values are translated as white, and below which all values are black (Hohl, 1998). Setting brightness and contrast levels can also affect the quality of images obtained, as can gamma correction (a method that looks at a histogram of the image and places points strategically along the histogram to isolate data types). Filtering methods may be used to selectively remove noise from a document. This may be particularly useful when scanning old maps.

Resolution

This is the density of the raster image produced by the scanning process. The resolution of scanners is usually measured in dots per inch (dpi) as a linear measurement

BOX 5.2 CONTINUED

along the scan line. In general, choose a resolution that matches the data being captured. Commonly-used guidelines are 200 dpi for text, 300 dpi for line maps and 400 dpi for high-quality orthophotos (Hohl, 1998). Interpolation may be used to increase the resolution of a scan. The values of pixels smaller than the scan head is physically capable of seeing are interpolated. This can create problems when scanning maps with sharp lines such as roads or contours as it tends to produce fuzzy rather than sharp lines in the scanned image.

Accuracy

The accuracy of the scanned image is important if the image is to be imported into a GIS database. The scanned document needs to be fit for its intended use in terms of its physical as well as its cartographic qualities. Remember that a scanner will not distinguish between coffee stains and real map data! The engineering tolerances of the scanner itself will also affect the accuracy of the output. For example, the rotating drum mechanism of drum scanners is much more accurate than the rollers employed to pull a map through a large-format feed scanner.

Vectorization

The output from scanned maps is often used to generate vector data. This process involves either automatic or interactive (user-controlled) raster to vector conversion. Problems occur with this process due to topographical effects at intersections, generalization of features smaller than the resolution of the scanner and the subsequent coding of attributes (these are dealt with further in Chapter 10).

Georeferencing

Ultimately the output from a scanner needs to be correctly referenced according to the co-ordinate system used in the GIS. Normally this process is controlled using linear transformation from the row and column number of the scanned image to the chosen geographic co-ordinate system. This requires a series of fixed reference points on the image for which the geographic co-ordinates are known. Distortion across scanned images can create a problem in this transformation when using cheap flat-bed scanners.

The accuracy of scanned output data depends on the quality of the scanner, the quality of the image-processing software used to process the scanned data, and the quality (and complexity) of the source document.

The resolution of the scanner used affects the quality, and quantity, of output data. Cheaper flat-bed scanners have resolutions of 50–200 dpi whereas more expensive drum scanners use resolutions of 500–2500 dpi. The higher the resolution, the larger the volumes of data produced. This can place a heavy burden on computing power and storage space, and increase the time spent editing the scanned data to make it usable by a GIS.

While digitizing tools may be provided within GIS software, scanning and image-processing software are seldom provided; therefore, users must consider the format of output files. If these are not compatible with the GIS being used, extra reformatting may be necessary. Image-processing software can be used to distinguish different features from a single scanned image. For example, data layers for roads, water and forestry may be produced if image-processing software is used to distinguish between map features on the basis of their colour.

Either ordinary 'off-the-shelf' maps or the colour separates used by publishers can be encoded using scanning. If colour separates are used the colour schemes used to distinguish features in a published map can be used to advantage. For example, on UK Ordnance Survey maps contour lines are represented as orange or brown lines made up of the primary colours yellow and magenta. Large volumes of scanned data require substantial computer power, occupy sizeable file storage space, and need more editing to make the data usable in a GIS. Thus, the scanning and subsequent editing of contour data are made much easier.

Chrisman (1997) considers that current scanning technology can produce data to a higher quality than manual digitizing, and that the combination of scanned graphic and line detection algorithms controlled on-screen by the user will help to make manual digitizing obsolete.

Another type of automatic digitizer is the automatic line follower. This encoding method might be appropriate where digital versions of clear, distinctive lines on a map are required (such as country boundaries on a world map, or clearly distinguished railways on a topographic map). The method mimics manual digitizing and uses a laser- and light-sensitive device to follow the lines on the map. Whereas scanners are raster devices, the automatic line follower is a vector device and produces output as (x,y) co-ordinate strings. The data produced by this method are suitable for vector GIS. Automatic line followers are not as common as scanners, largely due to their complexity. In addition, difficulties may be faced when digitizing features such as dashed or contour lines (contour lines are commonly broken to allow the insertion of height values). Considerable editing and checking of data may be necessary. However, automatic digitizing methods can be used effectively and are now widely employed by commercial data providers as they allow the production of data at a reasonable cost.

Electronic data transfer

Given the difficulties and the time associated with keyboard encoding, manual digitizing and automatic digitizing, the prospect of using data already in digital form is appealing. If a digital copy of the data required is available in a form compatible with your GIS, the input of these data into your GIS is merely a question of electronic data transfer. However, it is more than likely that the data you require will be in a different digital format to that recognized by your GIS. Therefore, the process of digital data transfer often has to be followed by data conversion. During conversion the data are changed to an appropriate format for use in your GIS.

Spatial data may be collected in digital form and transferred from devices such as GPS receivers, total stations (electronic distance-metering theodolites), and data loggers attached to all manner of scientific monitoring equipment. All that is required is a download cable and data communications software for a user to download the data to a file on their computer. In some cases it may be possible to output data from a collection device in a GIS format. For example, GPS receivers are available that will output data into the format required by a range of GIS packages, as well as into common spreadsheet and standard GPS formats.

Electronic data transfer will also be necessary if data have been purchased from a data supplier, or obtained from another agency that originally encoded the data. For example, the UK Ordnance Survey and US Geological Survey both provide data for the construction of digital terrain models in pre-prepared digital formats. In addition, both agencies, in common with many other national mapping agencies, provide detailed topographic data in a range of digital formats. Remotely sensed data are normally provided in electronic form, since this is the easiest way to transport the large files representing individual 'scenes' from the various sensors.

Thus, electronic data transfer is an appropriate method of data encoding where the data are already available in digital form (from a data collection device or another organization) in a format compatible with your GIS software. However, it may also be necessary to encode data using electronic transfer where the data are not in a format that is compatible with your GIS. To deal with this you will need to transform or convert your data to an appropriate format. Most GIS software packages will allow you to convert data from a number of different formats; however, there may still be times when it is necessary to use additional software or even write your own data transfer routine to alter formatting of data.

Finding out what data exist, how much they will cost, where they can be found and the format in which they are available are some of the most challenging stages in the development of a GIS application. Therefore, obtaining data from other sources requires users to address a range of important questions. These are outlined in Box 5.3.

As with so many aspects of GIS, Box 5.3 illustrates that there are not only technical issues to take into account when acquiring digital data from another source, but also human and organizational issues. However, if data can be successfully obtained and input to GIS, considerable time may be saved when electronic data transfer is possible, and duplication of effort may be avoided.

BOX 5.3

Obtaining spatial data from other sources

Users must address a number of questions if they wish to obtain data in digital form from another source:

1. *What data are available?* There are no data hypermarkets where you can go to browse, select and purchase spatial data. Instead, you must rummage around in the data marketplace trying to find what you need at an affordable price. Advertisements for data in digital format can be found in trade magazines, data can be obtained from national mapping agencies and a range of data is available from organizations via the Internet. However, since the search process can be time-consuming, and rather 'hit or miss', databases of data sources have been created to help users locate appropriate data. These databases are known as metadatabases. Two European examples of metadatabase projects, MEGRIN (Salge, 1996) and ESMI (Scholten, 1997), have also been designed to stimulate the data market and provide better access to data sources. The Internet is the nearest thing there is to a data hypermarket for GIS. Several organisations have set up data 'clearing houses' where you can browse for and purchase data online.
2. *What will the data cost?* Data are very difficult to price, so while some digital data are expensive, others are freely available over the Internet. The pricing policy varies depending on the agency that collected the data in the first place, and this in turn may be affected by national legislation. Because of the possibility of updating digital data sets, many are bought on an annual licensing agreement, entitling the purchaser to new versions and corrections. The cost of digital data may be an inhibiting factor for some GIS applications. In extreme cases, users may duplicate data by redigitizing in order to avoid the costs of purchase. In addition to the cost of purchasing data there may be copyright charges and restrictions to be taken into account.
3. *On what media will the data be supplied?* Data may be available in a number of ways. These range from magnetic media (diskette or tape) and optical disks (CD-ROM) to network transfers across internal local area networks (LAN) or the Internet. Even these methods of data input are not without their problems. Magnetic media may be damaged during transit or by faults in the copying process, while networks are often subject to faults or interruptions. These may lead to data being lost or corrupted. Problems may also be experienced with out-of-date media. Older data sets may still be stored on half-inch magnetic tape, top-loading disk packs or even punched cards. Finding a suitable reader for these old media formats is becoming increasingly difficult.
4. *What format will the data be in – will standards be adhered to?* In recent years, a great deal of effort has been put into the development of national and international data standards to ensure data quality and to improve compatibility and transfer of data between systems. Harding and Wilkinson (1997) list 27 different national and international organizations responsible for the development of standards related to geographical information. In addition, the GIS software vendors have developed their own standards. As a result, software is becoming increasingly compatible and there is a plethora of standards, developed by vendors, users and national and international geographic information agencies. Data standards and their implications for data quality are discussed further in Chapter 10.

Data editing

As a result of the problems encountered during data encoding, you cannot expect to input an error-free data set into your GIS. Data may include errors derived from the original source data, as well as errors that have been introduced during the encoding process. There may be errors in co-ordinate data as well as inaccuracies and uncertainty in attribute data. Data quality and data error will be considered in greater detail in Chapter 10. However, good practice in GIS involves continuous management of data quality, and it is normal at this stage in the data stream to make special provision for the identification and correction of errors. It is better to intercept errors before they contaminate the GIS database and go on to infect (propagate) the higher levels of information that are generated. The process is known as data editing or 'cleaning'. Data editing can be likened to the filter between the fuel tank and the engine that keeps the fuel clean and the engine running smoothly. Three topics are covered here: detection and correction of errors; re-projection, transformation and generalization; and edge matching and rubber sheeting.

Detecting and correcting errors

Errors in input data may derive from three main sources: errors in the source data; errors introduced during encoding; and errors propagated during data transfer and conversion. Errors in source data may be difficult to identify. For example, there may be subtle errors in a paper map source used for digitizing because of the methods used by particular surveyors, or there may be printing errors in paper-based records used as source data.

During encoding a range of errors can be introduced. During keyboard encoding it is easy for an operator to make a typing mistake; during digitizing an operator may encode the wrong line; and folds and stains can easily be scanned and mistaken for real geographical features. During data transfer, conversion of data between different formats required by different packages may lead to a loss of data.

Errors in attribute data are relatively easy to spot and may be identified using manual comparison with the original data. For example, if the operator notices that a hotel has been coded as a cafe, then the attribute database may be corrected accordingly. Various methods, in addition to manual comparison, exist for the correction of attribute errors. These are described in Box 5.4.

Errors in spatial data are often more difficult to identify and correct than errors in attribute data. These errors take many forms, depending on the data model being used (vector or raster) and the method of data capture. In our Happy Valley example, a spatial error may arise if a meteorological station has been located in the wrong place, if a forest polygon has been wrongly identified during image processing or if a railway line has been erroneously digitized as a road.

BOX 5.4

Methods of attribute data checking

Several methods may be used to check for errors in the encoding of attribute data. These include:

1. *Impossible values.* Simple checks for impossible data values can be made when the range of the data is known. Data values falling outside this range are obviously incorrect. For example, a negative rainfall measurement is impossible, as is a slope of 100 degrees.
2. *Extreme values.* Extreme data values should be cross-checked against the source document to see if they are correct. An entry in the attribute database that says the Mountain View Hotel has 2000 rooms needs to be checked. It is more likely that this hotel has 200 rooms and that the error is the result of a typing mistake.
3. *Internal consistency.* Checks can be made against summary statistics provided with source documents where data are derived from statistical tables. Totals and means for attribute data entered into the GIS should tally with the totals and means reported in the source document. If a discrepancy is found, then there must be an error somewhere in the attribute data.
4. *Scattergrams.* If two or more variables in the attribute data are correlated, then errors can be identified using scattergrams. The two variables are plotted along the x and y axes of a graph and values that depart noticeably from the regression line are investigated. Examples of correlated variables from Happy Valley might be altitude and temperature, or the category of a hotel and the cost of accommodation.
5. *Trend surfaces.* Trend surface analyses may be used to highlight points with values that depart markedly from the general trend of the data. This technique may be useful where a regional trend is known to exist. For example, in the case of Happy Valley most ski accidents occur on the nursery slopes and the general trend is for accidents to decrease as the ski piste becomes more difficult. Therefore, an advanced piste recording a high number of accidents reflects either an error in the data set or an area requiring investigation.

Table 5.2 Common errors in spatial data

Error	*Description*
Missing entities	Missing points, lines or boundary segments
Duplicate entities	Points, lines or boundary segments that have been digitized twice
Mislocated entities	Points, lines or boundary segments digitized in the wrong place
Missing labels	Unidentified polygons
Duplicate labels	Two or more identification labels for the same polygon
Artifacts of digitizing	Undershoots, overshoots, wrongly placed nodes, loops and spikes
Noise	Irrelevant data entered during digitizing, scanning or data transfer

Examples of errors that may arise during encoding (especially during manual digitizing) are presented in Table 5.2. Figure 5.3 illustrates some of the errors that may be encountered in vector data. Chrisman (1997) suggests that certain types of error can help to identify other problems with encoded data. For example, in an area data layer 'dead-end nodes' might indicate missing lines, overshoots or undershoots. The user can look for these features to direct editing rather than having to examine the whole map. Most GIS packages will provide a suite of editing tools for the identification and removal of errors in vector data. Corrections can be done interactively by the operator 'on-screen', or automatically by the GIS software. However, visual comparison of the digitized data against the source document, either on paper or on the computer screen, is a good starting point. This will reveal obvious omissions, duplications and erroneous additions. Systematic errors such as overshoots in digitized lines can be corrected automatically by some digitizing software, and it is important for data to be absolutely correct if topology is to be created for a vector data set. This is covered further in Chapter 10. Automatic corrections can save many hours of work but need to be used with care as incorrectly specified tolerances may miss some errors or correct 'errors' that never existed in the first place (see Chapter 10).

Errors will also be present in raster data. In common with vector data, missing entities and noise are particular problems. Data for some areas may be difficult to collect, owing to environmental or cultural obstacles. Trying to obtain aerial photographs of an area containing a major airport is one example, since it may be difficult to fly across the area using a flight path suitable for photography. Similarly, it may be difficult to get clear images of vegetation cover in an area during a rainy season using certain sensors. Noise may be inadvertently added to the data, either when they were first collected or during processing. This noise often shows up as scattered pixels whose attributes do not conform to those of neighbouring pixels. For example, an individual pixel representing water may be seen in a large area of forest. While this may be correct, it could

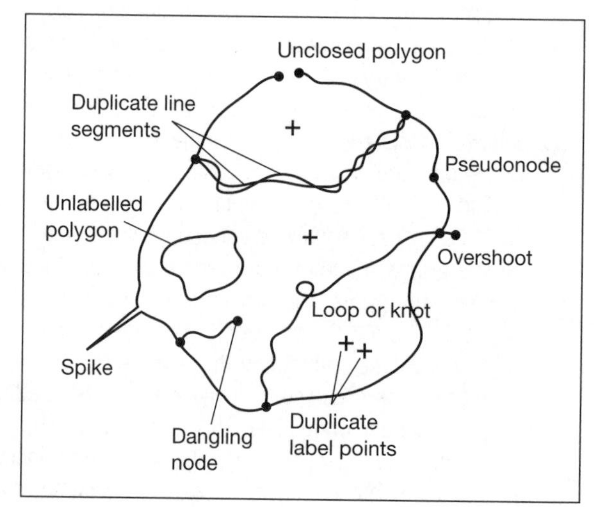

Figure 5.3 Examples of spatial error in vector data

also be the result of noise and needs to be checked. This form of error may be removed by filtering. Filtering is considered in this book as an analysis technique (Chapter 6) but, in brief, it involves passing a filter (a small grid of pixels specified by the user – often a 3 × 3 pixel square is used) over the noisy data set and recalculating the value of the central (target) pixel as a function of all the pixel values within the filter. This technique needs to be used with care as genuine features in the data can be lost if too large a filter is used. Figure 5.4 illustrates the effect of different-sized filters. Further details of errors that may be added to data during scanning and later processing are presented in Chapter 10.

Re-projection, transformation and generalization

Once spatial and attribute data have been encoded and edited, it may be necessary to process the data geometrically in order to provide a common framework of reference. Chapter 2 showed how the projection system of the source document may affect the positioning of spatial co-ordinates. The

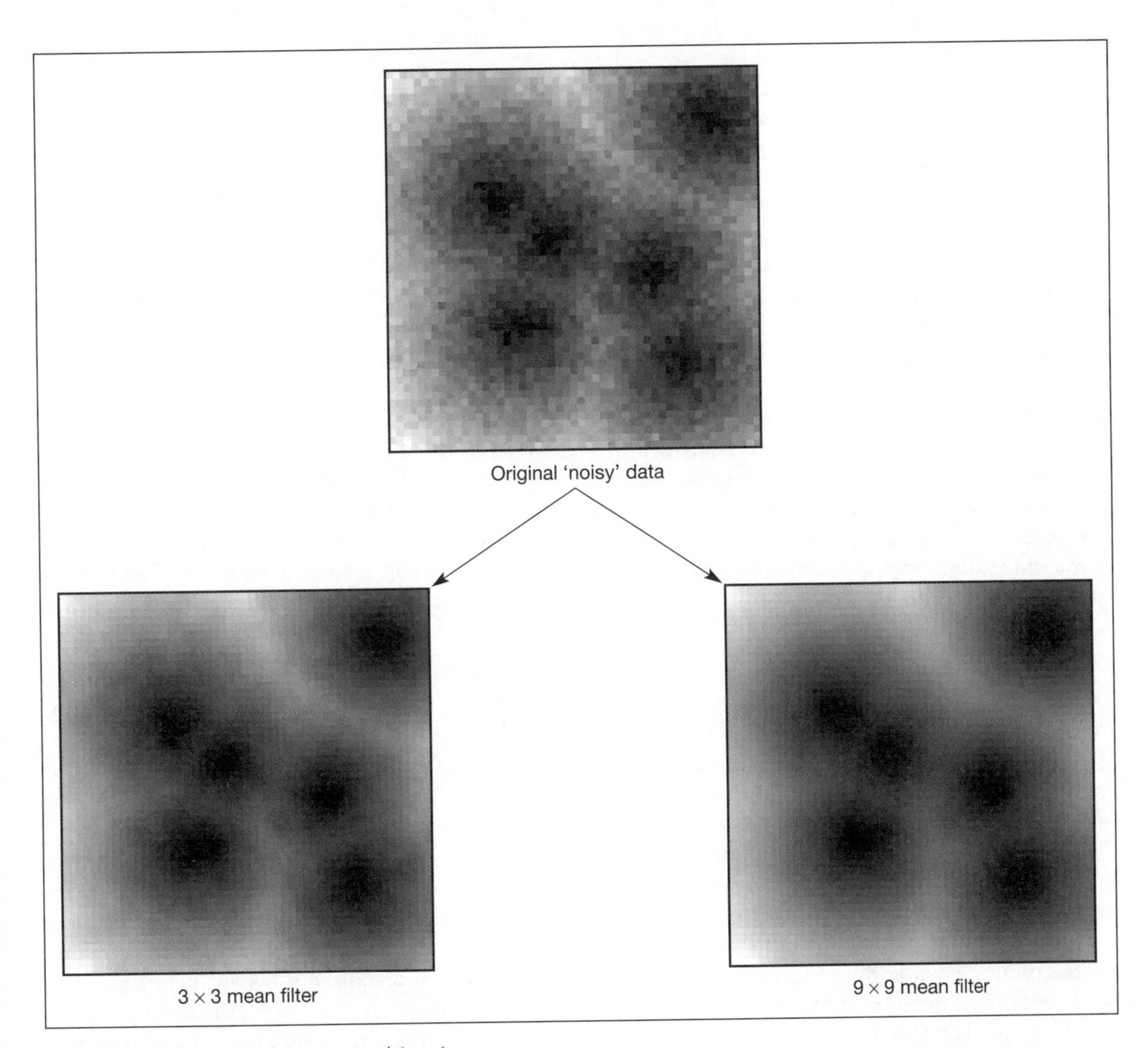

Figure 5.4 Filtering noise from a raster data set

scale and resolution of the source data are also important and need to be taken into account when combining data from a range of sources into a final integrated database. This section briefly considers the role of re-projection, transformation and generalization in the data stream.

Data derived from maps drawn on different projections will need to be converted to a common projection system before they can be combined or analysed. If not re-projected, data derived from a source map drawn using one projection will not plot in the same location as data derived from another source map using a different projection system. For example, if a coastline is digitized from a navigation chart drawn in the Mercator projection (cylindrical) and the internal census boundaries of the country are digitized from a map drawn using the Alber's Equal Area (conic) projection (as in the case of the US Bureau of Census) then the census boundaries along the coast will not plot directly on top of the coastline. In this case they will be offset and will need to be re-projected into a common projection system before being combined.

Data derived from different sources may also be referenced using different co-ordinate systems. The grid systems used may have different origins, different units of measurement or different orientation. If so, it will be necessary to transform the co-ordinates of each of the input data sets onto a common grid system. This is quite easily done and involves linear mathematical transformations. Some of the other methods commonly used are:

- *Translation and scaling*. One data set may be referenced in 1-metre co-ordinates while another is referenced in 10-metre co-ordinates. If a common grid system of 1-metre co-ordinates is required, then this is a simply a case of multiplying the co-ordinates in the 10-metre data set by a factor of 10.
- *Creating a common origin*. If two data sets use the same co-ordinate resolution but do not share the same origin, then the origin of one of the data sets may be shifted in line with the other simply by adding the difference between the two origins (dx,dy) to its co-ordinates.
- *Rotation*. Map co-ordinates may be rotated using simple trigonometry to fit one or more data sets onto a grid of common orientation.

Data may be derived from maps of different scales. The accuracy of the output from a GIS analysis can only be as good as the worst input data. Thus, if source maps of widely differing scales are to be used together, data derived from larger-scale mapping should be generalized to be comparable with the data derived from smaller-scale maps. This will also save processing time and disk space by avoiding the storage of unnecessary detail. Data derived from large-scale sources can be generalized once they have been input to the GIS. Routines exist in most vector GIS packages for weeding out unnecessary points from digitized lines such that the basic shape of the line is preserved. The simplest techniques for generalization delete points along a line at a fixed interval (for example, every third point). These techniques have the disadvantage that the shape of features may not be preserved (Laurini and Thompson, 1992). Most other methods are based on the Douglas–Peucker algorithm (Douglas and Peucker, 1973). This involves the following stages:

1. Joining the start and end nodes of a line with a straight line.
2. Examining the perpendicular distance from this straight line to individual vertices along the digitized line.
3. Discarding points within a certain threshold distance of the straight line.
4. Moving the straight line to join the start node with the point on the digitized line that was the greatest distance away from the straight line.
5. Repeating the process until there are no points left which are closer than the threshold distance.

Chrisman (1997) presents a practical example of the effects of line thinning. From a line consisting of 83 points he repeatedly thins the border between the states of Maryland and West Virginia to a line with only five points. Chrisman also highlights the risk of generating topological errors during line thinning where lines are close together.

When it is necessary to generalize raster data the most common method employed is to aggregate or amalgamate cells with the same attribute values. This approach results in a loss of detail which is often very severe. A more sympathetic approach is to use a filtering algorithm. These have been outlined above and are covered in more detail in

Chapter 6. If the main motivation for generalization is to save storage space, then, rather than resorting to one of the two techniques outlined above, it may be better to use an appropriate data compaction technique (see Chapter 3) as this will result in a volume reduction without any loss in detail.

Edge matching and rubber sheeting

When a study area extends across two or more map sheets small differences or mismatches between adjacent map sheets may need to be resolved. Normally, each map sheet would be digitized separately and then the adjacent sheets joined after editing, re-projection, transformation and generalization. The joining process is known as edge matching and involves three basic steps. First, mismatches at sheet boundaries must be resolved. Commonly, lines and polygon boundaries that straddle the edges of adjacent map sheets do not meet up when the maps are joined together. These must be joined together to complete features and ensure topologically correct data. More serious problems can occur when classification methods vary between map sheets. For example, different soil surveyors may interpret the pattern and type of soils differently, leading to serious differences on adjacent map sheets. This may require quite radical reclassification and reinterpretation to attempt a smooth join between sheets. This problem may also be seen in maps derived from multiple satellite images. If the satellite images were taken at different times of the day and under different weather and seasonal conditions then the classification of the composite image may produce artificial differences where images meet. These can be seen as clear straight lines at the sheet edges.

Second, for use as a vector data layer, topology must be rebuilt as new lines and polygons have been created from the segments that lie across map sheets. This process can be automated, but problems may occur due to the tolerances used. Too large a tolerance and small edge polygons may be lost, too small a tolerance and lines and polygon boundaries may remain unjoined. Finally, redundant map sheet boundary lines are deleted or dissolved. The process of edge matching is illustrated in Figure 5.5. Jackson and Woodsford (1991) note that although some quasi-automatic scanning edge matching is available, in practice the presence of anomalies in the data produced can require considerable human input to the process.

Certain data sources may give rise to internal distortions within individual map sheets. This is especially true for data derived from aerial photography as the movement of the aircraft and distortion caused by the camera lens can cause internal inaccuracies in the location of features within the image. These inaccuracies may remain even after transformation and re-projection. These problems can be rectified through a process known as rubber sheeting (or conflation). Rubber sheeting involves stretching the map in various directions as if it were drawn on a rubber sheet.

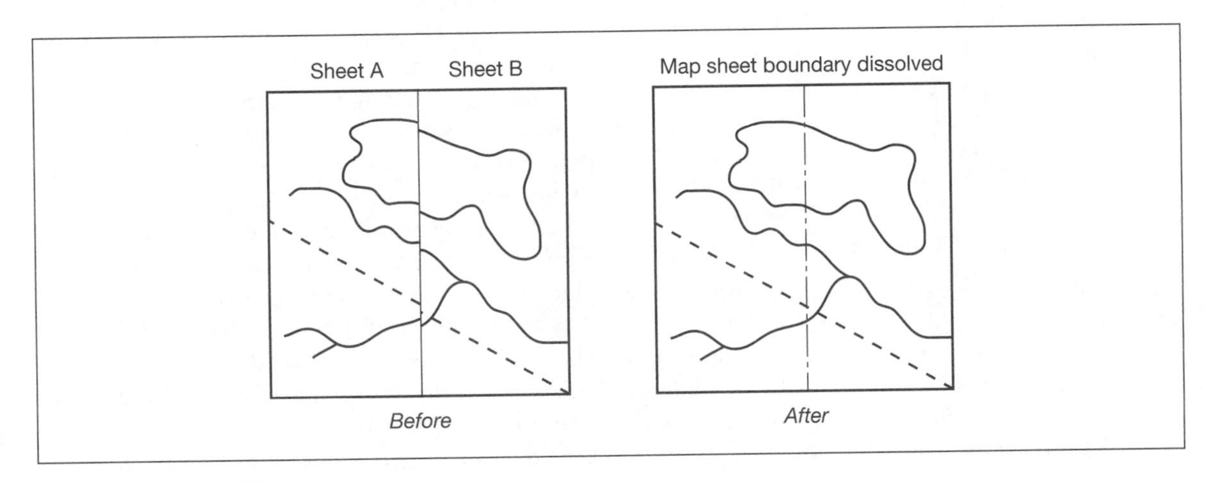

Figure 5.5 Edge matching

Objects on the map that are accurately placed are 'tacked down' and kept still while others that are in the wrong location or have the wrong shape are stretched to fit with the control points. These control points are fixed features that may be easily identified on the ground and on the image. Their true co-ordinates may be determined from a map covering the same area or from field observations using GPS. Distinctive buildings, road or stream intersections, peaks or coastal headlands may be useful control points. Figure 5.6 illustrates the process of rubber sheeting. This technique may also be used for re-projection where details of the base projection used in the source data are lacking. Difficulties associated with this technique include the lack of suitable control points and the processing time required for large and complex data sets. With too few control points the process of rubber sheeting is insufficiently controlled over much of the map sheet and may lead to unrealistic distortion in some areas.

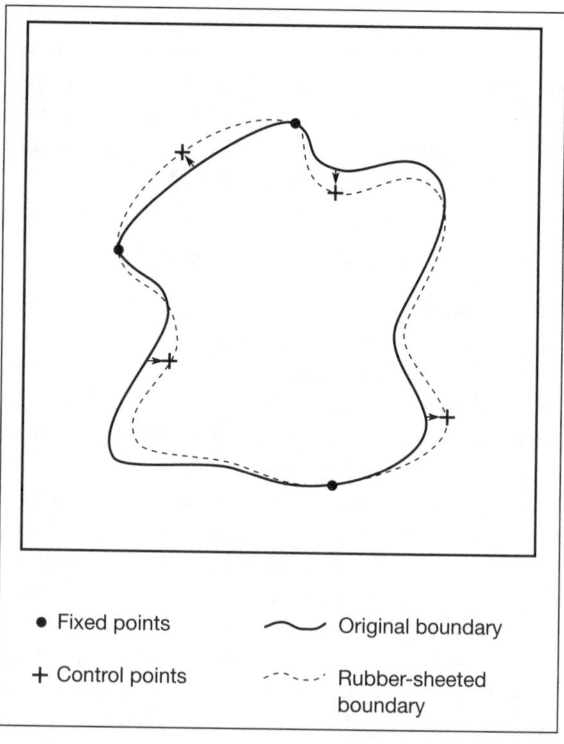

Figure 5.6 Rubber sheeting

Geocoding address data

Geocoding is the process of converting an address into a point location (McDonnell and Kemp, 1998). Since addresses are an important component of many spatial data sets, geocoding techniques have wide applicability during the encoding and preparation of data for analysis. Geocoding may be required to:

- turn hotel names and addresses collected during a questionnaire survey into a map of the distribution of holiday makers in Happy Valley;
- locate an incident in a street where an emergency response vehicle is required, based on details provided over the telephone by a witness at the scene;
- establish the home location of a store's credit card holders using postcodes collected on credit card application forms.

During geocoding the address itself, a postcode or another non-geographic descriptor (such as place name, property or parcel reference number) is used to determine the geographical co-ordinates of a location. UK postcodes can be geocoded with an Ordnance Survey grid reference. Several products are available that contain a single data record for each of the 1.6 million postcodes in the UK. In these files, each data record contains the OS Grid Reference (to 100 m resolution in England and Wales and to 10 m resolution in Scotland) and local government ward codes for the first address in each postcode. Many GIS software products can geocode US addresses, using the address, zip code or even place names.

Address matching is the process of geocoding street addresses to a street network. Locations are determined based on address ranges stored for each street segment.

Geocoding can be affected by the quality of data. Address data are frequently inconsistent: place names may be spelt incorrectly, addresses may be written in different formats and different abbreviations exist for words that appear frequently in addresses such as 'Street' and 'Avenue'. The use of standards for address data is particularly relevant to geocoding (see Chapter 2).

Towards an integrated database

As outlined in Chapter 3, most GIS adopt a layer view of the world. Each thematic layer in the database must be encoded, corrected and transformed to create a GIS ready for analysis. To explore how such an integrated database might be created in practice, the Happy Valley case study is considered in Box 5.5.

While Box 5.5 provides a hypothetical example of the work involved in the creation of an integrated GIS database, the variety and range of methods suggested is typical of real-world GIS applications. For the Zdarske Vrchy case study introduced in Chapter 1 over 30 different data sources were used to build the GIS database and an equal number of data encoding and editing techniques were necessary. For example, a map of ecological landscape zones was encoded by manual digitizing from a paper map. This map was itself a generalization of field surveys. A forest cover map was captured from remotely sensed data updated from more recent aerial photographs. The DTM used was interpolated (see Chapter 6) from contours which had been digitized from topographic maps. These data sets, and the others in the GIS, were re-projected to a common Transverse Mercator projection and transformed to a common scale of 1:50,000. The Zdarske Vrchy project lasted for four years in total, and the first two of these were almost entirely devoted to the data encoding and editing phase of the work (Petch *et al.*, 1995).

BOX 5.5

An integrated GIS for Happy Valley

The Happy Valley GIS team have a list of data requirements (Table 5.4). This includes details of the sources and uses of data and details of the entities, data models and attributes required. The data are from several different sources and require the use of a range of encoding and editing techniques. Some of the data are already in digital form while other data are from analogue sources.

Survey data in analogue form (the ski school and hotel surveys) are encoded by manual keyboard encoding. The roads are manually digitized in-house, and the team experiment with line-following software to automatically digitize the resort boundary. An external supplier of DTM data is sought and the digital data purchased at a scale compatible with the topographic map used for digitizing roads. An appropriate data format is documented in the licence agreement, which entitles the GIS team to any corrections to the data. Meteorological station locations are downloaded from a GPS receiver. After processing in the GPS software they are imported to the GIS. In this case attribute data are available as computer files downloaded from the automatic weather stations. However, only summary data are required in the GIS, so summary statistics are calculated and entered by keyboard encoding to an attribute database.

Stereoscopic pairs of aerial photographs are interpreted to produce maps of the ski pistes. These are then encoded by manual digitizing. Finally, the land use data are purchased as a classified satellite image from the local remote sensing data supplier. The team transfer the data electronically to their GIS, then check the data by undertaking some ground truthing of their own using GPS receivers to navigate to predefined sites.

After considerable effort, the Happy Valley GIS team now have a GIS database populated with data from a range of sources and in various formats and co-ordinate systems. There are vector data input by both manual and automatic digitizing; there are 10-metre resolution raster data sourced from satellite imagery; and there are a number of attribute files attached to vector point, line and area data. The task of the Happy Valley GIS team is to process these data so that they are free from errors (as far as is possible) and are in a common format and frame of spatial reference. The 1:25,000 data from which the DTM and valley infrastructure data originate are based on the Lambert Conformal Conic projection. This is chosen as the spatial frame of reference onto which all other data not based on this projection will be re-projected. The roads and boundary data were digitized from more than one map sheet. Thus, edge matching is required. Table 5.5 summarizes all the processes that were necessary for a selection of the data sets.

BOX 5.5 CONTINUED

Table 5.4 Data in the Happy Valley GIS

Name of data layer	*Source*	*Uses*	*Entity type*	*Data model*	*Attributes*
Infrastructure	Scanned from 1:5000 scale plans	Detailed resort planning and management	Mixed	Raster	None
Hotels	Accommodation survey carried out by GIS team, postcodes used to identify point locations for hotels	Customer service planning and management	Point	Vector	Name, address, category (basic, standard or luxury), number of rooms
Ski schools	GIS team survey – point locations as for hotels	Customer service planning and management	Point	Vector	Ski school name, address, number of instructors
Meteorological stations	Downloaded from GPS readings	Skiing forecasts and conditions reporting, ski run planning and management	Point	Vector	Meteorological Station reference number, snow cover, precipitation, wind speed and direction, temperature and sunshine hours
Roads	Digitized from 1:25,000 paper maps	Resort planning, traffic management	Lines (network)	Vector	Road name and number
Ski pistes	Digitized from aerial photographs	Ski run planning and management	Lines	Vector	Grade of run (green, blue, red, black)
Ski resort boundary	Digitized from 1:25,000 paper map	Resort planning and management	Area	Vector	Name of resort
Topography	Downloaded from purchased 1:25,000 DTM	Ski run and resort management	Surface	Vector/ Raster	Height
Land use	10-m resolution SPOT image, ground-truthed with GPS readings	Vegetation surveys and monitoring, environmental assessment, ski run and resort planning/ management	Areas	Vector/ Raster	Land use class (including agricultural, forestry, settlement, water)

BOX 5.5 CONTINUED

Table 5.5 Creating an integrated database for Happy Valley

Data set	*Encoding and editing processes required*
Survey data	• Keyboard encoding of co-ordinate data points into data file • Conversion of data file into GIS data layer • Design and implementation of attribute database • Linking of spatial and attribute data using unique identifiers
Meteorological stations	• GPS data downloaded into GPS software • Processed data transferred to GIS in compatible format • Point data layer created • Processing of attribute data in statistical package • Keyboard encoding of summary attribute data • Linking of spatial and attribute data using unique identifiers
Vector road data	• Manual digitizing • Automatic cleaning and topology generation • Manual on-screen editing to check errors missed and/or created by automatic procedure • Joining of map sheets and edge matching • Re-projection onto Lambert Conformal Conic projection • Creation of attribute database • Linking of spatial and attribute data
Ski pistes	• Photogrammetric interpretation of stereo pairs of aerial photographs • Manual digitizing of ski pistes • Automatic cleaning and topology generation • Manual on-screen editing to check errors missed and/or created by automatic procedures • Rubber sheeting • Creation of attribute database • Linking of spatial and attribute data

Conclusions

This chapter has given an overview of the process of creating an integrated GIS database using the data stream model (Figure. 5.1) to guide the process. This has taken us from source data through data encoding, to editing and on to manipulatory operations such as re-projection, transformation and rubber sheeting. The chapter has given some insights into how some of these processes are carried out, and highlighted issues relevant to data encoding and editing which will be covered again elsewhere in the book.

Much of the effort required in an individual GIS project is concerned with data encoding and editing. A commonly quoted figure suggests that 50–80 per cent of GIS project time may be taken up with data encoding and editing. While the amount of time and work involved is significant it is very important to ensure that the resulting database is of the best practical quality. The quality of the results of any analyses carried out on the GIS will ultimately depend on the quality of the input data and the procedures carried out in creating an integrated GIS database. This topic will be returned to in Chapter 10 in a more in-depth discussion of data quality issues, but first the book continues with details of what can be done with the data now integrated in your GIS.

REVISION QUESTIONS

- Describe the main phases of the data stream.
- How do source data type and nature affect data encoding?
- Describe the processes of point and stream mode digitizing.
- Why is data editing important?
- What methods are available for detecting and rectifying errors in:
  (a) attribute data, and
  (b) spatial data?
- What are re-projection, transformation and generalization?
- What is geocoding?

Further study

There are a number of GIS texts that cover data encoding and editing issues. Hohl (1998) devotes a whole book to the subject of GIS data conversion, while DeMers (1997) offers two relevant chapters – on data input and data storage and editing. Jackson and Woodsford (1991) present an excellent, if a little dated, review of hardware and software for data capture. Chrisman (1997) offers a reflective review of techniques and quality issues associated with data encoding in a chapter on 'Representation'. Chapter 5 of Martin (1996) and Openshaw (1995) contain overviews of aspects of data encoding issues relevant to population census data. For reference to some of the important research work in data encoding you need to seek out articles in journals such as the *International Journal of Geographical Information Systems*, *Cartographica* and *Cartography*, and papers in conference proceedings. For example, Chrisman (1987) and Walsby (1995) address issues of efficient digitizing, and the now famous line-thinning algorithm is presented in Douglas and Peucker (1973).

Burrough P A (1986) *Principles of Geographical Information Systems for Land Resources Assessment*. Clarendon Press, Oxford

Chrisman N R (1987) Efficient digitizing through the combination of appropriate hardware and software for error detection and editing. *International Journal of Geographical Information Systems* 1: 265–77

Chrisman N R (1997) *Exploring Geographic Information Systems*. Wiley, New York

DeMers M N (1997) *Fundamentals of Geographic Information Systems*. Wiley, New York

Douglas D H, Peucker T K (1973) Algorithms for the reduction of the number of points required to represent a digitized line or its caricature. *Canadian Cartographer* 10 (4): 110–2

Hohl P (ed.) (1998) *GIS Data Conversion: Strategies, Techniques, and Management*. Onward Press, Santa Fe.

Jackson M J, Woodsford P A (1991) GIS data capture hardware and software. In: Maguire D J, Goodchild M F, Rhind D W (eds) *Geographical Information Systems: Principles and Applications*. Longman, London, Vol. 1, pp. 239–49

Martin D (1996) *Geographical Information Systems and their Socio-economic Applications*, 2nd edn. Routledge, London

Openshaw S (ed.) (1995) *The 1991 Census User's Handbook*. Longman Geoinformation, London

Walsby J (1995) The causes and effects of manual digitizing on error creation in data input to GIS. In: Fisher P (ed.) *Innovations in GIS 2*. Taylor and Francis, London, pp. 113–24

6 Data analysis?

KEY QUESTIONS AND ISSUES

- What data analysis operations are available in GIS?
- How are features measured in GIS?
- How are spatial and attribute data queried in GIS?
- How are networks analysed in GIS?
- How are surfaces analysed in GIS?
- What are the differences between analysis carried out on raster and vector data?
- What are the problems faced when conducting data analysis in GIS?

Introduction

Chapters 1 to 5 have introduced some of the fundamental concepts behind GIS. The nature of the spatial data used has been reviewed, and the encoding and structuring of these data to produce a computer representation of the real world outlined. We could consider that our Martian friend has completed courses on how his car works, how it is built and what fuel to use. Now he is ready to take to the open road and use the car to get him from one place to another. This chapter begins a journey in GIS, which takes the user from data to information and ultimately to decision-making. It covers some of the options in GIS for data analysis. There is a wide range of functions for data analysis available in most GIS packages, including measurement techniques, attribute queries, proximity analysis, overlay operations and the analysis of models of surfaces and networks. To set these in an applied context, Box 6.1 considers what sort of data analysis might be required in the imaginary Happy Valley GIS. The questions presented in Box 6.1 will be returned to throughout the chapter and suggestions for methods of answering them will be presented.

BOX 6.1

Data analysis in Happy Valley

The Happy Valley GIS team has set up its system. Table 5.4 in Chapter 5 gave details of the data layers in the GIS. Using data analysis functions the team hope to answer questions about the tourist facilities in the valley and to identify a possible location for a new ski piste. Some of the questions they will address are:

- Which is the longest ski piste in Happy Valley?
- What is the total area of forestry in the valley?
- How many luxury hotels are there in the valley?
- Where are the luxury hotels with more than 20 bedrooms?
- Where are the areas of forestry in Happy Valley?
- Which hotels are within 200 m of a main road?
- In which land use types are the meteorological stations located?
- Which roads should be used for a scenic forest drive?
- What is the predominant land use type in the Happy Valley resort?
- Where could a new ski piste be located?
- What is the predicted snowfall for the new ski piste?
- What would be the quickest route by road from the centre of the resort to this new ski piste?
- What are the slope and aspect of the new ski piste?
- From where will the new ski piste be visible?

While Box 6.1 gives a substantial list of questions that the Happy Valley GIS could be used to answer, the list is by no means comprehensive. So too with this chapter – although a large number of data analysis functions will be examined, by no means all can be introduced and explained in a chapter of this size. Indeed, whole books have been devoted to the topic of spatial analysis in GIS (Fotheringham and Rogerson, 1994; Chou 1996). The aims of this chapter are to introduce the range of data analysis functions available, provide examples of how they might be applied and outline the problems faced in using them. The focus is on the practical application of data analysis functions in GIS, rather than a full explanation of the algorithms and concepts behind them. Reference is made to sources of further information throughout, and there are many suggestions for additional reading included at the end of the chapter.

The chapter begins by introducing methods for measurement and queries in GIS. Proximity, neighbourhood and reclassification functions are outlined, then methods for integrating data using overlay functions are explained. Interpolation techniques (used for the prediction of data at unknown locations) are introduced, and since height values are frequently predicted, analysis of surfaces follows. Finally, analysis of network data is considered. This chapter sets the scene for Chapter 7 (Analytical Modelling), which takes GIS through the next step, from a system capable of analysing models of spatial form to a system capable of building models of spatial process. There is also a strong link with Chapter 12 on project management, where issues such as how to plan and prepare for successful analysis are covered.

An important first step to understanding spatial data analysis in GIS is a knowledge of the terminology used. Finding standard terms is difficult since different GIS software packages often use different words to describe the same function, and individuals with backgrounds in different fields tend to prefer particular terms. The terminology used in this chapter is summarized in Table 6.1.

Measurements in GIS – lengths, perimeters and areas

Calculating lengths, perimeters and areas is a common application of GIS. Measuring the length of a ski piste from a digital map is a relatively straightforward task. However, it is possible that different measurements can be obtained depending on the type of GIS used (raster or vector) and the method of measurement employed. It is important to remember that all measurements from a GIS will be an approximation, since vector data are made up of straight line segments (even lines which appear as curves on the screen are stored as a collection of short straight line segments), and all raster entities are approximated using a grid cell representation.

In a raster GIS there is more than one answer to the question 'what is the distance between A and B?' where A and B are two ends of a straight line.

Table 6.1 Data analysis terminology

Term	*Definition*
Entity	An individual point, line or area in a GIS database
Attribute	Data about an entity. In a vector GIS attributes are stored in a database. For example, the street name for a line entity that represents a road may be stored. In a raster GIS the value of a cell in the raster grid is a numerical code used to represent the attribute present. For example, code '1' may be used for a motorway, '2' for a main road and '3' for a minor road. Further attributes of an entity (for instance street name) may be stored in a database linked to the raster image
Feature	An object in the real world to be encoded in a GIS database
Data layer	A data set for the area of interest in a GIS. For example, the Happy Valley GIS contains data layers on nine themes, including roads, ski runs, hotels and land use. Data layers in a GIS normally contain data of only one entity type (that is points, lines or areas). It would be unusual for a data layer to contain, for instance, roads and hotels; these would be stored as separate thematic data layers
Image	A data layer in a raster GIS. It should be remembered that each cell in a raster image will contain a single value that is a key to the attribute present there
Cell	An individual pixel in a raster image
Function or operation	A data analysis procedure performed by a GIS
Algorithm	The computer implementation of a sequence of actions designed to solve a problem

The answer varies according to the measurement method used. Normally the shortest path, or Euclidean distance, is calculated by drawing a straight line between the end points of a line, and creating a right-angled triangle so that Pythagorean geometry can be used (Figure 6.1a). In this case the distance AB is calculated using the formula:

$$AB^2 = AC^2 + CB^2$$

Therefore:

$$AB = \sqrt{AC^2 + CB^2}$$

Alternatively, a 'Manhattan' distance can be calculated. This is the distance along raster cell sides from one point to the other. The name for this method comes from the way in which you would get across a city, like Manhattan, consisting of dense 'blocks' of buildings. As it is impossible to pass diagonally through a block, you have to traverse the sides (Figure 6.1b). A third method of calculating distance in a raster GIS uses 'proximity' (Berry, 1993). In this method, concentric equidistant zones are established around the start point A (Figure 6.1c). The resulting image shows the shortest straight line distance from every point on the map (including end point B) to the location of interest (A). Thus, the distance from A to B can be ascertained. Some authors (Berry, 1987; Tomlin, 1990) use the term 'spread' for functions which act in this way to reflect the way the distance surface grows outwards from the feature of interest (Plate 10).

To obtain a perimeter measurement in a raster GIS, the number of cell sides that make up the boundary of a feature is multiplied by the known resolution of the raster grid. For area calculations, the number of cells a feature occupies is multiplied by the known area of an individual grid cell (Figure 6.1d). The area of forestry in Happy Valley (from a classified satellite image in a raster GIS) could be obtained in this way.

Area and perimeter calculations in raster data can be affected by the origin and orientation of the raster grid. These problems are discussed further in Chapter 10, and they can be avoided by orientation of grids with north–south alignment and use of consistent origins.

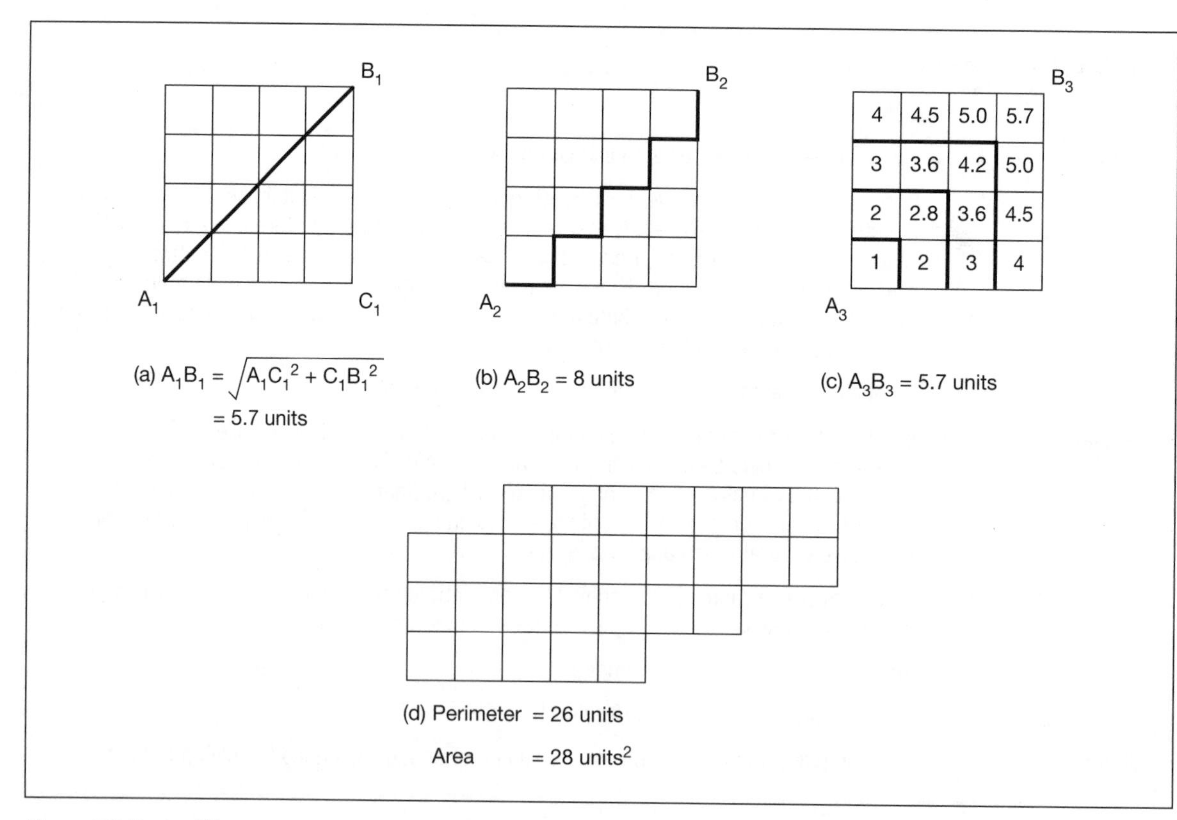

Figure 6.1 Raster GIS measurements: (a) Pythagorean distance; (b) Manhattan distance; (c) proximity distances; (d) perimeter and area

In a vector GIS distances are measured using Pythagoras's theorem to obtain the Euclidean distance (Figure 6.2). Geometry is also used to calculate perimeters and areas. Perimeters are built up of the sum of straight line lengths, and areas are calculated by totalling the areas of simple geometric shapes formed by subdividing the feature of interest (Figure 6.2). In vector GIS, length, perimeter and area data can be stored as attributes in a database, so these need to be calculated only once and then are permanently saved.

Ski pistes in Happy Valley are likely to bend and weave their way through the trees. To find the total length of an individual piste it would be necessary, therefore, to sum the lengths of several straight line segments that approximate the piste.

Queries

Performing queries on a GIS database to retrieve data is an essential part of most GIS projects. Queries offer a method of data retrieval, and can be performed on data that are part of the GIS database, or on new data produced as a result of data analysis. Queries are useful at all stages of GIS analysis for checking the quality of Figure 6.1 Raster GIS measurements: (a) Pythagorean distance; (b) Manhattan distance; (c) proximity distances; (d) perimeter and area data and the results obtained. For example, a query may be used if a data point representing a hotel is found to lie erroneously in the sea after data encoding. A query may establish that the address of the hotel had been wrongly entered into a database, resulting in the allocation of an incorrect spatial reference. Alternatively, queries may be used after analysis has been conducted. For instance, following exten-

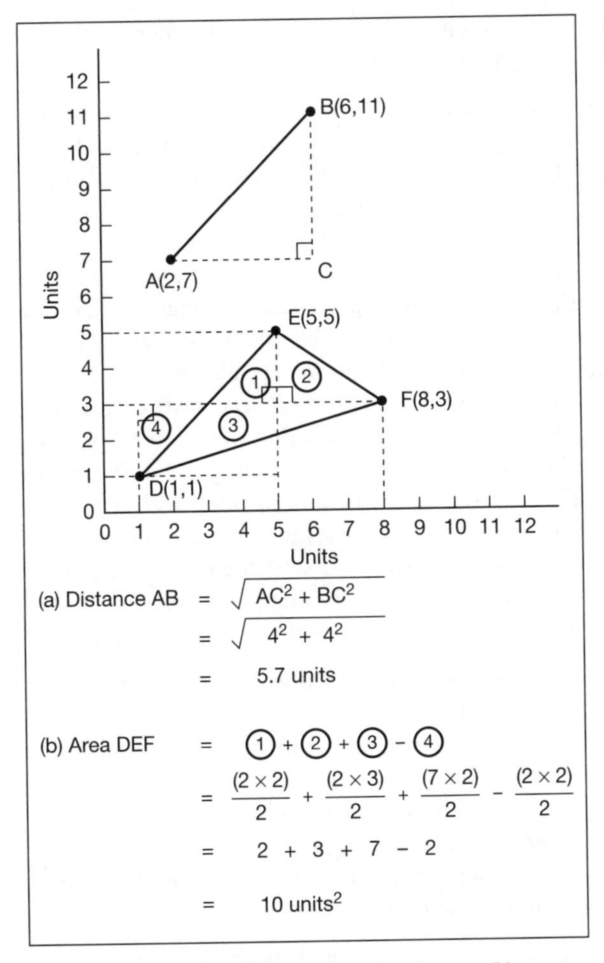

Figure 6.2 Vector GIS measurements: (a) distance; (b) area

sive searches using GIS for a suitable hotel (perhaps one with a small number of bedrooms, located a short distance from the ski slopes, and accessible by public transport), a query may be used to obtain the name and address of the hotel to allow a booking to be made.

There are two general types of query that can be performed with GIS: spatial and aspatial. Aspatial queries are questions about the attributes of features. 'How many luxury hotels are there?' is an aspatial query since neither the question nor the answer involves analysis of the spatial component of data. This query could be performed by database software alone. In Box 6.1 the question 'Where are the luxury hotels in Happy Valley?' was asked. Since this requires information about 'where' it is a spatial query. The location of the hotels will be reported and could be presented in map form.

The method of specifying queries in a GIS can have a highly interactive flavour. Users may interrogate a map on the computer screen or browse through databases with the help of prompts and query builders. A user may point at a hotel on the computer screen and click to obtain the answer to 'What is the name of *this* hotel?'. Queries can be made more complex by combination with questions about distances, areas and perimeters, particularly in a vector GIS, where these data are stored as attributes in a database. This allows questions such as 'Where is the longest ski run?' to be answered in one step, where a raster GIS might require two – one to calculate the lengths of all the ski runs, and the second to identify the longest.

Individual queries can be combined to identify entities in a database that satisfy two or more spatial and aspatial criteria, for example 'Where are the luxury hotels which have more than 20 bedrooms?' Boolean operators are often used to combine queries of this nature. These use AND, NOT, OR and XOR, operations that are also used for the combination of different data sets by overlay (Box 6.2). Boolean operators need to be used with care since 'Where are the hotels which are in the category 'luxury' AND have more than 20 bedrooms?' will yield a different answer from the question 'Where are the hotels which are in the category 'luxury' OR have more than 20 bedrooms?' The second query will probably identify more hotels.

Reclassification

Reclassification is an important variation on the query idea in GIS, and can be used in place of a query in raster GIS. Consider the raster land use image referred to in Table 5.4. If we wished to ask 'Where are all the areas of forestry?' an answer could be obtained using a query or by reclassifying the image. Reclassification would result in a new image. For example, if cells representing forestry in the original image had a value of 10, a set of rules for the reclassification could be:

> Cells with values = forestry (value 10) should take the new value of 1

BOX 6.2

Boolean operators

Boolean expressions used in GIS are AND, OR, NOT and XOR. These are explained best with the help of Venn diagrams, where each circle in the diagram represents the set of data meeting a specific criterion (Figure 6.3). In the diagrams, A is the set of hotels that are in the 'luxury' category, and B is the set of hotels that have more than 20 bedrooms.

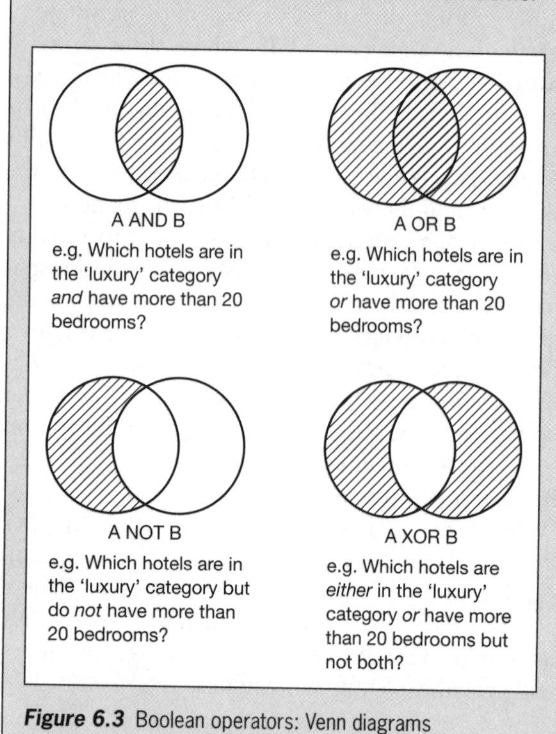

Figure 6.3 Boolean operators: Venn diagrams

Cells with values ≠ forestry should take the new value of 0

Reclassification would result in a new image with all areas of forestry coded with value 1, and all areas that were not forest coded with value 0. This is a Boolean image. Reclassification of this nature, producing a two-code image from a complex original image, is a very useful technique because it allows the resulting image (containing only values 1 and 0) to be used in further analysis.

Other reclassifications are also possible. In the Zdarske Vrchy case study reclassification was used to allocate a new value to different land use classes based on their ecological importance (Table 6.2). Forestry and agricultural land were considered to have a relatively high conservation value, while water and settlement had lower conservation value. A new image was produced after implementation of the following set of rules:

Cells with original value = 10 should take the new value of 5
Cells with original value = 11 should take the new value of 2
Cells with original value = 12 should take the new value of 1
Cells with original value = 13 should take the new value of 4

The resulting map helped the ecologist to identify areas of high conservation value. However, the new classes (1, 2, 4 and 5) are still simply labels (ordinal values) and care needs to be taken to ensure appropriate further analysis of the image.

Table 6.2 Reclassification values for the land use data layer

Land use	*Old value*	*New value after reclassification: Boolean example*	*New value after reclassification: Weighting example*
Forestry	10	1	5
Water	11	0	2
Settlement	12	0	1
Agricultural land	13	0	4

Buffering and neighbourhood functions

There is a range of functions available in GIS that allow a spatial entity to influence its neighbours, or the neighbours to influence the character of an entity. The most common example is buffering, the creation of a zone of interest around an entity. Other neighbourhood functions include data filtering. This involves the recalculation of cells in a raster image based on the characteristics of neighbours.

The question 'Which hotels are within 200 m of a main road?' could be approached in a number of ways. One option would be, first, to produce a buffer zone identifying all land up to 200 m from the main roads; and second, to find out which hotels fall within this buffer zone using a point-in-polygon overlay (see later in this chapter). Then a query would be used to find the names of the hotels. An alternative approach would be to measure the distance from each hotel to a main road, then identify those which are less than 200 m away. This example illustrates that in most GIS data analysis there is more than one method of achieving an answer to your question. The trick is to find the most efficient method, and the most appropriate analysis. In the example above, repeated measurement of distances from hotels to roads could be time-consuming and prone to human error. Thus, the first approach using buffering would be more appropriate.

Buffering, as already stated, is used to identify a zone of interest around an entity, or set of entities. If a point is buffered a circular zone is created. Buffering lines and areas creates new areas (Figure 6.4).

Buffering is very simple conceptually but a complex computational operation. Creating buffer zones around point features is the easiest operation; a circle of the required radius is simply drawn around each point. Creating buffer zones around line and area features is more complicated. Some GIS do this by placing a circle of the required radius at one end of the line or area boundary to be buffered. This circle is then moved along the length of the segment (Figure 6.4). The path that the edge of the circle tangential to the line makes is used to define the boundary of the buffer zone.

Figure 6.4 illustrates only the most basic set of buffer operations as there are many variations on this theme. For example, buffer zones may be of

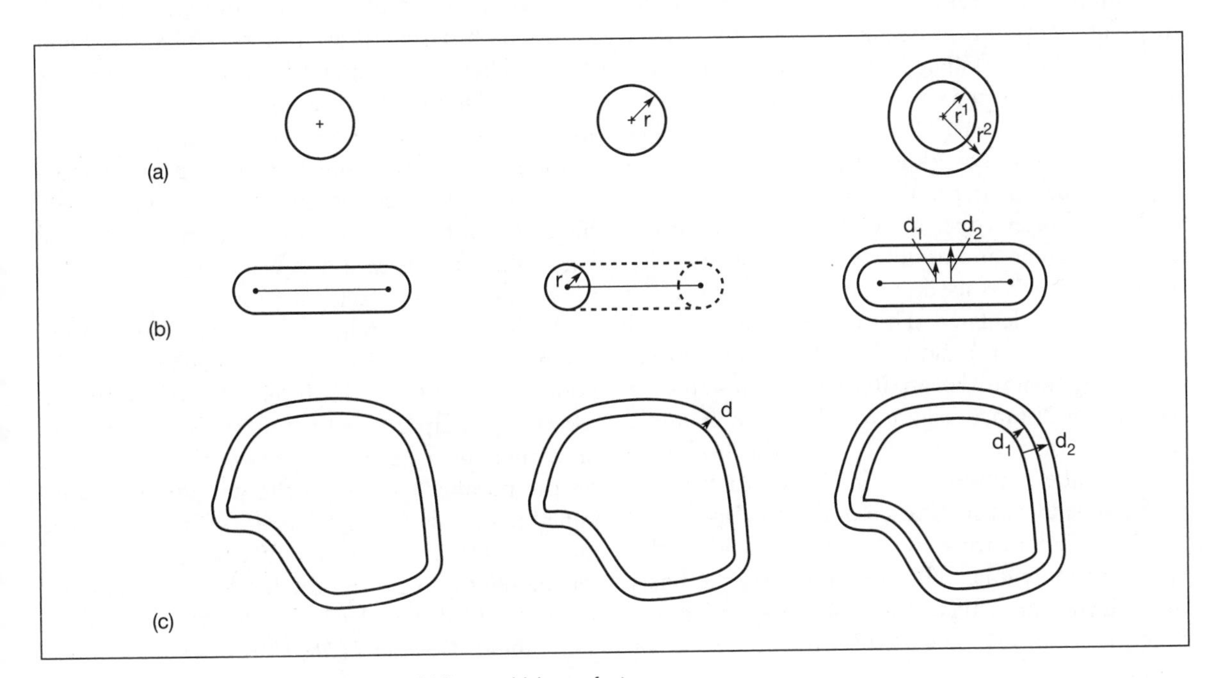

Figure 6.4 Buffer zones around (a) point, (b) line, and (c) area features

fixed or varying width according to feature attributes (Plate 11). When analysing a road network, wide buffer zones could be attached to motorways and narrower buffer zones to minor roads to reflect traffic densities. In the radioactive waste case study (introduced in Chapter 1) local accessibility of potential disposal sites was defined using buffering techniques. Buffer zones of varying widths were drawn around roads and railway lines. Motorways were buffered at 3 km (these being the main carriers of volume traffic) and primary routes at 1.5 km. Railways were also buffered at 3 km (Plate 3). Sites lying within any of these buffer zones were potentially feasible waste disposal sites as they were sufficiently accessible (Openshaw *et al.*, 1989).

Buffering has a whole series of uses when combined with other data layers. In the radioactive waste example the buffer zones were used as part of a process to identify the land use, population totals and conservation status of accessible land (Openshaw *et al.*, 1989). Andersson (1987) used buffering of bus stop data with population data to identify the best locations for bus stops. A buffer zone was calculated for each potential bus stop, using a value that reflected the distance an individual was prepared to walk to reach the stop. The population density within this buffer zone was then calculated. A set of bus stops was identified which maximized the overall population catchment areas. This type of operation relies on multiple layer overlays. These are considered in greater depth later in this chapter.

While buffer zones are often created with the use of one command or option in vector GIS, a different approach is used in many raster GIS. Here proximity is calculated. This method was outlined earlier (Figure 6.1c) and will result in a new raster data layer where the attribute of each cell is a measure of distance. Figure 6.5 shows an example of a proximity map produced for the hotels in Happy Valley, subsequently reclassed to create a buffer zone of 125 m for each of the hotels.

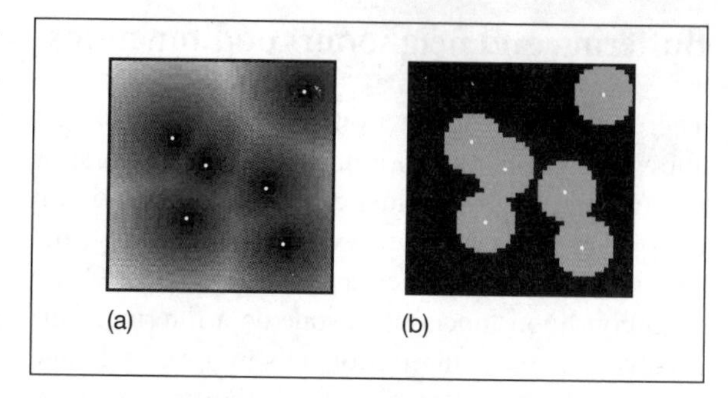

Figure 6.5 Proximity map for hotels in Happy Valley: (a) distance surface; (b) 125 m buffer zones

Other operations in raster GIS where the values of individual cells are altered on the basis of adjacency are called neighbourhood functions. Filtering is one example used for the processing of remotely sensed imagery. Filtering will change the value of a cell based on the attributes of neighbouring cells. The filter is defined as a group of cells around a target cell. The size and shape of the filter are determined by the operator. Common filter shapes are squares and circles, and the dimensions of the filter determine the number of neighbouring cells used in the filtering process. The filter is passed across the raster data set and used to recalculate the value of the target cell that lies at its centre. The new value assigned to the target cell is calculated using one of a number of algorithms. Examples include the maximum cell value within the filter and the most frequent value. These and other filtering techniques are shown in Figure 6.6. Filtering operations might be used in the preparation of the forestry data layer in the Happy Valley GIS. The raster data obtained from a classified satellite image may require filtering to 'smooth' noisy data caused by high spatial variability in vegetation cover or problems with the data collection device.

In the Zdarske Vrchy case study a filtering algorithm was used to create a forest edge index for the region. A 6 cell × 6 cell filter was passed over a forest edge map derived from satellite imagery. A new value for the central cell was calculated, based on the number of cells in the window that contained forest edge (Plates 7 and 8). The forest edge index was then used to identify suitable habitats for the reintroduction of the black grouse. For this species of bird the edge of the forest is an important habitat (Downey *et al.*, 1992).

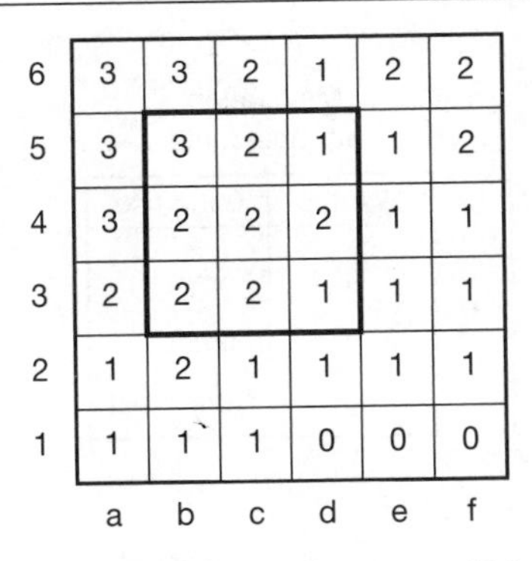

6	3	3	2	1	2	2
5	3	3	2	1	1	2
4	3	2	2	2	1	1
3	2	2	2	1	1	1
2	1	2	1	1	1	1
1	1	1	1	0	0	0
	a	b	c	d	e	f

Example: Forest data in Happy Valley GIS
- Applying 3×3 square filter to recalculate value for cell c4:

minimum filter c4 = 1
maximum filter c4 = 3
mean filter c4 = 1.89
modal filter c4 = 2
(most frequently occurring class)
diversity filter c4 = 3
(number of different classes)

Figure 6.6 Raster GIS filter operations

Integrating data – map overlay

The ability to integrate data from two sources using map overlay is perhaps the key GIS analysis function. Using GIS it is possible to take two different thematic map layers of the same area and overlay them one on top of the other to form a new layer. The techniques of GIS map overlay may be likened to sieve mapping, the overlaying of tracings of paper maps on a light table. Map overlay has its origins in the work of McHarg (1969), who used large and complex manual map overlays in landscape architecture (see Chapter 9).

Map overlay has many applications. At one level it can be used for the visual comparison of data layers. For example, the results of the hotel query 'Where are all the 'luxury' hotels?' may be plotted on top of, or overlaid, on a map of the road network to give some spatial context to the results. In this case no new data are produced. In fact, the location of ski pistes or land use layers could just as easily be used. This technique is used for the overlay of vector data (for example pipelines) on a raster background image (often a scanned topographic map).

Overlays where new spatial data sets are created involve the merging of data from two or more input data layers to create a new output data layer. This type of overlay may be used in a variety of ways. For example, obtaining an answer to the question 'Which hotels are within 200 m of a main road?' requires the use of several operations. First, a buffering operation must be applied to find all the areas of land within 200 m of a main road, then an overlay function used to combine this buffer zone with the hotel data layer. This will allow the identification of hotels within the buffer zone. Alternatively, the selection of a site for a new ski piste may require the overlay of several data sets to investigate criteria of land use, hotel location, slope and aspect.

As with many other operations and analyses in GIS there are differences in the way map overlays are performed between the raster and vector worlds. In vector-based systems map overlay is time-consuming, complex and computationally expensive. In raster-based systems it is just the opposite – quick, straightforward and efficient.

Vector overlay

Vector map overlay relies heavily on the two associated disciplines of geometry and topology. The data layers being overlaid need to be topologically correct (see Chapter 3) so that lines meet at nodes and all polygon boundaries are closed. To create topology for a new data layer produced as a result of the overlay process, the intersections of lines and polygons from the input layers need to be calculated using geometry. For complex data this is no small task and requires considerable computational power. Figure 6.7 shows the three main types of vector overlay; point-in-polygon, line-in-polygon and polygon-on-polygon. This figure also illustrates the complexity of the overlay operations. The overlay of two or more data layers representing simple spatial features results in a more complex output layer. This will contain more polygons, more intersections and more line segments than either of the input layers.

Point-in-polygon overlay is used to find out the polygon in which a point falls. In the Happy Valley

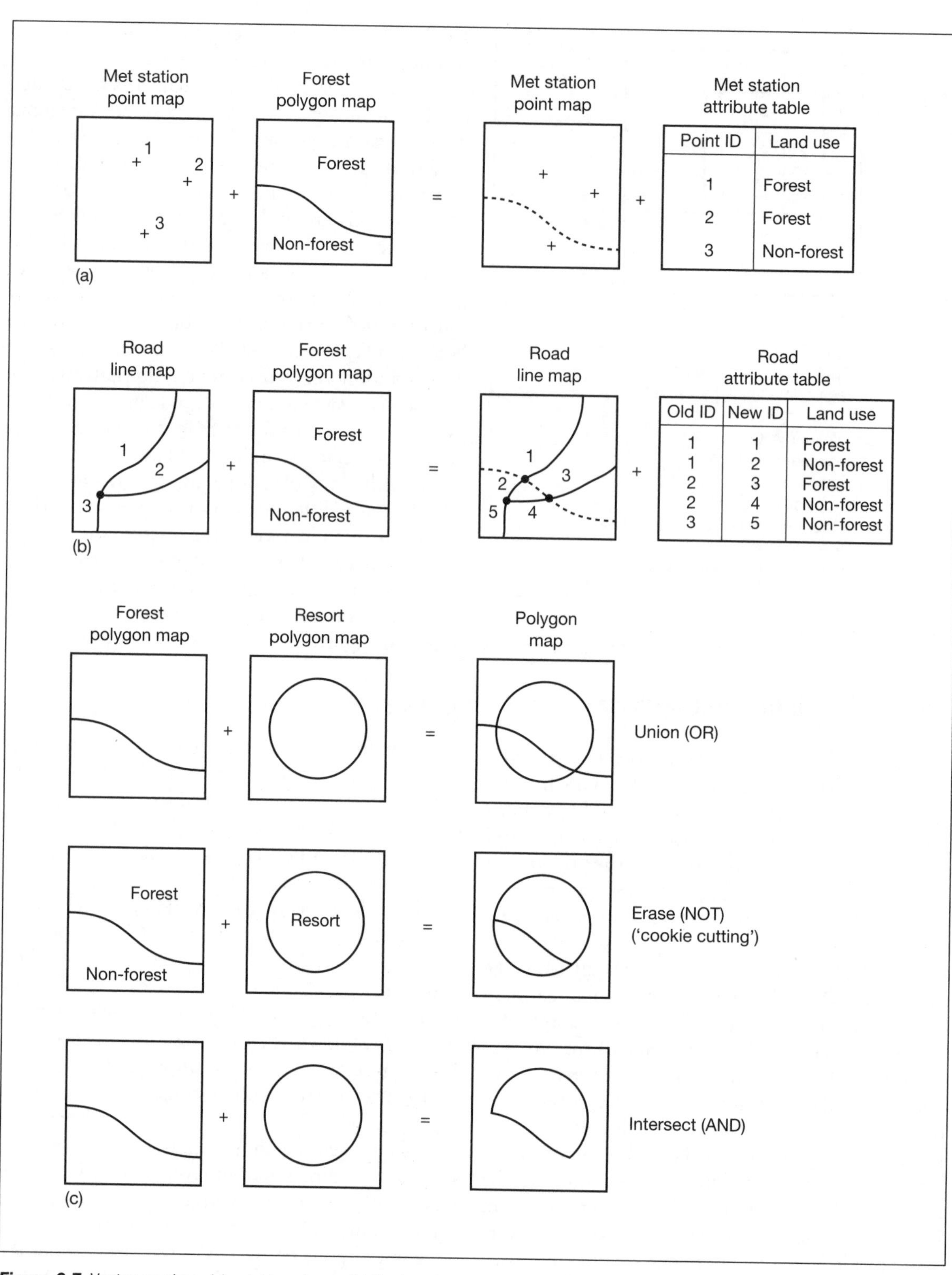

Figure 6.7 Vector overlays: (a) point-in-polygon; (b) line-in-polygon; (c) polygon-on-polygon

GIS meteorological stations are represented as points and land use as polygons. Using point-in-polygon overlay on these vector data layers it is possible to find out in which land use polygon each meteorological station is located. Figure 6.7 illustrates this overlay process. On the output map a new set of rain gauge points is created with additional attributes describing land use. The algorithm for point-in-polygon is explained in Box 6.3.

Line-in-polygon overlay is more complicated. Imagine that we need to know where roads pass through forest areas to plan a scenic forest drive. To do this we need to overlay the road data on a data layer containing forest polygons. The output map will contain roads split into smaller segments representing 'roads in forest areas' and 'roads outside forest areas'. Topological information must be retained in the output map (Figure 6.7b), therefore this is more complex than either of the two input maps. The output map will contain a database record for each new road segment.

Polygon-on-polygon overlay could be used to examine the areas of forestry in the Happy Valley resort. Two input data layers – a forest data layer

BOX 6.3

Point-in-polygon analysis

There are several methods of solving point-in-polygon problems but the most common technique used in GIS is the 'half line' or Jordan method (Laurini and Thompson, 1992; Martin, 1996). Consider a circular polygon with a point at its centre, and another point outside the circular polygon (Figure 6.8a). If a line (the 'half line') is extended from the central point to the edge of the data layer it will cross the polygon boundary once. A line extended from the point outside the circle will not cross the polygon boundary. Now consider a more complex polygon with crenellated edges (Figure 6.8b). A line extended from the central point will always cross the polygon boundary an odd number of times, whilst a line extended from the point outside the polygon will cross the boundary an even number of times. Thus, the Jordan algorithm is based on the rule that if the number of crossings is an odd number a point is inside the polygon, and if the number of crossings is an even number a point is outside the polygon. In practice, the 'half line' is drawn parallel to the x axis and the method works for even the most complicated polygons containing holes and islands (Laurini and Thompson, 1992).

A difficulty with this method arises when points lie exactly on the polygon boundary (Figure 6.8c). Are these considered in or out? The answer may depend on the decision rule implemented in the software you are using, and some systems will say in and others out. Laurini and Thompson (1992) suggest that where the point has been produced using a mouse, it can be moved slightly to facilitate the in or out decision. Another problem situation is where the 'half-line' is coincident with the polygon boundary, making the counting of crossings impossible (Figure 6.8c).

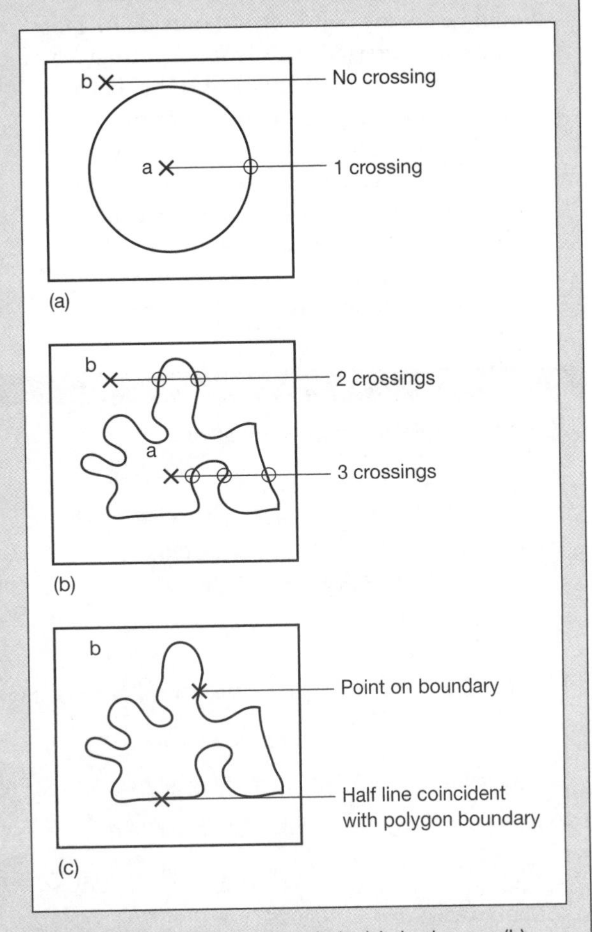

Figure 6.8 Point-in-polygon analysis: (a) simple case; (b) complex case; (c) problem cases

containing forest polygons, and the resort boundary layer – are required. Three different outputs could be obtained (Figure 6.7c):

1 The output data layer could contain all the polygons from both of the input maps. In this case the question posed is 'Where are areas of forestry OR areas which are within Happy Valley?' This corresponds to the Boolean OR operation, or in mathematical set terms, UNION. This may be useful if the Happy Valley management committee was interested in buying new forest areas to extend the scenic forest drive.
2 The output data layer could contain the whole of the resort area, and forest within this. The boundary of the resort would be used as the edge of the output map, and forest areas would be cut away if they fall outside it. This operation is referred to as 'cookie cutting'. It is equivalent to the mathematical IDENTITY operation and the identity of the resort boundary is retained in the output. The questions being answered are 'Where is the resort boundary, and where are areas of forest within this?' This overlay might be used in preparation for calculation of the percentage of the area of the resort covered by forest.
3 The output data layer could contain areas that meet both criteria: that is, areas that are both forest and within Happy Valley. An output map would be produced showing the whole of any forest polygons that are entirely within the resort boundary, and 'cut' away forest polygons which cross the resort boundary. This is the mathematical INTERSECT operation, and the output map shows where the two input layers intersect. 'Where are forest areas within Happy Valley?' is the question being answered. As a thematic data layer showing forestry in Happy Valley this may be useful for further analysis of the condition of the resort's forestry resources.

Overlay operations are seldom used in isolation. In practice, it is common to query a data layer first, then perform an overlay. To obtain areas of forestry used in the examples above, it would be necessary to extract these areas from the land use data layer first using a query. Box 6.4 discusses the use of polygon-on-polygon overlay in the nuclear waste case study (see also Plate 14).

One problem with vector overlay is the possible generation of sliver (or 'weird') polygons (Chrisman, 1997). These appear after the overlay of two data sets that contain the same spatial entities.

BOX 6.4

Siting a nuclear waste repository

Polygon-on-polygon overlay is often used for site selection or suitability analysis. In the nuclear waste case study the aim was to find those areas of Britain that simultaneously satisfied four siting constraints. The constraints are summarized in Table 6.3. Using a vector GIS to solve this problem the approach can be expressed in terms of GIS operations and Boolean overlays. This is illustrated in the flow chart shown in Figure 6.9.

Table 6.3 Siting criteria for a nuclear waste repository

Theme	*Criteria*
Geology	Chosen site must be in an area of suitable geology
Accessibility	Chosen site must be easily accessible
Population	Chosen site must be away from areas of high population density
Conservation	Chosen site must be outside any areas designated for conservation value (for example National Park, Site of Special Scientific Interest)

BOX 6.4 CONTINUED

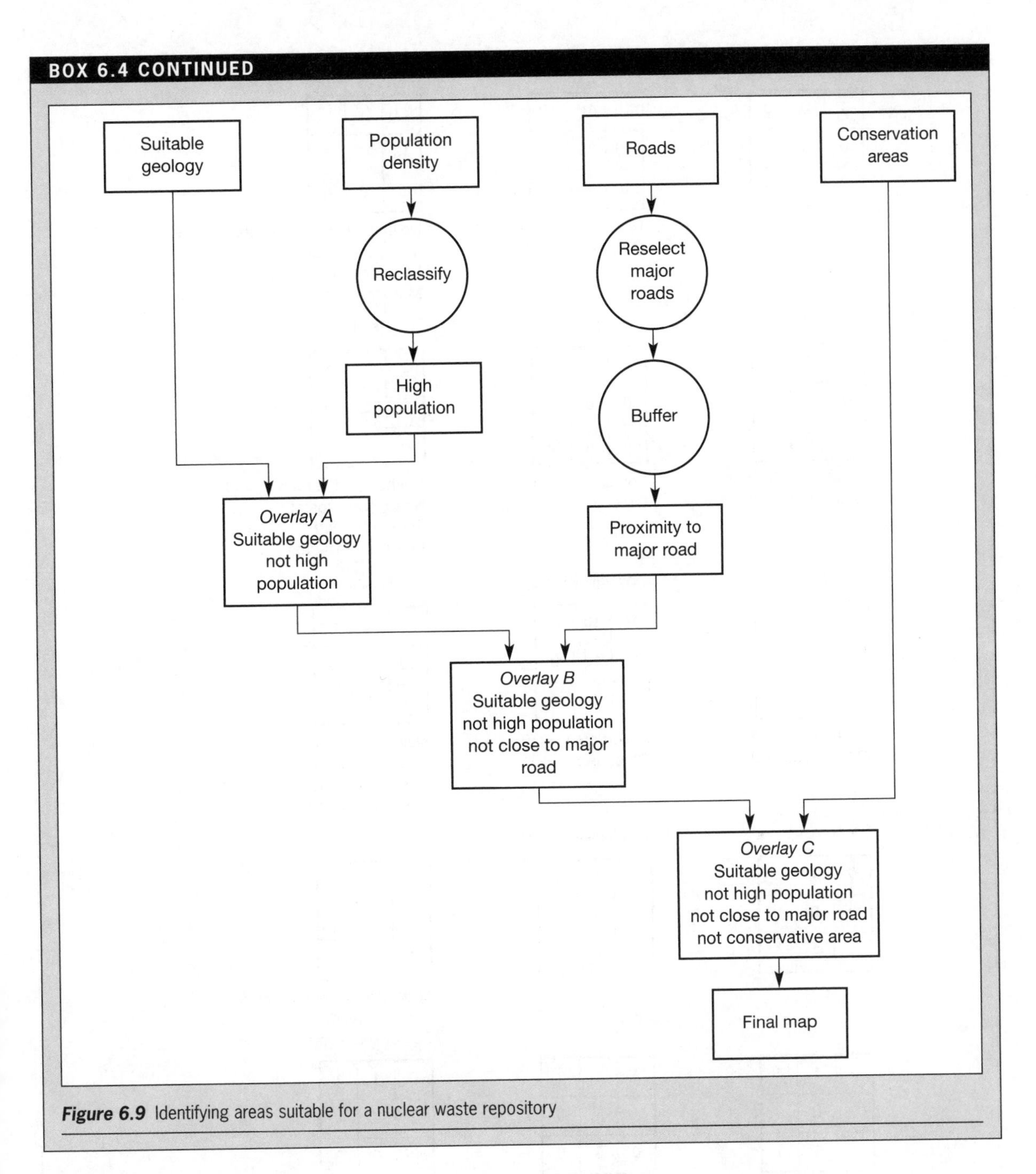

Figure 6.9 Identifying areas suitable for a nuclear waste repository

If the Happy Valley resort boundary were digitized by two different people two separate representations of the area would be created. If these were overlaid, long thin polygons would be seen along the boundary instead of a single line. These 'sliver' polygons arise from inconsistencies and inaccuracies in the digitized data. Frequently such errors go undetected but they can become apparent during vector overlay operations. Sliver polygons are considered further in Chapter 10. Other overlay problems, which relate to both vector and raster methods, are considered in Box 6.6.

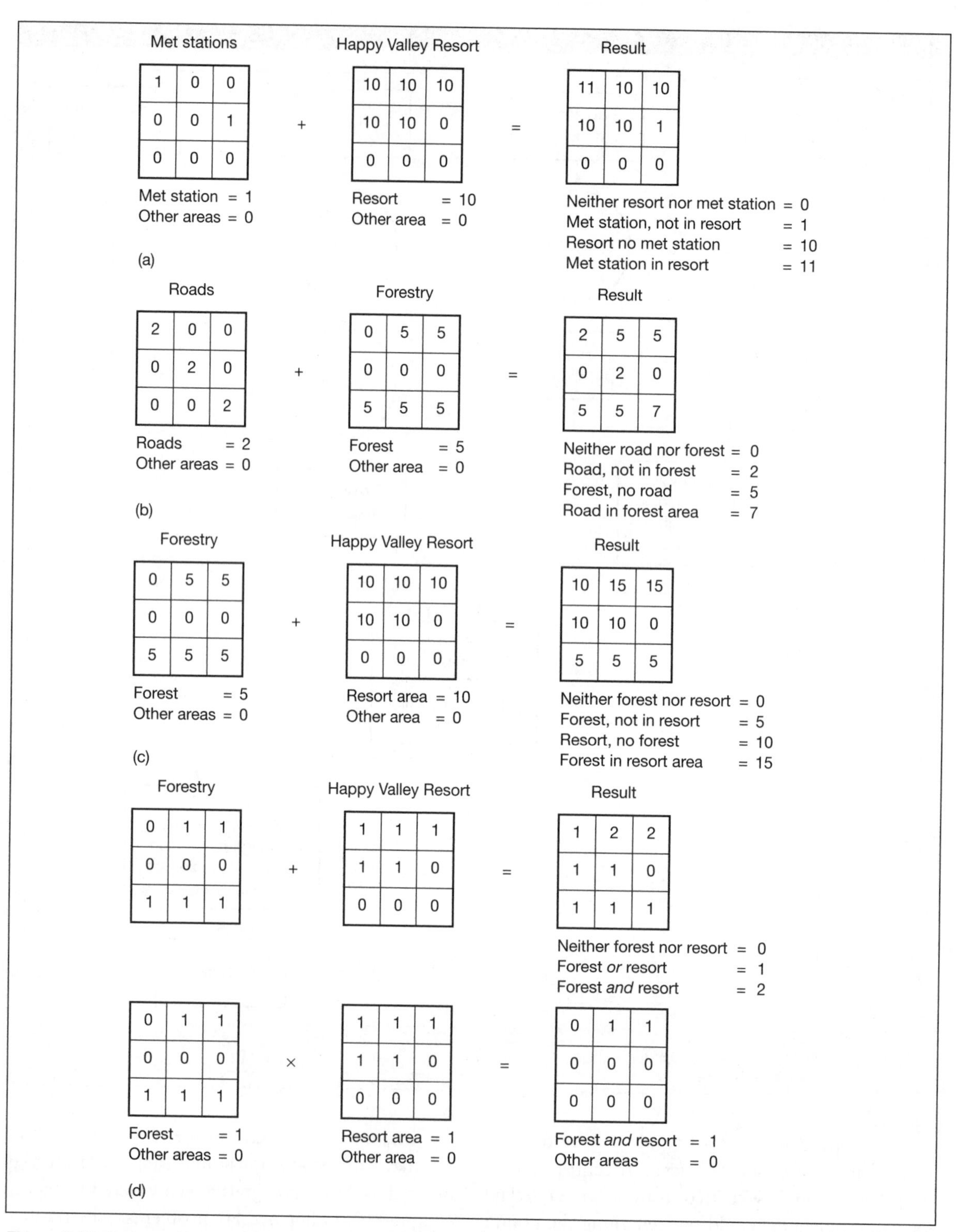

Figure 6.10 Raster overlays: (a) point-in-polygon (using add); (b) line-in-polygon (using add); (c) polygon-on-polygon (using add); (d) polygon-on-polygon (Boolean alternatives)

Raster overlay

In the raster data structure everything is represented by cells – a point is represented by a single cell, a line by a string of cells and an area by a group of cells. Therefore, the methods of performing overlays are different from those in vector GIS. Raster map overlay introduces the idea of map algebra or 'mapematics' (Berry, 1993). Using map algebra input data layers may be added, subtracted, multiplied or divided to produce output data. Mathematical operations are performed on individual cell values from two or more input layers to produce an output value. Thus, the most important consideration in raster overlay is the appropriate coding of point, line and area features in the input data layers.

Imagine that four of the Happy Valley data layers have been rasterized: the location of meteorological stations, the road network, the land use layer and the resort boundary. The meteorological stations are represented in a raster data layer where the value of '1' has been given to cells containing the stations. Roads are coded '2' in the roads raster layer and cells have been linked to create chains. Each cell in the land use layer has a value reflecting the type of land use present. Settlement is coded '1', water '2', agricultural land '4' and forestry '5'. The resort area has been given a value '10'. On all data layers '0' is the value given to cells that do not contain features of interest.

To find out which meteorological stations are contained within the Happy Valley resort an equivalent to the vector point-in-polygon overlay is required. To do this, one approach would be to add the two data layers (Figure 6.10a). The output map would contain cells with the following values:

- 0 for cells outside the resort boundary and without a meteorological station;
- 1 for cells containing meteorological stations but outside the resort boundary;
- 10 for cells inside the resort boundary but without a meteorological station;
- 11 for cells inside the resort boundary and containing a meteorological station.

In an operation equivalent to the vector line-in-polygon method (Figure 6.10b), the sections of roads that pass through forest areas could be obtained. This would require the roads data layer, and a reclassified version of the land use map that contained only forest areas. Again the two maps could be added. The output map would contain cells with the following values:

- 0 for cells with neither roads nor forest present;
- 2 for cells with roads, but outside forest areas;
- 5 for cells with forest present, but roads absent;
- 7 for cells with both forest and roads present.

If the value '2' for a road were added to land use codes, the new value for a cell could be the same as that for another land use type (for example a road value 2 + land use value 2 (water) = 4 (which is the same as the value here for agricultural land). Thus, the coding of raster images used in overlay is very important, and frequently users employ Boolean images (using only codes 1 and 0) so that algebraic equations will produce meaningful answers.

Polygon-on-polygon analysis is conducted in just the same way (Figure 6.10c). Again, the coding of input layers is the key to understanding the output produced by raster overlay. For example, adding the forest layer and the resort boundary would produce an output layer with the following codes:

- 0 for cells outside the resort boundary and with forest absent;
- 5 for cells outside the resort boundary and with forest present;
- 10 for cells inside the resort boundary and with forest absent;
- 15 for cells inside the resort boundary and with forest present.

The output map is equivalent to a union polygon-on-polygon overlay in vector GIS. Reclassification will produce variants on this, and other overlay operations are available, by multiplying, subtracting or dividing data layers, or by the implementation of a 'cover' or 'stamp' operation (which is equivalent to a 'cookie cutting' operation in vector GIS).

Box 6.5 illustrates some of the options available in raster overlay with reference to the nuclear waste siting case study.

The algebraic manipulation of images in raster GIS is a powerful and flexible way of combining data and organizing analysis. Equations can be written with maps as variables to allow the development of spatial models. Berry (1993) and

BOX 6.5

Raster analysis of the nuclear waste siting example

One solution to the process of selecting a nuclear waste site using raster data would be to create four binary data layers. These would represent geology, population density, a road and rail accessibility buffer and areas without conservation status. On these layers cells where the nuclear waste siting criteria are met are coded '1' and cells that do not meet the criteria are coded '0'. The solution to the problem can then be expressed as:

$$\text{Output cell value} = \text{geology} \times \text{population} \times \text{accessibility} \times \text{conservation}$$

Thus, for a cell in an area meeting all four criteria the value in the output layer will be:

$$\begin{aligned}\text{Output cell value} &= 1 \times 1 \times 1 \times 1 \\ &= 1\end{aligned}$$

Any cell coded as zero in any of the four input layers will cause the equation above to return a value of zero in the output layer. For example, if a cell falls in an area meeting the first three criteria, but failing the conservation criteria, the output value would be calculated by:

$$\begin{aligned}\text{Output cell value} &= \text{geology} \times \text{population} \times \text{accessibility} \times \text{conservation} \\ &= 1 \times 1 \times 1 \times 0 \\ &= 0\end{aligned}$$

An alternative approach to the overlay process would be to add the data layers. In this case the output cell values would reflect the number of siting criteria met by each cell. The method is summarized below for an output pixel meeting all four criteria:

$$\begin{aligned}\text{Output cell value} &= \text{geology} + \text{population} + \text{accessibility} + \text{conservation} \\ &= 1 + 1 + 1 + 1 \\ &= 4\end{aligned}$$

And, for a cell in an area meeting three criteria:

$$\begin{aligned}\text{Output cell value} &= \text{geology} + \text{population} + \text{accessibility} + \text{conservation} \\ &= 1 + 1 + 1 + 0 \\ &= 3\end{aligned}$$

Chrisman (1997) provide examples, and Chapter 12 looks at this approach in more detail in the context of designing GIS analysis.

There are two issues specifically affecting raster overlay that need to be considered by users: resolution and scales of measurement.

Resolution is determined by the size of the cell used. SPOT satellite data, for example, are collected at a resolution of 10 m. For some analyses you may wish to overlay a SPOT image with data collected at a different resolution, say 40 m. The result will be an output grid with a resolution of 10 m, which is greater than the resolution at which the second data set was collected. Since you cannot disaggregate data with any degree of certainty, a better approach to the overlay of these two data sets would be to aggregate cells in the SPOT image to match the resolution of the second data set.

The second issue is that of scales of measurement. Nonsensical overlays can be performed on map layers coded using nominal, ordinal, interval and ratio scales (see Chapter 2) if these scales are not sufficiently understood. For example, it is possible to add, subtract or multiply two maps, one showing land use coded using a nominal scale (where 1 represents settlement and 2 represents water) and another showing mean annual rainfall coded on a ratio scale. The result, however, is complete nonsense because there is no logical relationship between the numbers. A mean annual rainfall of 1000 mm minus land use type 1 is meaningless! Care is needed when overlaying raster data layers to determine whether the operation makes real sense according to the scales of measurement used.

There are other problems that affect both raster and vector overlay. Four of these are introduced in Box 6.6.

Spatial interpolation

Spatial interpolation is the procedure of estimating the values of properties at unsampled sites within an area covered by existing observations (Waters, 1989). In an ideal situation a spatial data set would provide an observed value at every spatial location. Satellite or aerial photography goes some way to providing such data; however, more

BOX 6.6

Problems affecting vector and raster overlay

1 *The modifiable areal unit problem.* The modifiable areal unit problem (MAUP) occurs when arbitrarily defined boundaries are used for the measurement and reporting of spatial phenomena (Openshaw, 1984). Data used in GIS are frequently continuous in nature. This means that the value of the entity recorded varies continuously over space. Population data are an example. Although individual people are measured in population censuses, data are normally reported for areas (for example, in the UK enumeration districts, wards, districts and counties are used). The boundaries of these units are arbitrary (from a statistical point of view) and subject to change. They do not necessarily coincide with breaks in the data. Thus, changing the boundaries of units and disaggregation or aggregation of units can affect the appearance of the data set.

   The overlay of two data sets, which both have modifiable units, should be undertaken with care, since overlay will almost inevitably lead to change in boundaries. The Department of the Environment (1987) suggests that results should be confirmed by repeating the analysis with different areal units.

2 *Ecological fallacy.* A problem associated with the MAUP is the ecological fallacy (Openshaw, 1984). This occurs when it is inferred that data for areas under study can be applied to the individuals within those areas. For example, if analysis of geodemographic data identifies the average income level in a postal zone, it is an ecological fallacy to assume that all the individuals within that postal zone have that level of income. Only when totally homogeneous data are aggregated into zonal data can the MAUP and the ecological fallacy be avoided. As geographical data are very rarely homogeneous it is important to recognize these problems when performing overlay analyses on polygon data. How aggregation systems are defined, zones constructed and boundaries drawn will affect the results of analysis (see Chapter 10). This problem has long been recognized, and there are many historical examples of boundaries being redrawn or small units aggregated in different ways to produce desired results (Monmonier, 1991).

3 *Selecting threshold criteria.* In the radioactive waste disposal case study four different map layers are combined in an overlay procedure to identify areas of the UK that simultaneously satisfy certain criteria. The problem is how to define the threshold criteria used for each constraint. For example, the choice of a population density limit of 490 persons per square kilometre is based on an adaptation of the Nuclear Installations Inspectorate (NII) nuclear power station siting guidelines (Openshaw *et al.*, 1989). Who is to say that 490 persons per square kilometre is a better threshold than 500 or even 491? Differences in the thresholds applied will produce differences in the output map. Fortunately, it is possible to investigate how much difference small changes in thresholds make to output maps. Thresholds can be re-applied to one or more of the input layers and overlay analysis repeated to view changes in the output. This form of investigatory analysis is termed 'what if?' modelling. Plate 5 shows results obtained from this type of 'what if?' modelling.

4 *Visual complexity.* Another map overlay problem is visual complexity. When two complex maps are overlain the output is likely to be more complex than either of the two input maps. The sheer complexity of some GIS outputs makes them very difficult to interpret. Generalization may be necessary, though with generalization comes a loss of detail.

often data are stratified (consisting of regularly spaced observations but not covering every spatial location) patchy (clusters of observations at specific locations) or even random (randomly spaced observation across the study area). The role for interpolation in GIS is to fill in the gaps between observed data points.

A common application of interpolation is for the construction of height contours. Contours on a topographic map are drawn from a finite number of height observations taken from surveys and aerial photographs. The height of the land surface between these points is estimated using an interpolation method and represented on a map using contours. Traditionally, contour maps were produced by hand, but today they are often drawn by computer. In the old hand–eye method, often referred to as line threading or eye-balling, contour lines were drawn between adjacent spot heights and divided into the chosen contour interval by assuming that the slope between adjacent spot heights remained constant. This technique suffered from a number of problems: the inaccurate assumption that slope is constant, human error, subjectivity and the amount of time needed if a large number of data points required interpolation. In the Happy Valley example interpolation could be used to estimate snow depth from observations obtained throughout the valley.

Most GIS packages offer a number of interpolation methods for use with point, line and area data. Box 6.7 presents a widely used classification of interpolation methods that is useful when deciding which method is appropriate for a particular data set (after Burrough, 1986; Lam, 1983).

Whichever interpolation technique is used, the data derived are only an estimate of what the real values should be at a particular location. The quality of any analysis that relies on interpolated data is, therefore, subject to a degree of uncertainty. It is essential for the user to appreciate the limitations of interpolated data when using the results in further GIS analysis. Interpolation techniques are reviewed in full in other books and papers

BOX 6.7

A classification of interpolation methods

1 *Local or global.* Global interpolation methods apply a single mathematical function to all observed data points and generally produce smooth surfaces. Local methods apply a single mathematical function repeatedly to small subsets of the total set of observed data points and then link these regional surfaces to create a composite surface covering the whole study area.
2 *Exact or approximate.* Exact methods of point interpolation honour all the observed data points for which data are available. This means that the surface produced passes exactly through the observed data points, and must not smooth or alter their values. Exact methods are most appropriate when there is a high degree of certainty attached to the measurements made at the observed data points. Approximate methods of interpolation do not have to honour the observed data points, but can smooth or alter them to fit a general trend. Approximate methods are more appropriate where there is a degree of uncertainty surrounding the measurements made at the sample points.
3 *Gradual or abrupt.* Gradual and abrupt interpolation methods are distinguished by the continuity of the surface produced. Gradual methods produce a smooth surface between sample points whereas abrupt methods produce surfaces with a stepped appearance. Terrain models may require both methods as it may be necessary to reproduce both gradual changes between observed points, for example in rolling terrain, and abrupt changes where cliffs, ridges and valleys occur.
4 *Deterministic or stochastic.* Deterministic methods of interpolation can be used when there is sufficient knowledge about the geographical surface being modelled to allow its character to be described as a mathematical function. Unfortunately, this is rarely the case for surfaces used to represent real-world features. To handle this uncertainty stochastic (random) models are used to incorporate random variation in the interpolated surface.

(Burrough, 1986; Davis, 1986; Lam 1983; Waters, 1989) so brief details of just three of the most frequently used methods – Thiessen polygons, TIN and spatial moving average – will be presented here. Plate 12 shows the results of applying these three methods to a random sample of 100 data points extracted from a digital elevation model.

Thiessen polygons (or Voronoi polygons) are an exact method of interpolation that assumes that the values of unsampled locations are equal to the value of the nearest sampled point. This is a local interpolator because the global characteristics of the data set exert no influence over the interpolation process. It is also an abrupt method of interpolation as sharp boundaries are present between the interpolated polygons.

Thiessen polygons are created by subdividing lines joining nearest neighbour points, drawing perpendicular bisectors through these lines, and then using these bisectors to assemble polygon edges (Laurini and Thompson, 1992). If observed data points are regularly spaced a regular lattice of square polygons will result. If the observed data points are irregularly spaced a surface of irregular polygons will be produced (Plate 12b).

The most common use of Thiessen polygons is to establish area territories for a set of points. Examples of applications include the transformation of point climate stations to watersheds and the construction of areas of influence around population centres (Chrisman, 1997; Laurini and Thompson, 1992). Although Thiessen polygons can be drawn around elevation observations this is not the most appropriate method to use because elevation data exhibit gradual rather than abrupt propeties.

A triangulated irregular network or TIN is an elegant way of constructing a surface from a set of irregularly spaced data points. This method of spatial interpolation is often used to generate digital terrain models (Chapter 3). An example 3D perspective TIN, created using the same data as in these interpopulation examples, is shown in Plate 13. The TIN model is an exact interpolation method based on local data points. In this method adjacent data points are connected by lines to form a network of irregular triangles (Plate 12c), hence the name. Because the value at each of the data points (forming the corners of the triangles) is known and the distance between these points can be calculated, a linear equation and trigonometry can be used to work out an interpolated value for any other point within the boundary of the TIN.

The spatial moving average is perhaps the commonest interpolation method used by GIS. It involves calculating a value for a location based on the range of values attached to neighbouring points that fall within a user-defined range. This neighbourhood filter is passed systematically across the region of interest, calculating new interpolated values as it goes. As such, the spatial average method produces a surface that suppresses the values of known data points to reveal global patterns in the data. This is an example of an approximate point interpolation method since it does not honour the values of known data points and recalculates these values as the filter passes over them.

A circular filter is the most obvious choice because known data points in all directions have an equal chance of falling within the radius of the interpolated location. In raster GIS a square or rectangular filter is often employed since this is easier to calculate. The size of the filter is based on assumptions about the distance over which local variability is deemed important. A distance weighting function may be applied to enhance the influence of closer data points (for example, a point 1 km away from the interpolated location will be considered more important than one 2 km away). Burrough (1986) provides details of a range of variations on this method. An interpolated surface, derived using a distance-weighted spatial moving average approach, is shown in Plate 12d.

The spatial moving average method is best suited to those instances where the values of the known data points are not exact and may be subject to measurement error, but where they nonetheless represent variation in a global pattern. Examples of suitable applications include the interpolation of census data, questionnaire responses and field survey measurements such as soil pH or rainfall infiltration rates.

Comparing the surfaces generated by these three different methods with the original DEM gives an appreciation of the difficulties involved in the interpolation of point data. Error maps of the absolute difference between estimated elevation and the original DEM can be produced to show where the errors are greatest (Plates 12 e–g). The TIN method shows the least error overall. All three methods show low levels of inaccuracy in the areas of low relief, but they perform worse in areas of high relief in the south-west of the DEM. The biggest errors occur around the edge of the image. In this example, the random sample of 100 points was taken from the central section of the DEM (Plate 12a) to demonstrate the difficulty of interpolation outside the sample region. Even within the sample an 'edge effect' can still be seen. For this reason it is recommended that sample points from outside the region of interest are collected with data that will later be interpolated to produce a surface. However, the TIN method minimizes edge effects because it does not allow interpolation outside the 'convex hull' or boundary of the sample points (Plate 12f).

Analysis of surfaces

Consideration of techniques available in GIS for surface analysis follows on logically from the discussion of interpolation, since interpolation techniques will invariably have been used to create a surface for analysis. Since the Earth is three-dimensional, it would seem that all GIS applications should include some element of three-dimensional analysis. However, software packages able to handle and analyse three-dimensional data are limited, and, as Chapter 3 outlined, analysis in GIS is more likely to be 2.5D, since the surfaces that are produced are simply that – surfaces. There is no underlying or overlying information. This prevents the analysis of, for example, geological or atmospheric data, and even to add realistic-looking trees with height to a GIS terrain model, a CAD or other design package may be necessary.

Despite these limitations, there are still many analysis functions in GIS that operate on surfaces. These range from draping functions that allow the results of other analyses to be overlain on a surface, to slope and aspect calculation, and visibility analysis. The selection of a new ski piste site gives some ideas of how these might be applied. Consider first the value of draping the results of the site selection analysis over a terrain surface compared with presentation of the results on a 'flat' map (see Plates 1 and 2). The visualization of the piste will be much enhanced when it can be seen in context with the terrain it descends. Slope calculations may help with the designation of the piste as an easy or difficult ski run. Aspect calculations will help to analyse the snow retention characteristics. Visibility analysis could be applied to examine what a skier would see from the piste, and to determine from which other locations the ski piste would be visible.

Slope, aspect and visibility are considered here as these are the most commonly used applications of terrain models used in GIS. Other applications of DTMs, for example hydrological modelling, are considered further in Chapter 7.

Calculating slope and aspect

Slope is the steepness or gradient of a unit of terrain, usually measured as an angle in degrees or as a percentage. Aspect is the direction in which a unit of terrain faces, usually expressed in degrees from north. These two variables are important for many GIS applications. As suggested above, both may be required in the planning of a new ski piste in Happy Valley. Slope values may be important for the classification of the slope for skiing. Aspect is important to ensure that the piste selected will retain snow cover throughout the ski season (a completely south-facing slope may not be suitable, since snow melt would be more pronounced than on a north-facing slope).

Slope and aspect are calculated in two ways according to the type of DTM being used. In raster DTMs slope and aspect are calculated using a 3×3 window that is passed over a database to determine a 'best fit tilted plane' for the cell at the centre of the window. This allows the calculation of constants for the equation

$$z = a + bx + cy$$

where z = height at the point of interest (the centre of the window), (x,y) = co-ordinates of the point at the centre of the window, and a, b, c = constants.

Slope and aspect for the centre cell can then be calculated (S = slope, A = aspect) using the formulae

$$S = b^2 + c^2$$

$$A = \tan^{-1}\left(\frac{c}{b}\right)$$

In the vector TIN model slope and aspect variables are usually calculated using a series of linear equations when the TIN is generated. The equations calculate the slope and aspect of the individual triangles formed by the TIN. Figure 6.11 shows an example of a shaded slope and aspect image.

Slope and aspect may also be calculated as a first step to more complex terrain analysis. Parameters such as the rate of change of slope, convexity or curvature (Weibel and Heller, 1991) may be necessary for landform analysis or classification.

Visibility analysis

One of the common uses of DTMs is in visibility analysis, the identification of areas of terrain that can be seen from a particular point on a terrain surface. This is used in a variety of applications ranging from locating radio transmitters for maximum coverage to minimizing the impact of commercial forestry in protected areas (DeMers, 1997; Fishwick and Clayson, 1995). In Happy Valley, visibility analysis could be used to determine the areas of the resort that would be visible from the top of the proposed new ski piste. Conversely, it could be used to determine other locations in the valley that could see the top of the new ski piste, or any other point along the piste.

DeMers (1997) explains how these analyses work. The location of the observer is connected to each possible target location in the terrain (i.e. a line is drawn from the observer at the top of the ski piste to all other locations in the area). The line, or ray, is followed from each target back to the observer, looking for locations that are higher (Figure 6.12). Higher points will obscure what is behind them. Thus, through repeated 'ray tracing' a viewshed map can be built (Clarke, 1990). Plate 14 shows the viewsheds calculated for both point and line features placed within the DEM data used in Plates 12 and 13.

The method for calculating a visibility surface is similar in both raster and vector GIS and in each case the results of analysis would be presented as a viewshed map. This is a Boolean polygon map that indicates which areas of terrain are visible from the location of interest. This map may be more easily interpreted and understood by draping it over the top of a terrain surface for visualization.

Some GIS packages offer more sophisticated visibility analyses which allow the user to specify the height of the viewing position as an offset above the terrain surface height at that location. In this manner it is easy to calculate from where the tops of tall structures such as ski lift pylons, or tall chimneys, can be seen.

Some GIS packages allow the incorporation of barrier features, such as buildings and trees, into the visibility analysis to produce a more realistic result, or the computation of 'relief shadows' (areas which lie in the shadow of the sun). Some GIS even allow you to 'walk' or 'fly' through a terrain model to visualize what the view would be like at various points on or above

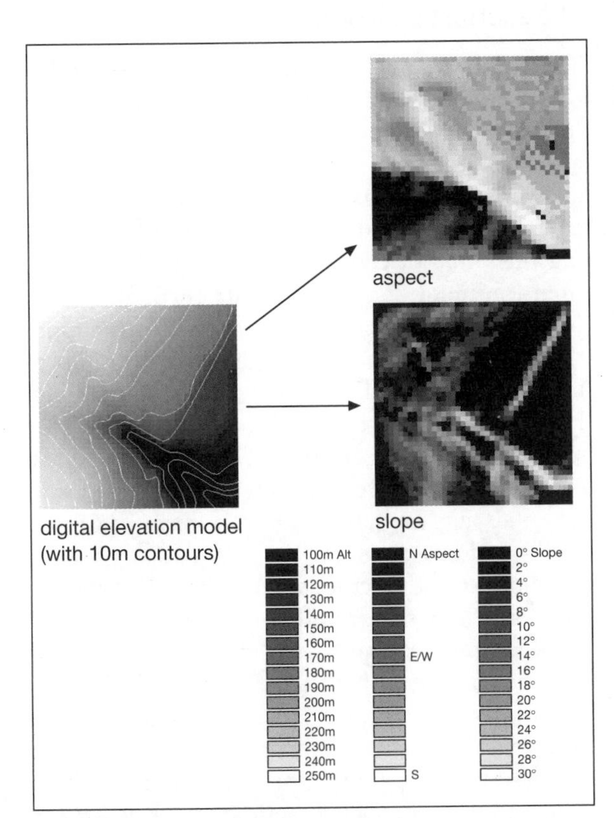

Figure 6.11 Three-dimensional analysis: shaded slope and aspect image

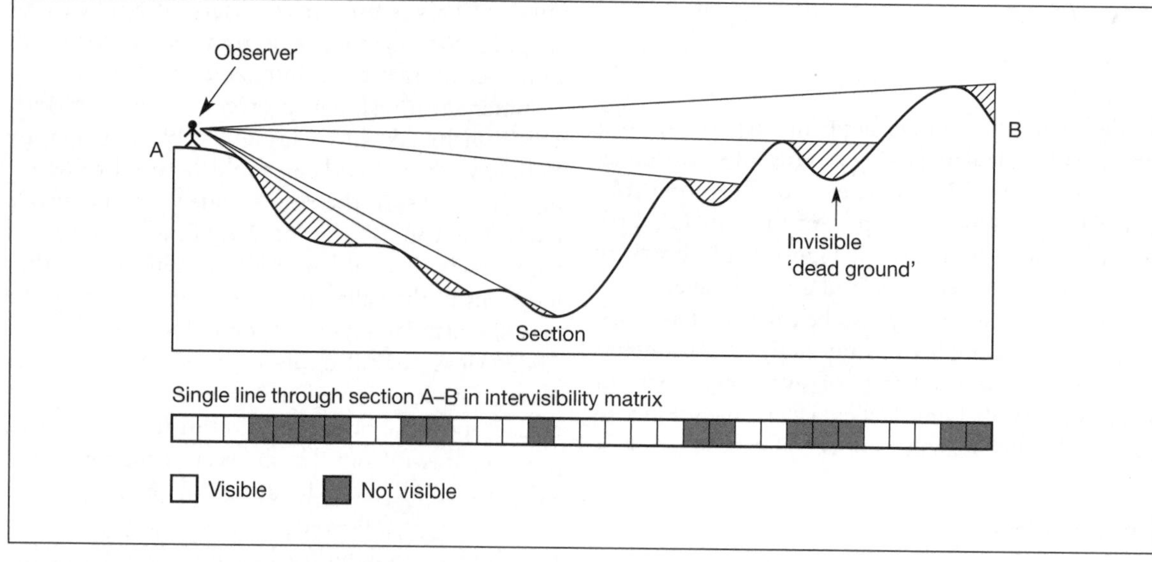

Figure 6.12 Ray tracing for visibility analysis

the DTM. This is enhanced by the ability to drape other data onto the surface of a DTM (such as a land use map or aerial photograph) and adds texture effects. Thus, the realism of views is increased (see Plate 2).

Network analysis

A network is a set of interconnected lines making up a set of features through which resources can flow. Rivers are one example, but roads, pipelines and cables also form networks that can be modelled in GIS. Chapter 3 considered the structuring of network data in vector GIS. Here we consider the applications and analysis of network data and how answers to networking problems could be obtained using a raster GIS. There are several classic network-type problems, including identifying shortest paths, the travelling salesperson problem, allocation modelling and route tracing. In the Happy Valley context, network analysis applications could be used to find the shortest route from one hotel to another, to plan a circular forest drive, to route waste collection vehicles or to allocate hotel guests to their nearest ski school.

The shortest path problem

The shortest path between one point and another on a network may be based on shortest distance, in which case either raster or vector GIS could attempt a solution. A raster GIS could provide an answer from a proximity analysis. Impediments to travel can be added to a raster grid by increasing the value of cells that are barriers to travel, then finding a 'least cost' route through a grid. Networks structured in vector GIS offer more flexibility and a more thorough analysis of impediments such as traffic restrictions and congestion. However, the shortest path may not be defined simply in terms of distance. For example, for an emergency vehicle to reach an accident the quickest route may be needed and this may require the traverse of less congested minor roads.

Shortest path methods work by evaluating the links and turns required to traverse a network between required stops. Several potential paths may be considered before the route with the least cumulative impedance is constructed from the intervening network. This process is repeated for all required stops until the whole journey path has been defined. Such a method might be used to identify the quickest route from the centre of a resort to a new ski piste.

The travelling salesperson problem

The travelling salesperson problem is a common application of network analysis. The name arises from one application area where a salesperson needs to visit a specific set of clients in a day, and to do so by the best route (usually the quickest). The waste collection vehicle has the same problem – it needs to visit all the hotels in Happy Valley, then return to base. In each case the question is 'In which order should the stops be visited, and which path should be taken between them?' This is a complex computing task. Imagine a situation where the delivery van has to visit just 10 customers. The possible number of combinations in which the 10 customers could be visited is 9 factorial (9! or 9 × 8 × 7 × 6 × 5 × 4 × 3 × 2 × 1 = 362,880). Add to this the difficulties of determining the route with the least cumulative impedance and this turns into a hefty computing problem. In GIS network analysis the ordering of the stops can be determined by calculating the minimum path between each stop and every other stop in the list based on impedance met in the network. A heuristic (trial and error) method can then be used to order the visits so that the total impedance from the first stop to the last is minimized. In the same way, the calculation of a route for a scenic drive could be aided by the inclusion of specified stops that need to be visited in a particular order.

Location-allocation modelling

Network analysis may also be used for the allocation of resources by the modelling of supply and demand through a network. To match supply with demand requires the movement of goods, people, information or services through the network. In other words, supply must be moved to the point of demand or vice versa. Allocation methods usually work by allocating links in the network to the nearest supply centre, taking impedance values into account. Supply and demand values can also be used to determine the maximum catchment area of a particular supply centre based on the demand located along adjacent links in the network. Without regard for supply and demand limitations, a given set of supply centres would service a whole network. If limits to supply and demand levels are indicated then situations can arise where parts of the network are not serviced despite a demand being present. Consider an example in which five ski schools service the network shown in Figure 6.13. If we assume that clients always enrol at the nearest school (based on link and turn impedance values) and there are no limits to the number of places at each school (i.e. unlimited supply) then each link in the network is assigned to the nearest school, and clients staying at hotels along those links will be allocated with the link. If the supply of ski school places is limited at one or all of the schools then the pattern of allocated network links changes according to the nearest ski school with free places. It is possible to have a situation where all the places are filled leaving some clients without a ski school place. In this case a new ski school needs to be built in the unserviced area, or the number of places increased at one or more of the existing ski schools.

Allocation modelling in network data is the basis for more advanced analyses such as location-allocation modelling. This can be used to identify the optimum location for a new centre to service a shortfall in supply relative to demand. In the above example, demand for ski school places outstrips supply such that parts of the network are left unserviced. If it is not possible to increase supply at

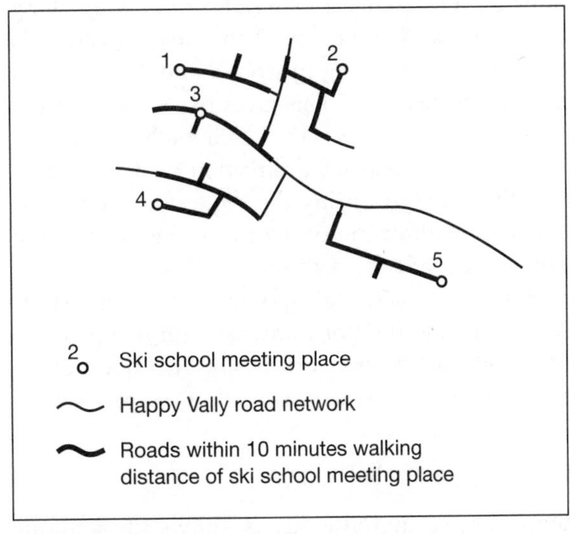

Figure 6.13 Network data analysis: allocation of ski school to clients on a network

existing ski schools, a new school must be built. The obvious location is within the unserviced area identified above, but identifying the optimum location becomes another network problem. This is because, once built, the new school will radically alter the pattern of ski school catchment areas throughout the whole network. This is a more complex modelling problem that is revisited in Chapter 7.

Route tracing

The ability to trace flows of goods, people, services and information through a network is another useful function of network analysis. Route tracing is particularly useful for networks where flows are unidirectional, such as stream networks, sewerage systems and cable TV networks. In hydrological applications route tracing can be used to determine the streams contributing to a reservoir or to trace pollutants downstream from the site of a spillage. Route tracing can be used to find all the customers serviced by a particular sewer main or find those affected by a broken cable.

Connectivity, the way network links join at network nodes, is the key concept in route tracing. Without the correct connectivity in a network, route tracing and most other forms of network analysis would not work. Directionality is also important for route tracing as this indicates the direction in which the materials are moving along the network. Knowledge of the flow direction is critical to establishing upstream and downstream links in the network. This gives rise to the concept of a directed network in which each link in the system has an associated direction of flow. This is usually achieved during the digitizing process by keeping the direction of digitizing the same as the flow direction in the network.

Once a directional network has been established, tracing the links downstream or upstream of a point on the network is simply a question of following or going against the flow, respectively.

Quantitative spatial analysis

Quantitative spatial analysis allows ideas about spatial processes and patterns to be tested and is used to help find meaning in spatial data. Quantitative analysis methods can (after Fotheringham *et al.*, 2000):

- Reduce large data sets to smaller amounts of more meaningful information.
- Explore data to suggest hypotheses or examine the distribution of data. Exploratory data analysis (EDA) techniques are used for this.
- Explore spatial patterns, test hypotheses about these patterns and examine the role of randomness in their generation.

Quantitative methods for spatial data analysis developed distinct from other statistical methods used for non-spatial data because of the distinguishing features of spatial data. Observations of spatial data are not independent, but it is assumed that features that are close together in geographical space are in some way related (this is Tobler's First Law of Geography). Different results can also be obtained when applying the same technique to the same distribution of data, simply by varying the spatial units used (this problem is known as the Modifiable Areal Unit Problem – see Box 6.6).

In some ways GIS has helped to revive the quantitative tradition in the spatial sciences, but it is interesting to note how few statistical analysis tools there are in GIS software packages. As a result, many users turn to external statistical packages to perform even relatively simple quantitative analyses. However, all GIS can provide useful data for statistical analyses simply by outputting spatially referenced data in a form that can be read by another package.

There are three main types of quantitative spatial methods:

1. *Exploratory and descriptive statistics*. These can be used to describe the distribution of spatial phenomena. For example, a histogram may be created to show the distribution of land parcels of different sizes. For exploratory and descriptive statistical models a GIS can provide: attribute data; distance from one point observation to another point, line or area feature; distances to nearest observations for input into nearest neighbour statistics or cluster analysis; and location of point

Table 6.4 Happy Valley GIS questions and possible solutions.

Happy Valley GIS question	*Suggested GIS functions required to address this question*
Which is the longest ski piste in Happy Valley?	Measurement of length or Query of attribute database (if length is included as an attribute)
What is the total area of forestry in the valley?	Area measurement or Reclassification of forestry map, then area calculation
How many luxury hotels are there in the valley?	Attribute (aspatial) data query
Where are the luxury hotels with less than 20 bedrooms?	Attribute query followed by spatial query or Combinatory AND query
Which hotels are within 200 m of a main road?	Buffering, then point-in-polygon overlay, then query or Proximity analysis, then reclassification, then point-in-polygon overlay and query
In which land use types are the meteorological stations located?	Point-in-polygon overlay
Which roads should I use for a scenic forest drive?	Line-in-polygon overlay
What is the predominant land use type in the Happy Valley resort?	Polygon-on-polygon overlay (identity), then query
Where could a new ski piste be located?	A map algebra overlay, depending on siting criteria selected
What is the predicted snowfall for the new ski piste?	Interpolation, then overlay
What are the slope and aspect of the new ski piste?	Three-dimensional analysis – slope and aspect calculation
From where will the new ski piste be visible?	Three-dimensional analysis – visibility analysis

observations relative to area data for chi-squared analyses.

2 *Predictive statistics*. These are used to look at relationships between spatial phenomena. For example, regression analysis might be used to look at the relationship between altitude and vegetation type. In this instance, geographical datasets of altitude and vegetation type can be overlaid in the GIS to provide a scatterplot of paired observations that show the relationship between the dependent variable (vegetation) and the independent variable (altitude). The scatterplot can be used for regression analysis and the subsequent model used to predict the vegetation type for areas where there are no data. Multiple input layers of dependent and independent variables can be used as the basis for multiple regression models.

3 *Prescriptive statistics*. These are used to ask 'what if?' questions. They help with the prediction of what might happen in a particular set of circumstances. For example, optimization methods could be used to look at possible market analysis scenarios to inform retail decision-making.

Each analysis technique may be applied at a local or global level. Global application to a whole dataset (for example, calculating average rainfall for a whole country) will mask regional and local variations. Local application to parts of a dataset (for example, calculating average rainfall for each state or county) provides statistics that can easily be displayed with GIS and that can be used to search for exceptions and 'hotspots'. It is also possible to geographically weight a regression analysis so that greater emphasis is placed on the relationship between local observations.

Conclusions

This chapter has covered the most common methods of data analysis available in current GIS packages. The next chapter will delve further into some of the more sophisticated modelling capabilities of GIS. However, it is now possible to see how analysis functions might be used singly or together to produce information that can be used to assist decision makers. The chapter started with some suggested questions for the Happy Valley GIS team. It is useful to revisit these questions, and consider the GIS functions needed to address them (Table 6.4). Table 6.4 indicates that there is a choice of methods in many cases, giving more than one way of obtaining an answer to a question. This is true for many GIS problems, and users need to be flexible in their approach to analysis and able to mix and match the functions available to obtain a suitable solution to their problem. In addition, users need to be aware of the strengths and weaknesses of different types of GIS (raster and vector) for data analysis, and the limitations of some of the functions they routinely apply (for example, length measurements). However, an informed GIS analyst should be able to produce useful answers by combining data analysis functions into an analysis schema. How to do this will be covered further in Chapter 12.

Data analysis is an area of continuing development in GIS; as users demand more sophisticated analysis, the vendors attempt to satisfy them. Specialist software packages have emerged which concentrate on only a few of the functions described in this chapter (there are, for example, several examples of packages geared towards network analysis). New analysis tools frequently start out in the academic domain. For example, the book by Fotheringham and Rogerson (1994), which laments the lack of spatial analysis functions (even exploratory data analysis functions like the calculation of means and production of charts and graphs), suggests some areas for development. There are continued calls for improved GIS functionality, particularly from those interested in spatial analysis. National research agencies have placed spatial analysis on their research agendas (National Research Council, 1997; Harding and Wilkinson, 1997). However, the range of functions currently available is sufficient to meet the needs of the majority of GIS users.

REVISION QUESTIONS

- Using examples, discuss the different types of queries used in GIS.
- Suggest 10 applications for buffering.
- Outline the ways in which overlay is used for the integration of data.
- Discuss the advantages and disadvantages of adopting either a vector or a raster approach to data analysis.
- What is interpolation? Why are interpolation techniques included in GIS?
- How can surface data be analysed in GIS?
- Propose a methodology using data analysis techniques for the selection of a site for a new ski resort. You have the following data available: land use (classified into rural and urban), roads (motorways, highways and minor roads), meteorological data (point met stations with average snowfall) and terrain (a 2.5D model). Your proposed site should be in a rural area, with good accessibility by car and bus, in an area of high snowfall and with good scenic potential.

Further study

There are many opportunities to follow up material covered in this chapter. Many GIS textbooks devote useful space to data analysis (for example Aronoff, 1989; Burrough, 1986; Laurini and Thompson, 1992; DeMers, 1997; Worboys, 1995; Chrisman 1997) and there are a few books devoted completely to spatial analysis and GIS (Fotheringham and Rogerson, 1994; Chou, 1996, Mitchell, 1999).

Map overlay is a technique that has enthralled cartographers and geographers for some time. There are several texts that will help you understand the origins and basics of the method. These include Ian McHarg's now famous book *Design with Nature* (1969) and the writings of geographers such as Waldo Tobler and Mark Monmonier (Tobler, 1976; Monmonier, 1991). The difficulties of automating map overlay operations occupied some of the finest geographical and computer minds in the 1970s, including Nick Chrisman and Mike Goodchild. More recent texts covering the subject include Berry (1993) and Chrisman (1997). The modifiable areal unit problem (MAUP) is described in detail by Openshaw (1984) and a good example of applied GIS overlay technique is given by Openshaw *et al.* (1989) in their pursuit of a solution to the nuclear waste disposal example that we have been using in this book.

For an excellent review of spatial interpolation methods see Lam (1983). Other useful texts include Burrough (1986) and Davis (1986).

A readable review of GIS applications in network analysis relevant to transportation problems is provided by Vincent and Daly (1990). Short sections on network analysis can also be found in Laurini and Thompson (1992) and Chrisman (1997). Spatial interaction modelling is covered by Wilson (1975) and Fotheringham and O'Kelly (1989).

Surface analysis in GIS is a popular subject on which there are many papers and texts. A good starting point for further reading is the book *Mountain Environments and GIS* edited by Price and Heywood (1994), which contains many examples of DTMs at work. This book also contains a concise overview of DTMs by Stocks and Heywood (1994). Alternatively, Maguire *et al.* (1991) contains two relevant chapters on DTMs and three-dimensional modelling (Weibel and Heller, 1991; Raper and Kelk, 1991).

Longley *et al.* (2001) offer an excellent chapter on advanced spatial analysis, which covers descriptive statistics, optimization methods and hypothesis testing. Fotheringham *et al.* (2000) explain the role of GIS in spatial analysis and explore current issues in spatial analysis.

Aronoff S (1989) *Geographic Information Systems: A Management Perspective*. WDL Publications, Ottawa

Berry J K (1993) *Beyond Mapping: Concepts, Algorithms and Issues in GIS*. GIS World Inc., Colorado

Burrough P A (1986) *Principles of Geographical Information Systems for Land Resources Assessment*. Clarendon Press, Oxford

Chou Y-H (1996) *Exploring Spatial Analysis in Geographic Information Systems*. OnWord Press, USA

Chrisman N R (1997) *Exploring Geographic Information Systems*. Wiley, New York

Davis J C (1986) *Statistics and Data Analysis in Geology*. Wiley, New York

DeMers M N (1997) *Fundamentals of Geographic Information Systems*. Wiley, New York

Fotheringham A S, Brunsdon C, Charlton M (2000) *Quantitative Geography: Perspectives on Spatial Data Analysis*. Sage, London

Fotheringham A S, O'Kelly M E (1989) *Spatial Interaction Models: Formulations and Applications*. Kluwer Academic, Dordrecht

Fotheringham A S, Rogerson P (eds) (1994) *Spatial Analysis and GIS*. Taylor and Francis, London

Lam N S (1983) Spatial interpolation methods: a review. *American Cartographer* 10: 129–49

Laurini R, Thompson D (1992) *Fundamentals of Spatial Information Systems*. Academic Press, London

Longley P A, Goodchild M F, Maguire D J, Rhind D W (2001) *Geographic Information Systems and Science*. Wiley, Chichester

Maguire D J, Goodchild M F, Rhind D W (eds) (1991) *Geographical Information Systems: Principles and Applications*. Longman, London

McHarg I L (1969) *Design with Nature*. Doubleday, New York

Mitchell A (1999) *The ESRI Guide to GIS Analysis: Vol. 1: Geographic Patterns and Relationships*. ESRI Press, Redlands

Monmonier M (1991) *How to Lie with Maps*. University of Chicago Press, Chicago

Openshaw S (1984). *The Modifiable Areal Unit Problem: Concepts and Techniques in Modern Geography*, Vol. 38. GeoBooks, Norwich, UK

Openshaw S, Carver S, Fernie J (1989) *Britain's Nuclear Waste: Safety and Siting*. Belhaven, London

Price M F, Heywood D I (eds) (1994) *Mountain Environments and Geographical Information Systems*. Taylor and Francis, London

Raper J F, Kelk B (1991) Three-dimensional GIS. In: Maguire D J, Goodchild M F, Rhind D W (eds) *Geographical Information Systems: Principles and Applications*. Longman, London, Vol. 1, pp. 299–317

Stocks A M, Heywood D I (1994) Terrain modelling for mountains. In: Price M F, Heywood D I (eds) *Mountain Environments and Geographical Information Systems*. Taylor and Francis, London, pp. 25–40

Tobler W R (1976) Spatial interaction patterns. *Journal of Environmental Systems* 6: 271–301

Vincent P, Daly R (1990) GIS and large travelling salesman problems. *Mapping Awareness* 4(1): 19–21

Weibel R, Heller M (1991) Digital terrain modelling. In: Maguire D J, Goodchild M F, Rhind D W (eds) *Geographical Information Systems: Principles and Applications*. Longman, London, Vol. 1, pp. 269–97

Wilson A G (1975) Some new forms of spatial interaction model: a review. *Transportation Research* 9:167–79

Worboys M F (1995) *GIS: A Computing Perspective*. Taylor and Francis, London

7 Analytical modelling in GIS

KEY QUESTIONS AND ISSUES

- What are models of spatial processes?
- What types of process model exist?
- How are process models implemented in GIS?
- How has GIS been used in the modelling of physical and human processes?
- What is diffusion modelling and where can it be used?
- What is multi-criteria evaluation and how is it implemented in GIS?
- What are the main problems associated with the use of GIS to build process models?
- How can public input be incorporated in GIS analysis?

Introduction

Chapters 3 and 6 explored how models of spatial form are represented and analysed using GIS. These models can be used in many ways in data analysis operations; however, they tell us nothing about the processes responsible for creating or changing spatial form. To help us understand these processes a different type of model is necessary – a process model. If we return to the Martian and the car analogy, now the Martian realizes that there is more to driving than just switching on the ignition and turning the steering wheel. He needs to be

able to estimate how far he can go on a tank of fuel or predict what will happen if he hits the brakes too hard on an icy road.

This chapter provides a summary of process models before considering how they can be implemented in GIS. These models are then approached from an applications perspective, and three examples examined: physical process models, human process models and decision-making models. To conclude, the chapter considers some of the advantages and disadvantages of using GIS to construct spatial process models.

Process models

A process model simulates real-world processes. There are two reasons for constructing such a model. From a pragmatic point of view decisions need to be made and actions taken about spatial phenomena. Models help this process. From a philosophical point of view a process model may be the only way of evaluating our understanding of the complex behaviour of spatial systems (Beck *et al.*, 1995). In GIS, process models may be used in either role. For example, in the nuclear waste and house-hunting case studies GIS has been used to help the decision-making process. In the Zdarske Vrchy case study GIS was used to aid understanding of ecological and environmental processes such as the effects of acid rain on forest ecosystems, and the effects of habitat change on the endangered black grouse.

There are many different approaches to process modelling. To decide which approach is appropriate in a particular situation, an understanding of the range of models available and their strengths and weaknesses is required. Thus, a classification of process models is a good first step.

It is a relatively simple task to classify process models into one of two types: *a priori* or *a posteriori* (Hardisty *et al.*, 1993). *A priori* models are used to model processes for which a body of theory has yet to be established. In these situations the model is used to help in the search for theory. Scientists involved in research to establish whether global warming is taking place would use *a priori* models, as the phenomenon of global warming is still under investigation.

A posteriori models, on the other hand, are designed to explore an established theory. These models are usually constructed when attempting to apply theory to new areas. An *a posteriori* model might be developed to help predict avalanches in a mountain area where a new ski piste has been proposed. Avalanche formation theory is reasonably well-established, and several models already exist which could be applied to explore the problem.

It is possible for a model to change from *a posteriori* to *a priori* and vice versa. In the avalanche prediction example, the existing model may fail to explain the pattern of avalanche occurrences. If so, it would be necessary to adapt the model to explore the new situation and thus contribute to theory on the formation of avalanches.

Beyond the *a priori* / *a posteriori* division, developing a further classification of process models becomes quite complex. However, two useful classifications (Hardisty *et al.*, 1993; Steyaert, 1993) have been integrated here to provide a starting point for examining the different types of models. The classification includes natural and scale analogue models, conceptual models and mathematical models.

Natural and scale analogue models

A natural analogue model uses actual events or real-world objects as a basis for model construction (Hardisty *et al.*, 1993). These events or objects occur either in different places or at different times. A natural analogue model to predict the formation of avalanches in the previously unstudied area of a new ski piste might be constructed by observing how avalanches form in an area of similar character. The impact that avalanches would have on the proposed ski piste could also be examined by looking at experiences of ski piste construction in other areas.

There are also scale analogue models (Steyaert, 1993) such as topographic maps and aerial photographs, which are scaled down and generalized replicas of reality. These are exactly the sort of analogue models that GIS might use to model the avalanche prediction problem.

Conceptual models

Conceptual process models are usually expressed in verbal or graphical form, and attempt to describe

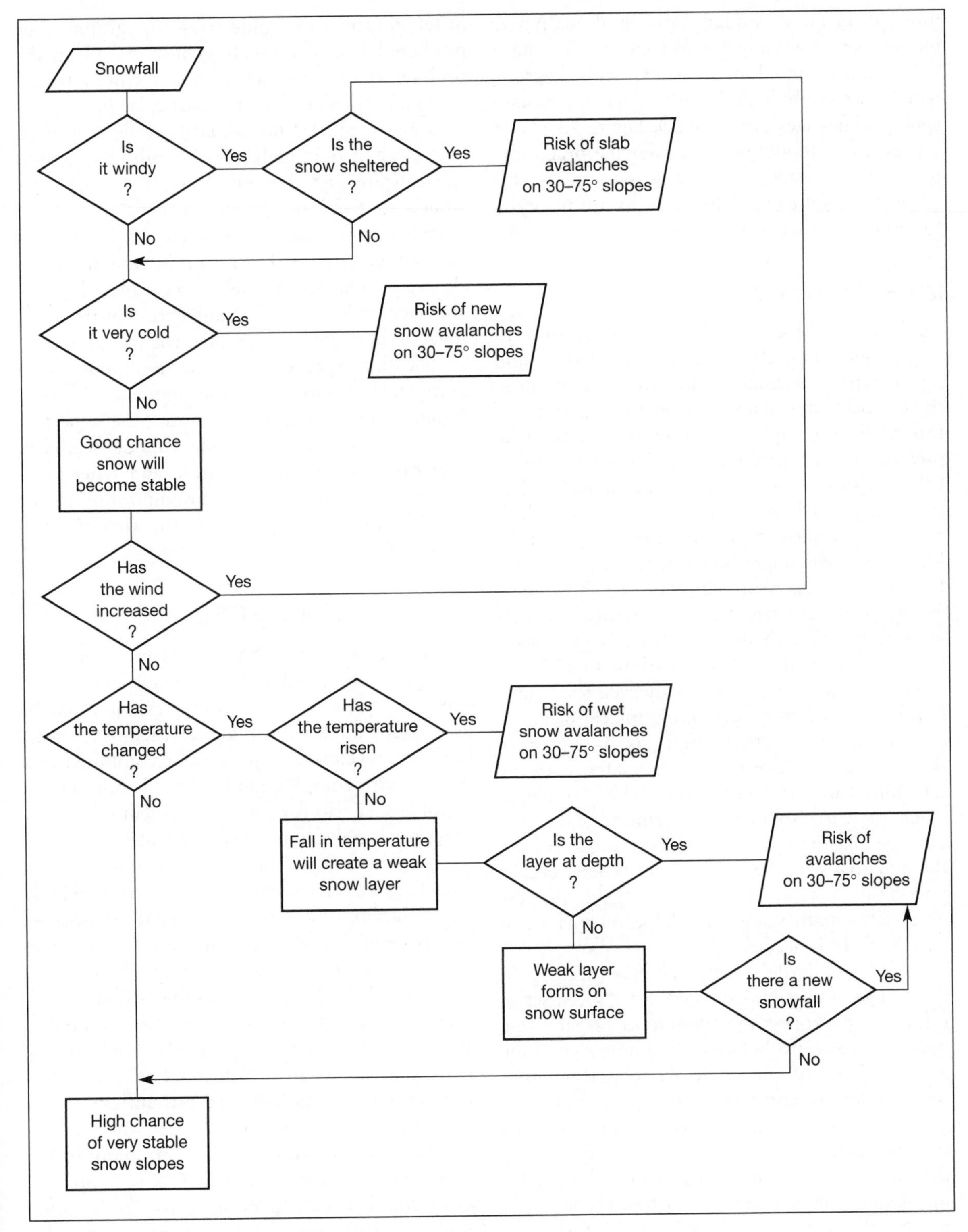

Figure 7.1 Simplified conceptual model of avalanche prediction

in words or pictures quantitative and qualitative interactions between real-world features. The most common conceptual model is a systems diagram, which uses symbols to describe the main components and linkages of the model. Figure 7.1 shows a conceptual model of the avalanche prediction problem as a systems diagram. Conceptual modelling is returned to in Chapter 12 as a framework for project management in GIS.

Mathematical models

Mathematical process models use a range of techniques including deterministic, stochastic and optimization methods. In deterministic models, there is only one possible answer for a given set of inputs. For example, a deterministic avalanche prediction model might show a linear relationship between slope angle and size of avalanche. The steeper the slope, the smaller the avalanche which results, since snow build-up on the slope will be less. This model might be created by developing a least-squares regression equation for slope angle against avalanche size. Such deterministic models work well for clearly defined, structured problems in which a limited number of variables interact to cause a predictable outcome. However, few simple linear relationships exist in geographical phenomena. In most situations there is a degree of randomness, or uncertainty, associated with the outcome. This is true in the avalanche example, as slope angle is only one factor among many that must be taken into account when trying to predict the potential size of an avalanche. Where there is uncertainty about the nature of the process involved, a mathematical model known as a stochastic model is needed.

Stochastic models recognize that there could be a range of possible outcomes for a given set of inputs, and express the likelihood of each one happening as a probability. We know that slope angle and size of avalanche are related but that the problem is much more complex than suggested by our deterministic model. In the deterministic model the assumption was made that as slopes get steeper there will be less build-up of snow and smaller avalanches. However, in reality other variables will be involved, for example direction of slope, exposure to wind, changes in temperature and underlying topography. The predicted size of an avalanche is based on the probability of a number of these factors interacting in a particular place at a particular time.

Steyaert (1993) notes that both deterministic and statistical models can be subdivided into steady-state and dynamic models. Steady-state models are fixed for a particular time and dynamic models have at least one element which changes over time. Our avalanche prediction model is clearly a dynamic model, since time plays an important role in the changing nature of the snow cover and hence when an avalanche will occur.

The final type of mathematical model is an optimization model. These models are constructed to maximize or minimize some aspect of the model's output. Assuming a model has been developed which shows where and when avalanches are likely to occur, an optimization model could be used to help identify the area of minimum avalanche risk at a given time.

Process modelling and GIS

In GIS all three approaches – natural and scale analogue, conceptual and mathematical modelling – are used to model spatial processes. They may be used in isolation, substituted for each other in an iterative development process or combined in a larger, more complex model. Box 7.1 presents a possible scenario for the development of a snow cover prediction model in Happy Valley.

The example in Box 7.1 shows how different modelling techniques can be used together to build up complex models of spatial processes. Unfortunately, proprietary GIS software provides few, if any, process models as part of the standard set of functions. In many ways this is understandable since process models are heavily adapted to meet the requirements of particular applications. Thus, generic models, which could be made available in GIS, would be far too inflexible for widespread use. However, there are some modelling functions in GIS. In addition, many of the analytical functions found in GIS (Chapter 6), when coupled with functions provided by other modelling software, provide an environment for constructing application-specific models (Box 7.2).

BOX 7.1

The development of a snow cover model for Happy Valley

The Happy Valley GIS team has attempted to construct a model of snow cover in the valley to assist the siting of a new ski piste. The ski piste must be in an area where snow cover is reliable throughout the winter. To assist the decision where to site the new piste the following models were developed:

1. The first model of seasonal snow cover variation was developed using GIS. A series of digitized aerial photographs was used to establish how the snow cover in the valley had changed over time. This sequence of photographs was used as a scale analogue model to predict future snow cover patterns. However, the results were unsatisfactory, as other factors clearly needed to be considered. The valley's meteorological expert suggested that altitude and slope angle were important factors to include in any model.
2. The next model developed was also a scale analogue model. Extra layers of data were added to the GIS. A DTM provided surface data, and analysis undertaken on this allowed the prediction of snow melt at different altitudes and on slopes with different aspects.
3. A stochastic mathematical model was developed next to compare data on snow melt patterns over a number of years. This predicted the probability of a particular slope having full snow cover at any time during the ski season.
4. Finally, a conceptual model of the whole modelling process was developed by summarizing the stages required in a systems diagram. This conceptual model was used to facilitate revisions to the modelling process following validation of results by field checking.

The remainder of this chapter explores how GIS can be used to help construct spatial models for physical, human and decision-making processes. These will be examined by reference to specific examples: forecasting and diffusion; location-allocation and spatial interaction models; and multi-criteria evaluation techniques. The range of applications considered is not comprehensive; however, the examples illustrate key issues in the model-building process.

Modelling physical and environmental processes

All the different types of models introduced in the earlier part of this chapter have been applied to the modelling of physical processes. Jakeman *et al.* (1995) provide many examples of the application of these techniques for modelling change in environmental systems. Goodchild *et al.* (1993) take a closer look at how environmental and physical models can be constructed in or coupled with GIS. Table 7.1 provides a summary of some of the application areas being dealt with in GIS.

BOX 7.2

Linking GIS with other spatial analysis and modelling software

The GIS toolbox can provide only a limited number of spatial analysis and modelling tools. Therefore, as Maguire (1995) suggests, there is considerable value in integrating GIS with other specialist systems to provide for the needs of the process modeller. There are two main ways in which GIS software can be linked with other spatial analysis and modelling tools, and the approach adopted will depend on the nature of the application:

1. The spatial analysis or modelling software can be integrated within a GIS environment, or the GIS can be integrated within the modelling or analysis environment. In this case one software environment will dominate and therefore dictate the structure of the model.
2. The second approach involves either tightly or loosely coupling the GIS to the other software environment. In tightly coupled models the linkage between the two products will be hidden from the user by an application interface. In a loosely coupled model the user may well have to perform transformations on the data output from the GIS, spatial analysis or modelling software before they can be used in another environment.

Table 7.1 Applications of spatial models in GIS

Application	*Source*
Physical and environmental applications	
GIS analysis of the potential impacts of climate change on mountain ecosystems and protected areas	Halpin, 1994
GIS analysis of forestry and caribou conflicts in the transboundary region of Mount Revelstoke and Glacier National Parks, Canada	Brown *et al.*, 1994
Simulation of fire growth in mountain environments	Vasconcelos *et al.*, 1994
Integration of geoscientific data using GIS	Bonham-Carter, 1991
Spatial interaction models of atmosphere–ecosystem coupling	Schimel and Burke, 1993
Geographical innovations in surface water flow analysis	Richards *et al.*, 1993
Human applications	
Techniques for modelling population-related raster databases	Martin and Bracken, 1991
Use of census data in the appraisal of residential properties within the UK: a neural network approach	Lewis and Ware, 1997
Design, modelling and use of spatial information systems in planning	Nijkamp and Scholten, 1993
Selecting and calibrating urban models using ARC/INFO	Batty and Xie, 1994a,1994b
An introduction to the fuzzy logic modelling of spatial interaction	See and Openshaw, 1997
An investigation of leukaemia clusters by use of a geographical analysis machine	Openshaw *et al.*, 1988

In a spatial context, perhaps one of the most challenging tasks for physical and environmental modellers is *forecasting* what may happen in the future under a given set of conditions. For example, predicting when and where a flood may occur or where an avalanche may take place is a difficult task. However, if the nature of the system under investigation is reasonably well understood then it can be modelled and future trends, conditions, changes and outcomes forecast.

Forecasting is inherently time-dependent. Unfortunately GIS data models handle the time dimension poorly (Chapter 3); therefore, most applications that involve making predictions or forecasts are based around some non-GIS model. Often these models are not spatial (aspatial). They may deal with a single site or assume large homogeneous areas. However, when these models are linked with GIS they can be expanded to include a range of spatial inputs and outputs. Linking existing aspatial models with GIS can be difficult because the models do not necessarily include important key spatial phenomena such as spatial interaction and spatial autocorrelation when making their predictions. Other models are inherently spatial and are particularly well suited for integration with GIS. The linking of existing forecasting models with GIS provides access to a larger database and tools to visualize how a spatial phenomenon diffuses over time. Different methods of visualizing output from forecasting models are presented in Chapter 8.

Forecasting models tend to be dynamic, with one or all of the input variables changing over time. For example, in the case of a forest fire, as time elapses from the start of the fire the prevailing weather conditions will change, as will the amount of forest still left to burn. Simulating the geographical spread of a dynamic phenomenon such as a forest fire requires a special type of forecasting model known as a *diffusion model*. These are best explained by reference to examples.

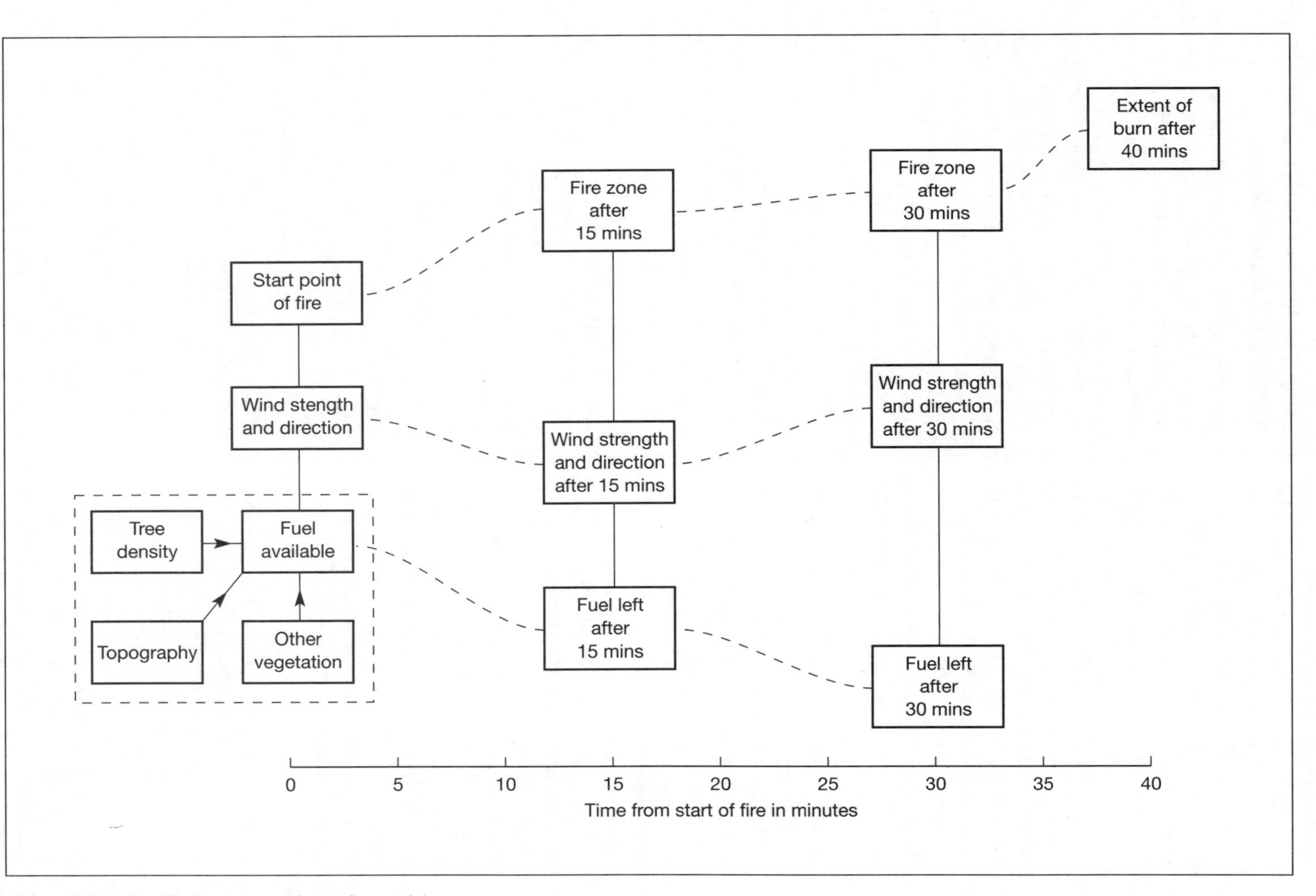

Figure 7.2 A simplified conceptual forest fire model

Imagine that the Happy Valley GIS team wishes to explore how GIS can be used to predict the rate of spread of forest fires in the area. The team has access to a conceptual fire model as shown in Figure 7.2. This shows that the rate of diffusion of the fire from its starting point is based on the spatial characteristics of the neighbourhood. The density of trees, the nature of the topography of an area, the presence of previously burnt areas and dynamic variables such as wind direction all affect the spread of fire. The spatial diffusion problem faced by the fire model is how to use this knowledge to simulate the movement of the fire over time. One way to achieve this is through close coupling with GIS.

Vasconcelos *et al.* (1994) describe one approach to modelling fire behaviour. Their method involved three stages:

1 A raster approach was used to create a spatial database of the criteria influencing fire behaviour. These criteria were determined from an aspatial model and included data on fuel, topography and weather.
2 The value associated with each cell in the raster images was then fed into an established aspatial fire-modelling program known as BEHAVE (Burgan and Rothermel, 1984). The output was a map showing the status of the fire after a given time.
3 The modelling capabilities of GIS were then used to 'spread' the fire to adjacent cells. The extent of the spread depended on the amount of fuel present, the topography of the land and recent weather. This diffusion process was continued until a specified time limit had been reached or all fuel sources for the fire had been consumed.

Using this approach, the GIS was used to produce three summary maps that showed the rate of spread of the fire, the spatial intensity of the fire and the direction of maximum spread.

In the forest fire example above, the focus has been on how the GIS can be used to provide input data, facilitate spatial diffusion and assist with the visualization of results. Another example of how GIS can be used to provide additional data for input to a forecasting model is discussed in Box 7.3.

BOX 7.3

Using GIS to provide terrain variables for hydrological modelling

The hydrological cycle is an example of a dynamic natural system that can be modelled with the help of GIS. Not all hydrological models involve a spatial element, even though the inputs and controls of the hydrological cycle are inherently spatial (for example, rainfall is never distributed evenly across a catchment). GIS can provide variables for input to hydrological modelling using spatial data input, management, analysis and display functions that may be lacking in other modelling systems. This is particularly true for terrain variables, such as slope, aspect and watershed analysis.

Terrain is fundamental to hydrology, since water always flows down a slope. The shape of the ground surface determines how water flows downhill – at what speed, in what direction, how much infiltrates into the soil and how flows accumulate. Therefore, DTMs are important in hydrological modelling. Analysis of a DTM in GIS provides data on slope and aspect, height, catchment boundaries and direction and accumulation of flow. Other digital information may be added to a DTM (such as soil type and land cover) to determine likely infiltration rates and the hydrologic response to given precipitation inputs. This is more realistic than assuming that all the water falling as rain onto a DTM surface will run off as surface flow. Data on infiltration rates, evapotranspiration, interception, groundwater percolation and subsurface storage can be derived in a GIS using DTM data in combination with data on soils, land use, vegetation, climate and human action. DTMs can also be used to determine the hydrological network in an area.

Thus, GIS can contribute to hydrological forecasting models by providing a suite of tools for the manipulation and generation of terrain-related variables.

A recent development in forecasting models has been to couple GIS with artificial intelligence (AI) techniques such as neural networks (Hewitson and Crane, 1994). Neural networks are essentially classification systems that look for pattern and order in complex multivariate data sets. If the data are longitudinal records of inputs and outputs to a system they can be used to 'train' a neural network to recognize complex relationships between variables. The resulting neural network can then be used to predict system outputs from a given set of inputs and a knowledge of preceding conditions.

Several problems make accurate forecasting very difficult. It is generally accepted that the accuracy of forecasts decreases with the length of the prediction. A classic example is the weather forecast. With improved understanding of weather systems and better data, weather forecasters can predict what the weather is going to do in the next 24 to 48 hours fairly accurately. Beyond 48 hours the accuracy of their predictions begins to fall to such an extent that long-term five-day forecasts are only about 40% correct (Tyler, 1989). The introduction of random or stochastic processes, as advocated by proponents of chaos theory (Stewart, 1989), may play an increasingly important role in improving the accuracy of longer-term predictions, especially models of natural systems like the weather and climate.

Modelling human processes

The complete range of modelling techniques introduced earlier in this chapter has also been applied to the modelling of human processes (Table 7.1). However, perhaps the most difficult aspect of modelling humans is simulating their spatial behaviour. *Spatial interaction models* offer one method of modelling spatial behaviour, and the links between these and GIS are the focus of this section. Spatial interaction models are used to help understand and predict the location of activities and the movement of materials, people and information (Birkin *et al.*, 1996). They can be used, for example, to predict the flow of people from their home to shops. In this case, the assumption is made that, as long as the number and characteristics of shops and people remain the same, so does the predicted behaviour of people. It is also assumed that people will use the shops closest to their home. If one parameter in the model is changed (for example, a new shopping centre may be added to increase the choice of shops available) the predicted behaviour of people will change. Spatial interaction models rely on the fact that the nature of this change can be predicted since it can be compared with the behaviour of people elsewhere. In this respect, spatial interaction models are a form of natural analogue model. Time does not have a role in the simplest spatial interaction models, therefore these models can be implemented in GIS. To explore how spatial interaction models work and identify the benefits of coupling them with GIS it is useful to consider an example.

Siting a new supermarket in Happy Valley will influence the shopping patterns of residents of Happy Valley and visitors to the area. Skiers staying in Happy Valley will not normally travel many miles to buy groceries but will tend to choose the nearest store that satisfies their requirements for value for money, choice, quality and convenience. Given that their primary reason for visiting Happy Valley is skiing, convenience may be the most important factor. In this case the most convenient supermarket may be one closest to their apartment. If the new supermarket is located close to the main apartment complex there is a good chance that a large number of skiers will make use of the facility. The amount of trade done by other stores in the area may be reduced as a result. Therefore, the distance between the origin of a journey (the apartment) and the destination (the supermarket) will affect the amount of 'interaction' that takes place between the two locations. If the new supermarket is in a poor location, for example if it is too far away from the main ski village, the skiers will continue to use existing stores that are closer. The only way a new supermarket situated away from the apartment blocks may be able to attract customers is to offer additional services and facilities to induce customers to travel further. The 'attractiveness' of a store can be represented in a spatial interaction model as an attractiveness index. This can be estimated from surveys of consumer opinion and by counting the number of people visiting the destination and determining how far they have travelled.

The concepts of supply and demand are also important for spatial interaction modelling. Demand occurs at an origin (for example, the apartment block), while supply is located at destinations (the supermarket). Production, the capacity of the origin to produce a trip for a particular activity, is also important, and is generally some function of the population of the origin. If the population of the Happy Valley ski village is over 2000 people, its capacity to produce trips to the supermarket will be greater than a nearby farming hamlet that is home to three families.

Distance between an origin and a destination does not have a fixed effect on spatial interaction. The effect of distance depends on the type of activity and customers. For example, skiers would use the nearest supermarket if all they needed was bread and milk, but for a good bottle of wine they may be prepared to travel further.

The effects of distance and attractiveness can be studied using *gravity models* (Birkin *et al.*, 1996). These use a distance-decay function (derived from Newton's law of gravitation) to compute interactions given the relative attractiveness of different destinations. The distance-decay function is used to exaggerate the distance between origins and destinations. In the supermarket example the distance-decay function applied should be large as skiers are prepared to travel only short distances to use a particular store. Other activities such as shopping for skis, clothes or presents demand a smaller distance-decay function because skiers will be more willing to travel greater distances for these items. Choosing the correct distance-decay function to use for a particular activity is essential in developing a sensible model. Gravity models can be calibrated using data on trip lengths derived from customer surveys since the function is related to the average trip length for the activity.

The basic data required to develop spatial interaction or gravity models are a set of origins and destinations. These can be provided by a GIS as point entities. In the case of our supermarket example, one data layer would represent the proposed site for the new supermarket and a second data layer the distribution of the apartments in Happy Valley. In addition, the GIS could be used to provide a further data layer representing the transport networks (including streets and footpaths) along which goods and people move. GIS can be used to calculate real distances between origins and destinations. In addition, the GIS database can be used to store attribute information about individual locations. In the case of the new supermarket these attributes could include the type of goods and services the store offers. Attributes for the apartments could reflect the socio-economic characteristics of skiers who rent them. This additional information could be used to determine the demand, supply and attractiveness characteristics for the model. GIS can also be used to display the results of the modelling process as a series of accessibility contours for both origin and destination locations, or as interaction probability surfaces (showing the probability of skiers in a given location making trips to the proposed supermarket). Birkin *et al.* (1996) provide numerous examples of how spatial interaction and gravity models have been coupled with GIS. Harder (1997) and Mitchell (1997) offer other examples of the application of GIS to the modelling of human processes in areas such as health care, schools and recycling.

Modelling the decision-making process

Outputs from process models of both human and physical systems are the raw information that will be required to assist various types of decision making. Output from a location-allocation model might show potential locations for a new supermarket within Happy Valley. This would need to be integrated with other regional planning data to ensure that the resort is being developed in line with regional strategic planning objectives. Moreover, it may be that a choice has to be made between the development of a supermarket or the construction of a new ice rink. As each development will have different spatial impacts on the environment, it may be necessary to evaluate what these will be to help decide which of the two development options should be pursued. In the same way, areas of high fire risk output from the Happy Valley forest fire model may need to be integrated with other environmental data to determine priority areas for nature conservation.

Map overlay is the traditional technique for integrating data for use in spatial decision making. For example, in siting nuclear waste facilities, the criteria defining geological suitability and conservation area status can be combined using the overlay procedures outlined in Chapter 6. However, overlay analyses suffer from certain limitations when dealing with decision-making problems of a less well-defined nature. These are summarized by Janssen and Rievelt (1990):

- Digital map overlays are difficult to comprehend when more than four or five factors are involved.
- Most overlay procedures in GIS do not allow for the fact that variables may not be equally important.
- When mapping variables for overlay analysis, decisions about threshold values (see Box 6.6) are important (the outcome of polygon overlay depends strongly on the choice of threshold values).

One way to address these problems is to use *multi-criteria evaluation* (MCE) techniques to either supplement or replace standard map overlay in GIS. MCE techniques allow map layers to be weighted to reflect their relative importance and, unlike polygon overlay in vector GIS, they do not rely on threshold values. Therefore, MCE provides a framework for exploring solutions to decision-making problems, which may be poorly defined.

MCE is a method for combining data according to their importance in making a given decision (Heywood *et al.*, 1995). At a conceptual level, MCE methods involve qualitative or quantitative weighting, scoring or ranking of criteria to reflect their importance to either a single or a multiple set of objectives (Eastman *et al.*, 1993). It is not the intention here to introduce the full range of MCE techniques, since these are documented elsewhere (Eastman *et al.*, 1993; Janssen, 1992; Voogd, 1983). However, an understanding of the key concepts is important to appreciate how MCE can be applied in a GIS context.

In essence, MCE techniques are numerical algorithms that define the 'suitability' of a particular solution on the basis of the input criteria and weights together with some mathematical or logical means of determining trade-offs when conflicts arise. These techniques have been available since the early 1970s and a whole literature has developed that is devoted to MCE and related fields (Nijkamp, 1980; Voogd, 1983). It is the overlap with GIS that is of interest, since the application of MCE techniques within a GIS framework permits the limitations of standard map overlay (defined above) to be addressed. An example of how MCE is implemented in GIS is given in Box 7.4. MCE, in common with other scientific disciplines, has developed its own specialized terminology. The examples below inevitably contain some of this 'jargon', but this is kept to a minimum and defined as necessary.

The first step in MCE is to define a problem. In the house-hunting case study this is where to look for a new home. A range of criteria that will influence the decision must then be defined. The criteria can be thought of as data layers for a GIS, and in the house-hunting example these included proximity to schools and roads, and quality of neighbourhood. The criteria selected should reflect the characteristics of the neighbourhood in which the decision maker wishes to live. For example, a decision maker with a young family may feel it is important to be near a school, though in a rural area. A method of weighting or scoring the criteria to assess their importance (a decision rule) must then be constructed. This is done by adding weights (often expressed as percentages) to reflect the importance of each criteria (Figure 7.3). A high level of importance (80%) may be attached to proximity to a school and a low level of importance (20%) for proximity to the road network. Application of an MCE method will result in a list of neighbourhoods in which the home buyer could choose to live.

MCE could also have been applied in the nuclear waste case study. The problem was defined and a solution developed in Chapter 6 using polygon overlay. The results of this overlay procedure gave a set of areas that satisfy certain criteria relating to geology, population, accessibility and conservation. The number of potentially suitable sites within the suitable area identified may total several thousand. Once the suitable area is defined, the question then becomes how to identify a limited number of best sites from the thousands available.

In the case of nuclear waste disposal it is possible to identify a whole range of factors that may be important in determining a site's suitability, including population density within a 50 kilometre

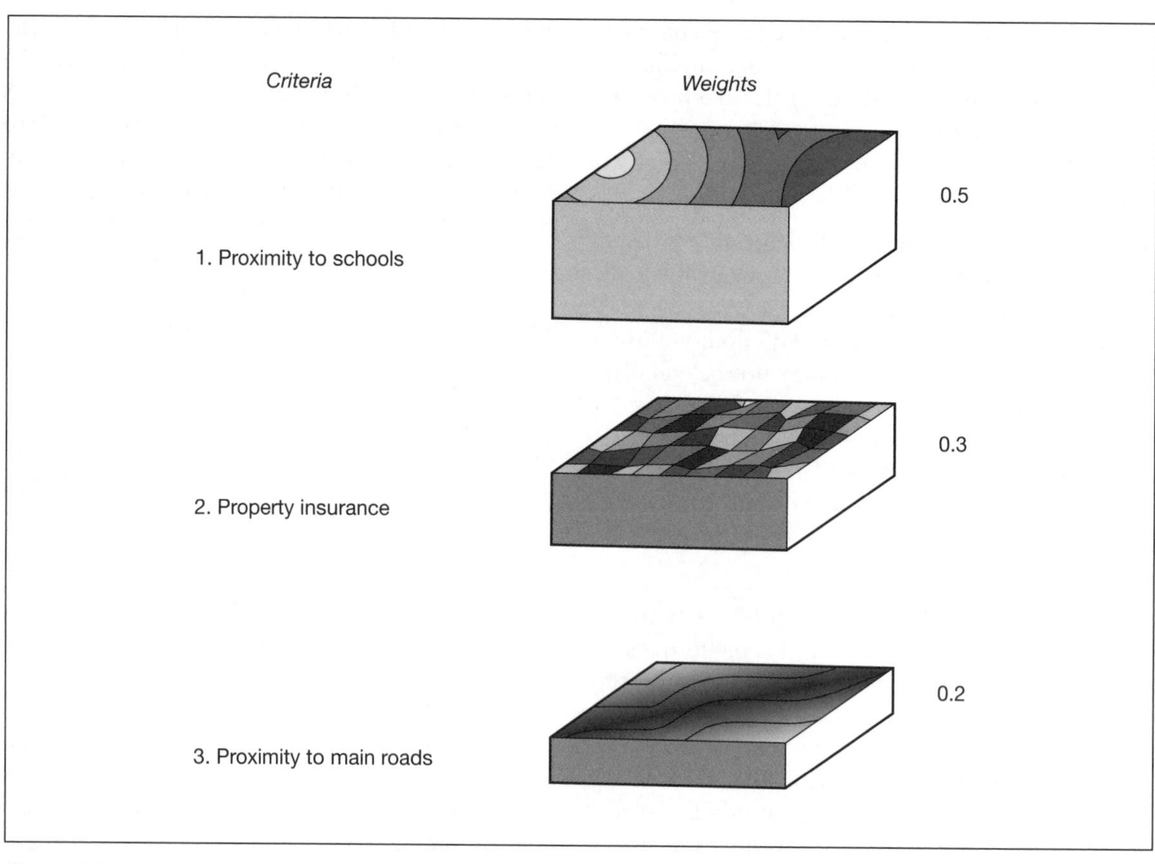

Figure 7.3 Weighting data layers in the house hunting case study

BOX 7.4

Implementing MCE in GIS

Perhaps the simplest MCE algorithm is the weighted linear summation technique. The steps involved in applying this model within a raster GIS are illustrated in Figure 7.4 and can be described as follows:

1 *Selection of criteria.* Most MCE analyses in GIS start with the identification of the data layers that are important to the problem. These criteria are represented as separate raster data layers in the GIS.
2 *Standardization of criterion scores.* Most MCE analyses, especially those using quantitative and mixed data sources, require some form of standardization of the scales of measurement used by the data layers. This is necessary to facilitate the comparison of factors measured using different units and scales of measurement (it would be nonsense to directly compare population density with distance from the nearest railway line). Therefore, standardization of the scales will enable meaningful comparisons to be made between the data layers. Standardization can be done in a number of ways, but it is normal to apply a linear stretch routine to re-scale the values in the raster map between the maximum and minimum values present. Care needs to be taken to observe 'polarity' such that beneficial factors are represented on a scale that gives a high value to high benefit and a low value to low benefit, while cost factors are represented on a scale that gives a low value to high cost and a high value to low cost. Another common method is to bring all the values on a data layer to an interval value between 0 and 1. For example, if the values on the proximity to an urban map range

BOX 7.4 CONTINUED

between 0 and 10 miles, using a standardized scale a value of 0 miles will equate to 0 and a value of 10 miles to 1. A value of 5 miles would score 0.5.

3 *Allocation of weights.* Weights are allocated which reflect the relative importance of data layers. A weighting of 80 % may be expressed as 0.8.

4 *Applying the MCE algorithm.* An MCE algorithm may then multiply these standardized scores by the weights for each of the data layers in stage 1 and sum these to allocate a score to each pixel on the output map. The map produced will be a surface with values ranging from 0 to 1. In the case of the house-hunting example, areas with a value close to 1 represent those areas where the home buyer should start looking for a house. Further evaluation of the results may be carried out by ranking the values in the results map and reclassifying the ranked map to show the top-ranked sites.

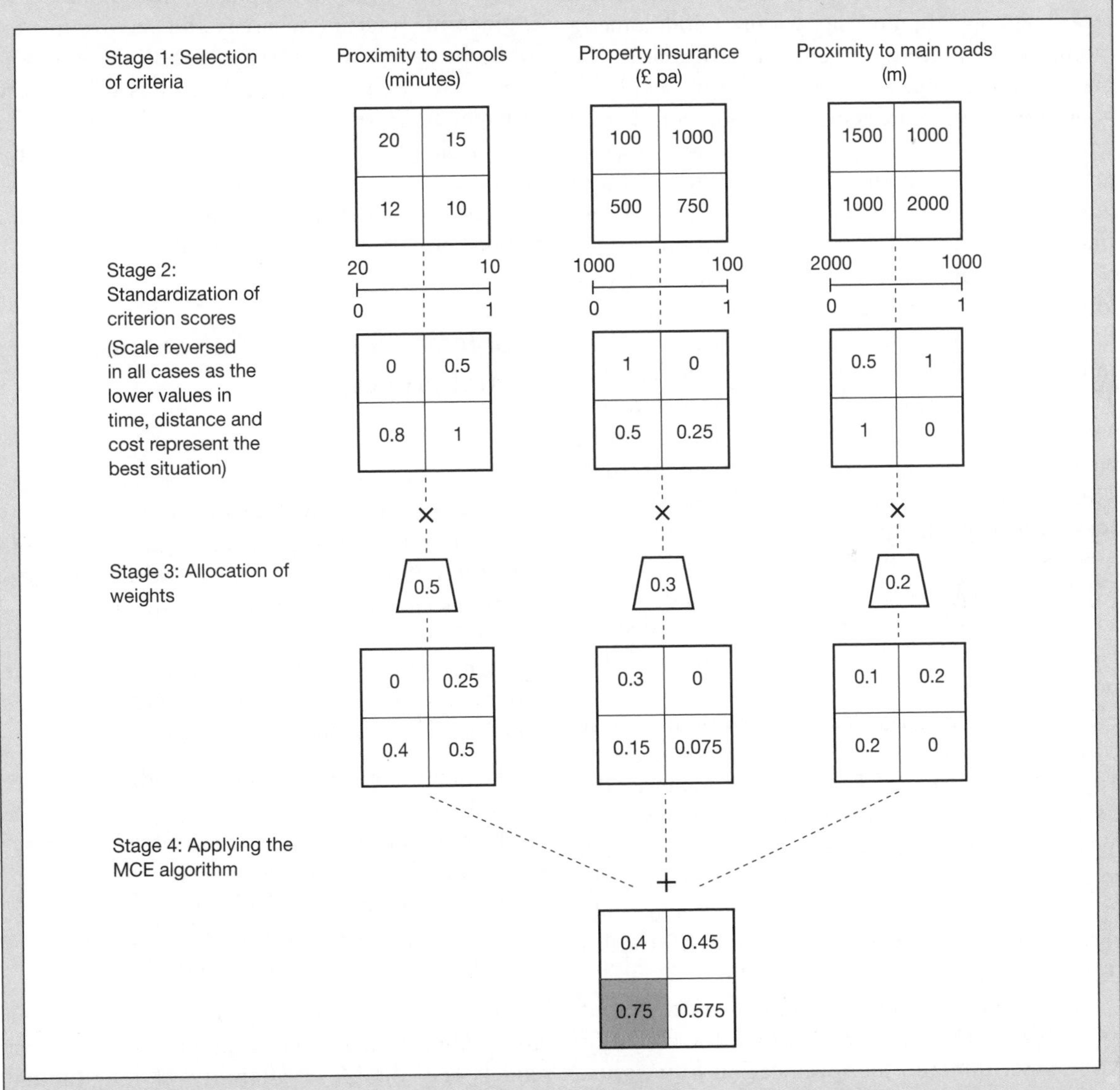

Figure 7.4 Applying a simple linear weighted summation model in raster GIS

radius, quality of on-site access, distance from conservation area(s), distance from nearest railway line, accessibility to waste producers and even marginality of the local parliamentary constituency! Satisfying certain deterministic criteria applied to these factors is not essential at this stage, but we may wish to optimize them such that the best sites are those which possess all the best qualities. This is not the simple ranking and sorting procedure that it may at first sound, since many of the factors specified may be of differing importance and may be conflicting. For example, consider the factors regarding population density within a 50-kilometre radius and distance to the nearest railway line. We may wish to optimize the siting solution by identifying those sites in areas of low population density within the 50-kilometre radius, and at the same time minimize the distance from the nearest railway line. In many areas these are conflicting criteria since railways usually run through and link highly populated areas. It is likely that the sites with very low population densities will be a long way from the nearest railway line, while the sites near to a railway line will have higher population densities. The solution to this problem is therefore to use MCE techniques, which allow the factors to be weighted according to their relative importance. Thus, the trade-offs necessary to determine which sites are actually the best available are simplified.

GIS is an ideal framework in which to use MCE to model spatial decision-making problems as it provides the data management and display facilities lacking in MCE software. In return MCE provides GIS with the means of evaluating complex multiple criteria decision problems where conflicting criteria and objectives are present. Together, GIS and MCE-based systems have the potential to provide the decision maker with a more rational, objective and unbiased approach to spatial decision making and support than hitherto. However, MCE techniques have only recently been incorporated into the GIS toolkit (Carver, 1991b; Eastman *et al.*, 1993).

MCE, like any other technique, is not without its problems. The main problems are the choice of MCE algorithm and the specification of weights. Different MCE algorithms may produce slightly different results when used with the same data and the same weights. This is due to the differences in the way the trade-offs are performed. The most intractable problem, however, is in the specification of (factor) weights since these strongly determine the outcome of most analyses. Different people will view the problem differently, and so specify different weighting schemes. In the context of the nuclear waste example, it is possible to imagine that the weighting schemes specified by the nuclear industry and the environmental lobby will be entirely different. The nuclear industry is more likely to place greater emphasis on economic and engineering factors whereas the environmental lobby is more likely to stress social and environmental factors. In this case, the results will be very different and further analysis would be required to identify least-conflict solutions (Carver, 1991b). This has led to developments in public participation GIS in which public inputs are the key to responsible GIS applications (see Box 7.5).

Problems with using GIS to model spatial processes

The use of GIS to facilitate the development of spatial process models is not without problems. We have already alluded to some of these, such as the availability of appropriate modelling tools with the GIS toolbox. However, while technical problems can be solved (by integrating or coupling GIS with appropriate software tools), there are other, more conceptual problems that the process modeller wishing to use GIS should be aware of. These include:

- the quality of source data for model calibration;
- the availability of data for model validation;
- the implementation within a GIS;
- the complexity of modelling reality; and
- the conceptual and technical problems of building multi-dimensional models.

The data requirements of process models can be large and problematic. For example, many forecast models rely on the availability of longitudinal records to calculate past trends and relationships. These records are used to predict future conditions.

BOX 7.5

Public participation GIS: information is power

One of the many criticisms levelled at GIS is that it is an elitist technology accessible only to those organizations with sufficient resources to pay for systems, data and technical staff (Pickles, 1995). GIS has also been criticized for its lack of knowledge-based input (Taylor, 1990). As GIS moves into the public sphere via the Internet, it is now possible to counter these two claims by giving the public much greater access to GIS and spatial datasets.

The public is an enormous data resource that can be of tremendous benefit to decision makers. Participatory approaches use this resource to help populate GIS databases with information on local knowledge and grassroots opinion. In return, decision makers are able to make decisions that are more in tune with local feelings and needs.

Early public participation GIS (PPGIS) involved groups of decision makers sitting around a single GIS workstation and asking them to explore a problem with the aid of relevant GIS datasets. An example of this was the Zdarske Vrchy case study where local resource managers used GIS to explore decisions about recreational management (Petch *et al.*, 1995). Shiffer (1995a, 1995b) extended the PPGIS model to include multimedia to draw the public further into the decision-making process. He used video and audio representations of aircraft during take-off and landing linked to a GIS database of residential areas around a proposed airport site.

Recent PPGIS have used Internet-based approaches because of the advantages this offers in terms of access for the public 24 hours a day, seven days a week. Open access systems via the Internet may also help break down some of the psychological barriers to participation present in face-to-face meetings that lead to 'us and them' type stand-offs between the decision-making authorities and the public.

A model of public participation in spatial decision-making facilitated by Internet-based PPGIS is that of 'explore, experiment and formulate' (Kingston *et al.*, 2000). In this model the participating public can *explore* the decision problem using the GIS database and *experiment* with possible alternative solutions before *formulating* their own informed decision as to the best course of action. The information accessed and the choices made in reaching this decision are fed back to the decision makers via the Internet along with the decision made. This allows much greater insight into public opinion than is possible from more traditional methods such as local meetings and consultative documents.

The nuclear waste case study is an example of a spatial decision problem that would benefit significantly from greater public input, perhaps via Internet-based PPGIS. It is clear that nobody wants to live next door to a nuclear waste disposal site because of the possible health risk, however real or imagined. We have seen in previous chapters how GIS can be used to identify possible sites for such a facility based on geographical criteria in a map overlay model. We have also seen how MCE methods may be integrated with GIS to model the suitability of individual sites based on the trade-offs between weighted factor maps. By incorporating MCE approaches into Internet-based PPGIS it should be possible to survey the opinions of a large number of people across the country concerning the general geographical factors (and their weights) relevant to the problem of identifying a suitable site for a nuclear waste repository. A link to just such a website can be found on the companion website to this book.

PPGIS using the Internet does have problems. These include difficulties regarding the representativeness and validity of responses gained. At present not everyone has equal access to the Internet nor do they have the necessary training in order to use it. Distinguishing valid responses from those made by people just 'playing around' or wishing to bias the results one way or another is difficult. Perhaps the biggest problem is the willingness of those people already in positions of power (politicians, government ministers, planners and local officials) to place greater decision-making powers in the hands of the public through such systems. Governments around the world are placing increased emphasis on enhancing services and public accountability by electronic means. It will be interesting to see just how far down this route to 'cyber-democracy' they are prepared to go.

Suitably long records in many cases may not exist. Similarly, models that have been developed in a non-spatial context may require detailed data inputs which, again, do not exist as spatial data sets. An example of this can be seen in hydrological modelling – where do you obtain detailed spatial data on evapotranspiration rates or soil moisture conditions? Another example is the Universal Soil Loss Equation, which has often been applied within a GIS context. The equation is written as:

$$A = R \times K \times LS \times C \times P$$

where: A = average annual soil loss; R = rainfall factor; K = soil erodibility factor; LS = slope length–steepness factor; C = cropping and management factor; and P = conservation practices factor.

Finding the relationship between the factors in the soil loss equation and making the necessary measurements is simple enough for small experimental soil plots. However, extending the measurements to a large study area, where the controlling factors can vary by orders of magnitude, is difficult. Often surrogate measures and interpolated data are used. For example, spatial variations in evapotranspiration may be estimated by combining previous field studies on different vegetation types and weather with land cover information derived from satellite imagery.

All process models need to be validated or checked to make sure that they work and that the results they produce make sense. The easiest way to validate a model is to compare its predictions with a real event (for example, the predictions of a forest fire model could be compared with a real fire situation). Starting a fire for model validation would not, however, be a feasible validation method. Equally, if we were trying to model the potential effects of global warming we would have a long time to wait before we could compare the results of our model with reality.

In the case of some processes, such as the forest fire example, we can use historical events to validate our model, if the historical records of the event provide us with enough detail. Therefore, we could use our fire model to simulate a fire that happened in 1950, and compare our results with a survey done after the fire had burnt out. However, this approach is unable to solve all problems. In some cases the best option is to produce a range of scenarios showing what the likely outcome may be and hope that reality is somewhere between the worst and best cases. The main GIS problem associated with validating spatial process models is therefore lack of appropriate spatial data.

Implementation of process models within GIS enforces a spatial context. When models were originally developed for non-spatial applications this can create conceptual and operational problems. The original model may not consider geographical processes such as spatial interaction, diffusion or spatial autocorrelation because it was designed for single, non-spatial applications. As a result, the extension of models into a spatial context must be undertaken with care.

A recent development in GIS software has been the introduction of visual modelling tools that allow users to construct spatial models within a modelling window (Plate 15). Data and process icons can be combined using onscreen 'drag and drop' techniques to build up a model flow chart. Once a model has been constructed in this manner, it is interpreted by the GIS and processed to generate the required output. These visual model builders have made spatial modelling accessible to a wider number of GIS users, but at the same time raise important questions about the users' ability to understand exactly what they require from the GIS and how to achieve this.

The human and physical worlds are extremely complex and interlinked. Theories of environmental determinism suggest that human development is strongly influenced (if not determined) by the physical environment. More anthropocentric theories stress the increasing level of control that the human race has over the natural environment and how the latter has been irrevocably changed by human activity. Whichever theory you subscribe to, there is no escaping that the two systems, human and natural, cannot really be considered in isolation from each other. Agriculture is a good example. A crop may not grow naturally in an area that receives a low rainfall, so a more drought-tolerant crop may be grown instead. Thus, the environment determines human activity. Alternatively, it may be possible to irrigate the land to allow less drought-tolerant crops to be grown. Thus, human action overcomes environmental factors. Therefore, complex models are

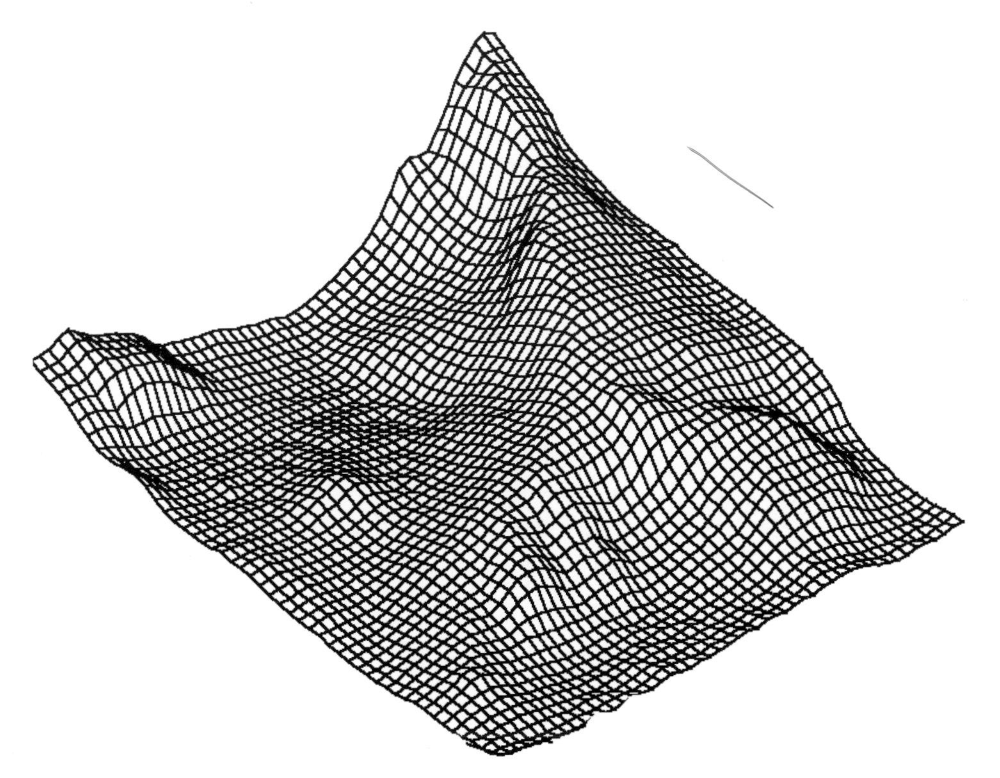

Plate 1. A wire frame diagram for part of the Snowdonia National Park, Wales (the grid interval is 20m).

Plate 2. The wire frame model in Plate 1 draped with orthorectified aerial photography of the same area.

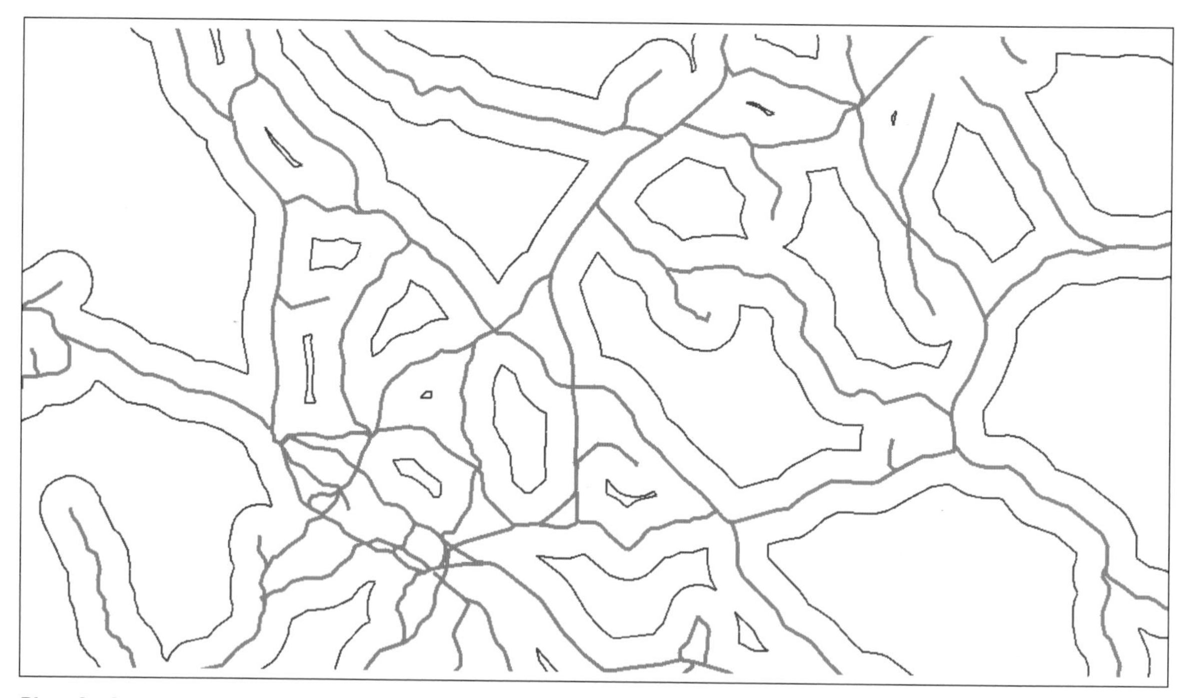

Plate 3. Radioactive waste case study: 3km buffer zones around the railway network.

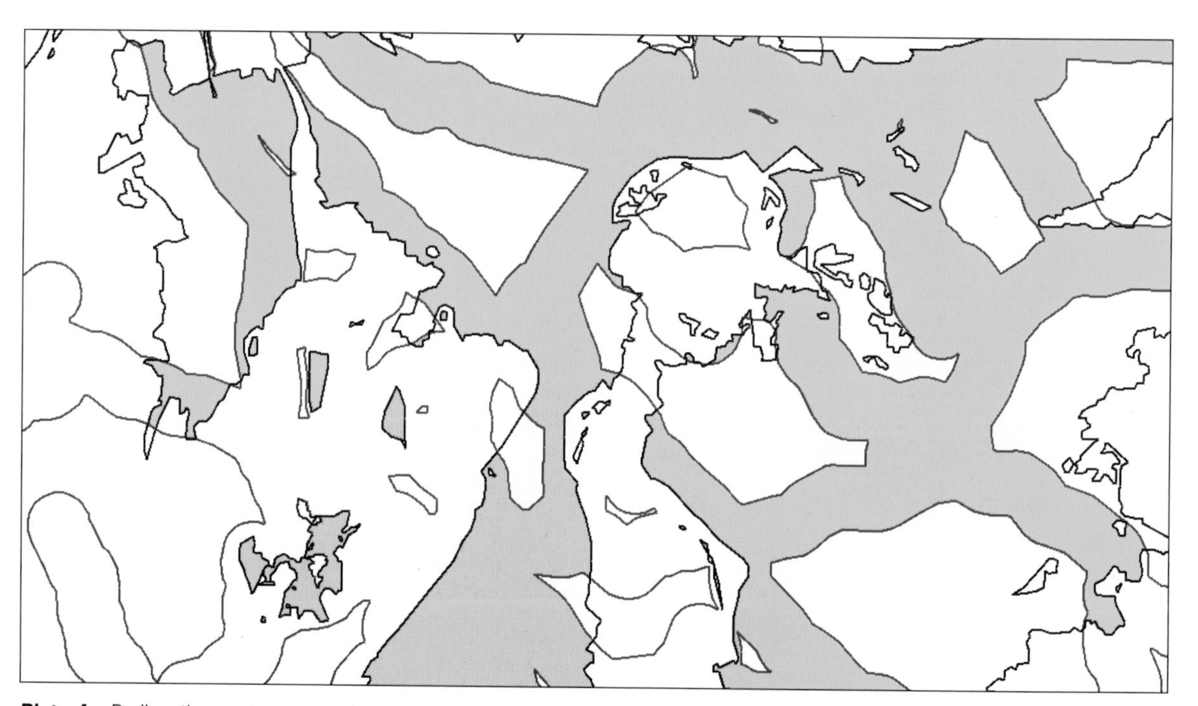

Plate 4. Radioactive waste case study: results of vector overlay showing the intersect of the rail buffer zone and clay geology.

Intersect of rail buffer and clay geology
Railway lines
3 km railway buffer zone
Surface clay geology

Plate 5. Radioactive waste case study: results from different siting scenarios.

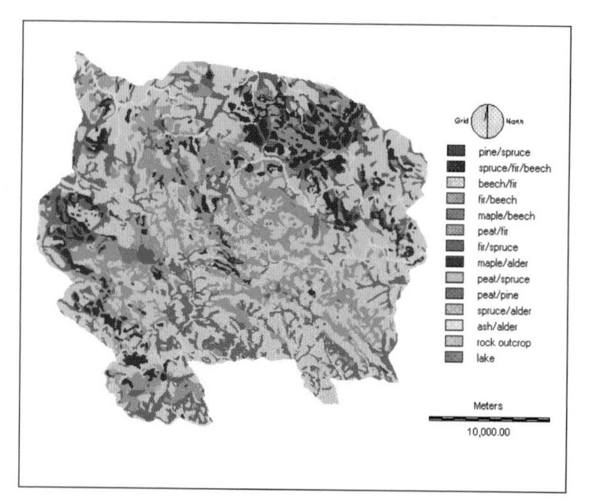

(a) Ecological survey

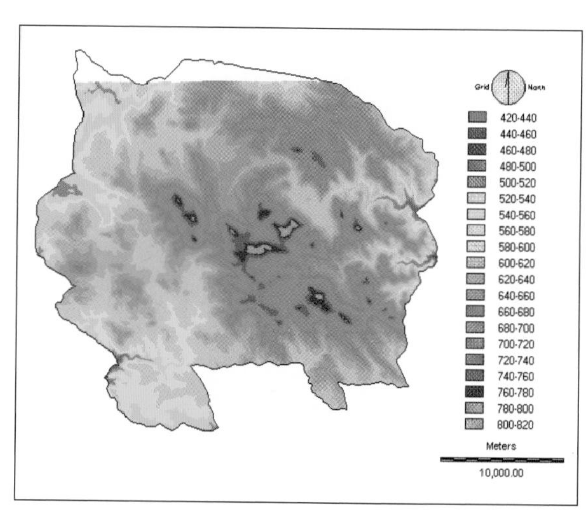

(b) Relief (20m contours)

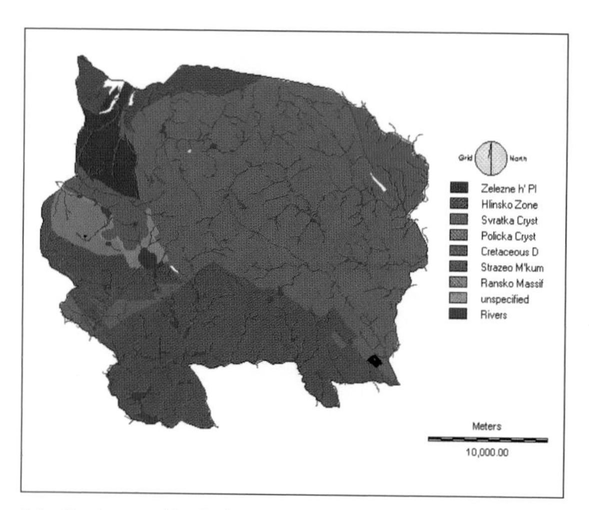

(c) Geology and hydrology

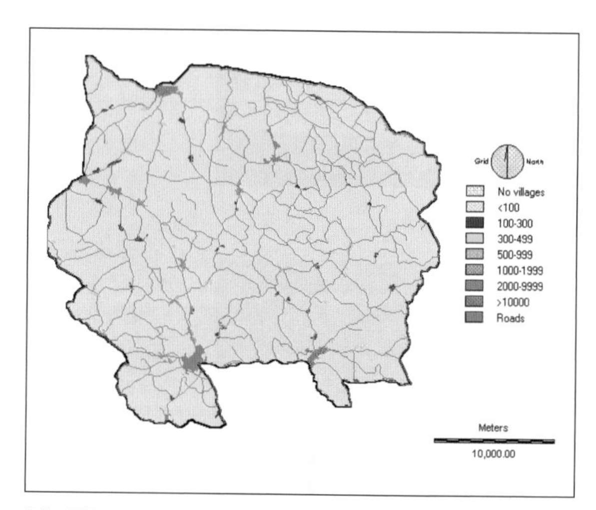

(d) Village population

(e) Recreational load: this map has been produced by combining the four maps shown above with several other data layers for the region.

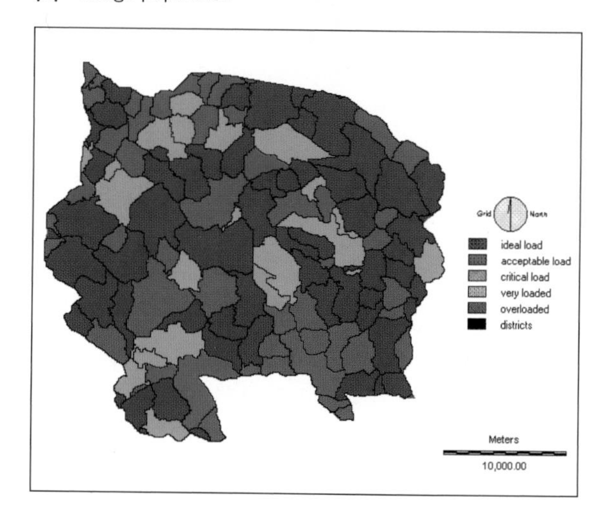

Plate 6. Zdarske Vrchy case study.

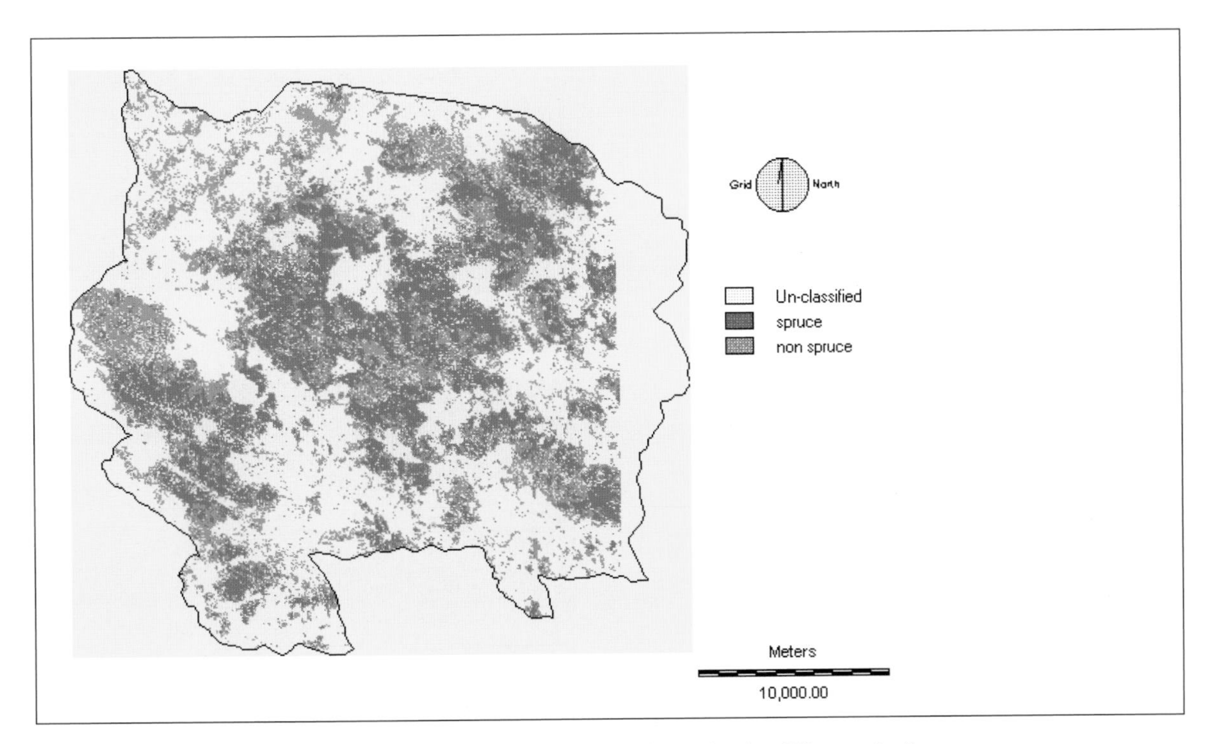

Plate 7. Zdarske Vrchy case study: forest cover, created by reclassifying a Landsat TM scene for the area.

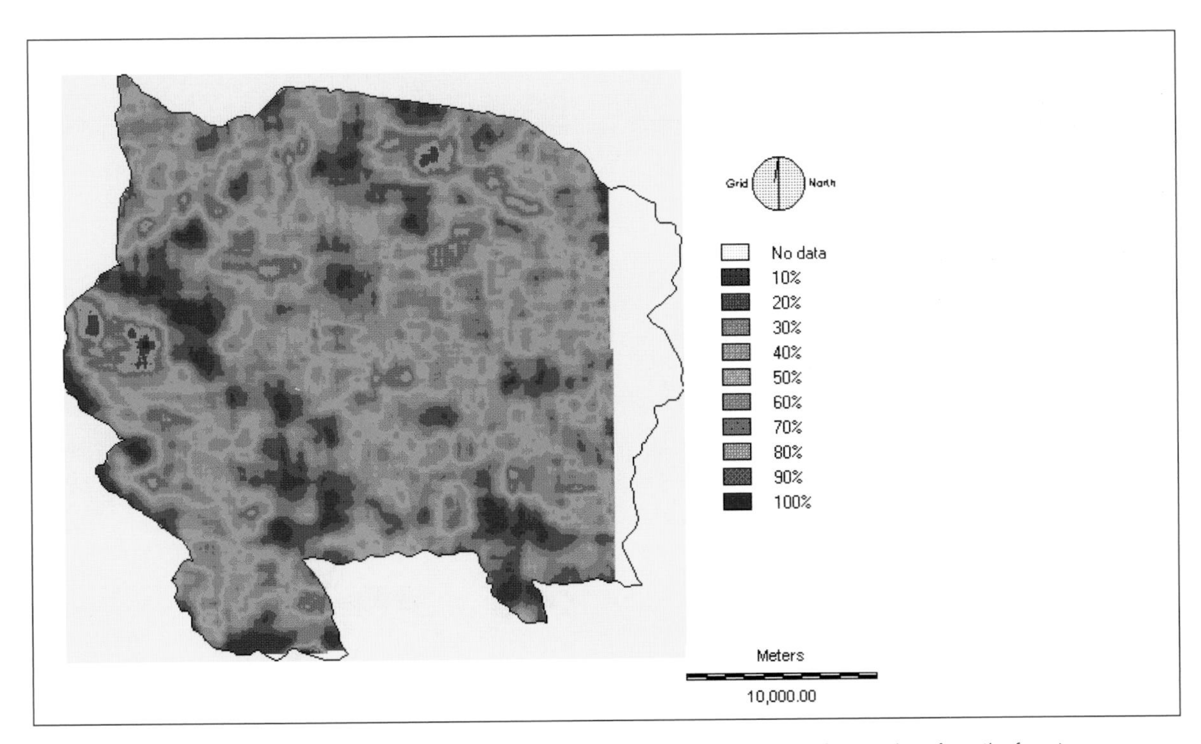

Plate 8. Zdarske Vrchy case study: a forest edge index calculated by first identifying the forest edges from the forest cover map (Plate 7) and then passing a 6 by 6 cell filter over the edges identified (if all cells in the filter contained forest edge the index value = 100%).

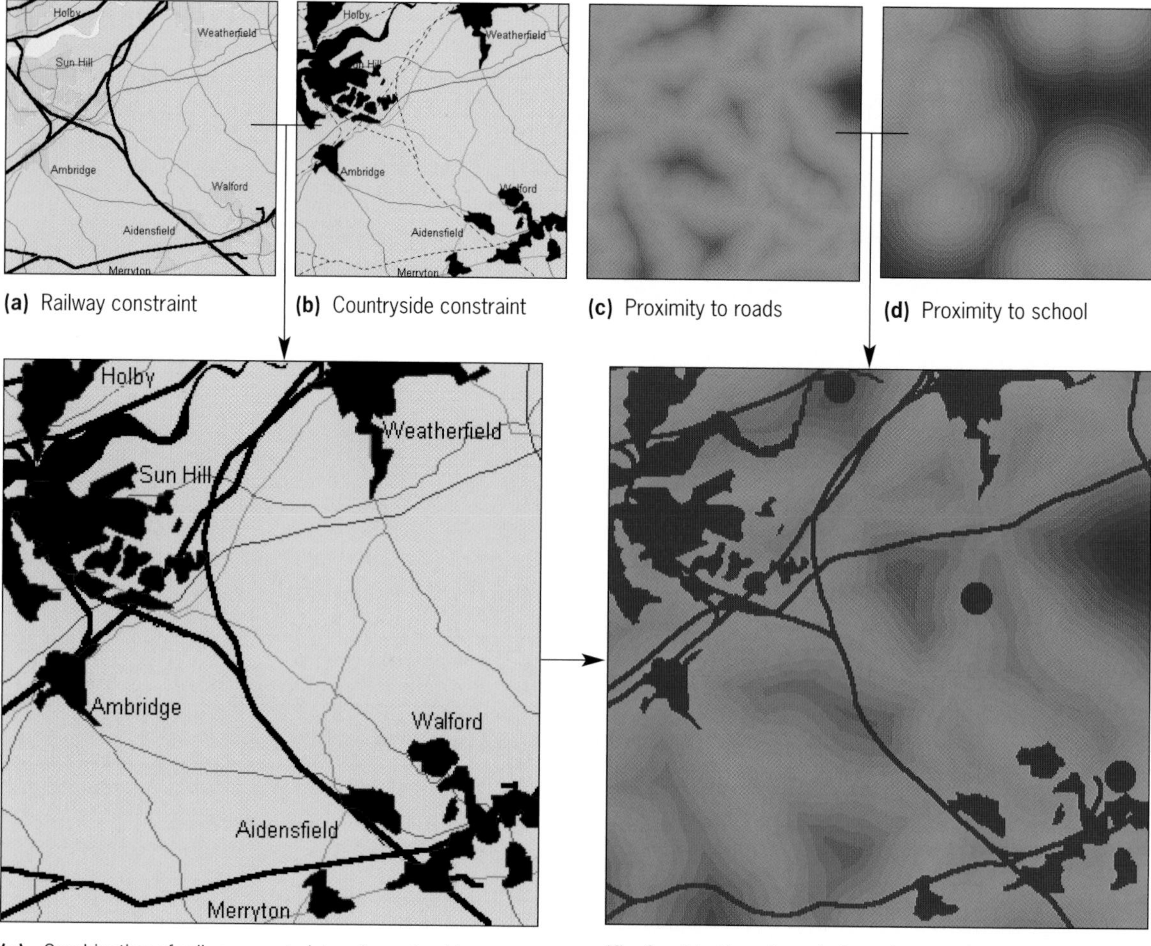

(a) Railway constraint

(b) Countryside constraint

(c) Proximity to roads

(d) Proximity to school

(e) Combination of railway constraint and countryside constraint.

(f) Combination of proximity and constraint maps (Plates 9 c, d & e) with proximity to road used as the most important factor.

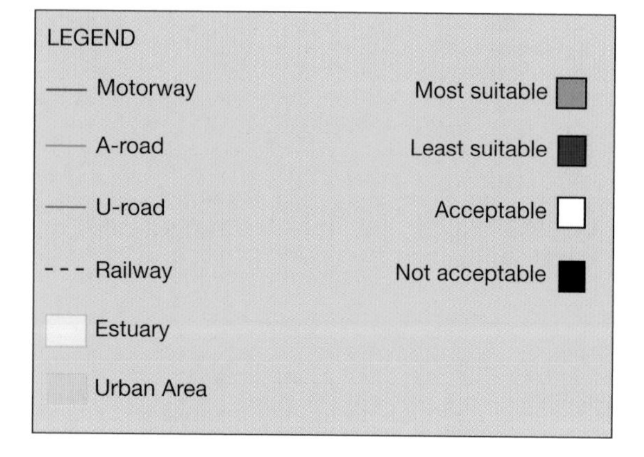

Plate 9. The house hunting case study.
Source: GeographyCal®

(g) Optimal sites (reclassified from plate 9 (f)).

Plate 10. House hunting case study: distance from office calculated using proximity method.

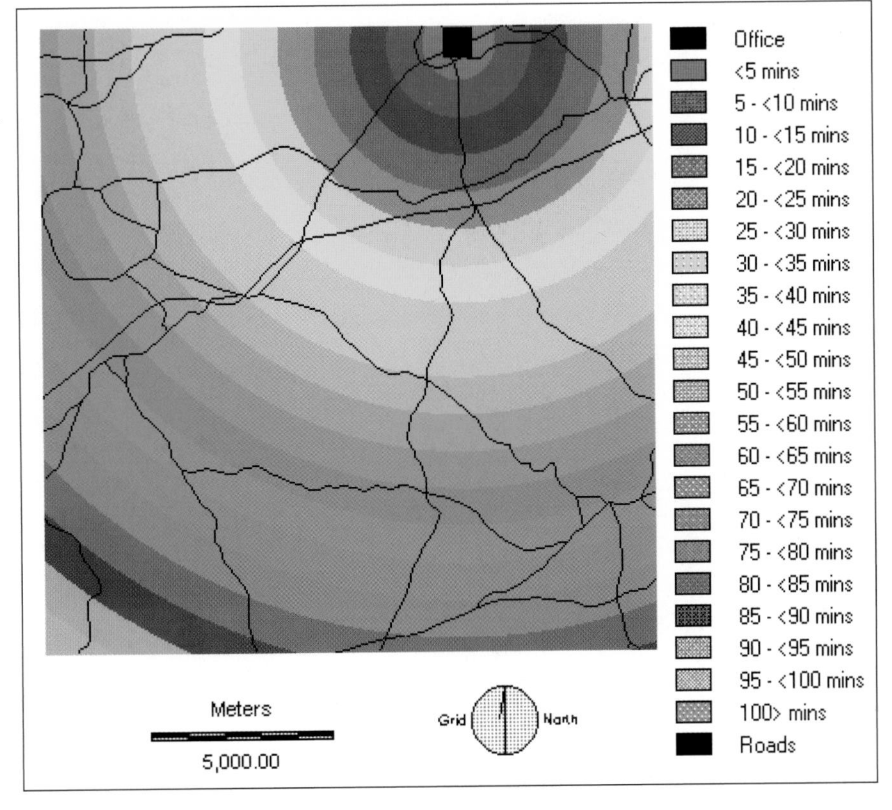

Plate 11. House hunting case study: distance from office adjusted for road network.

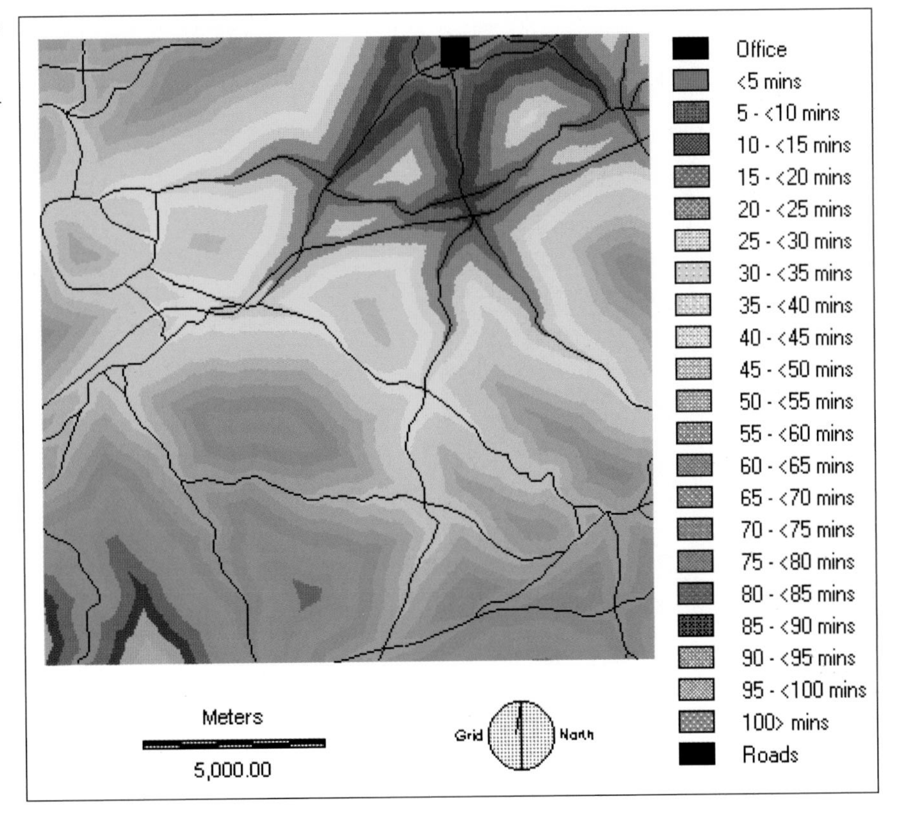

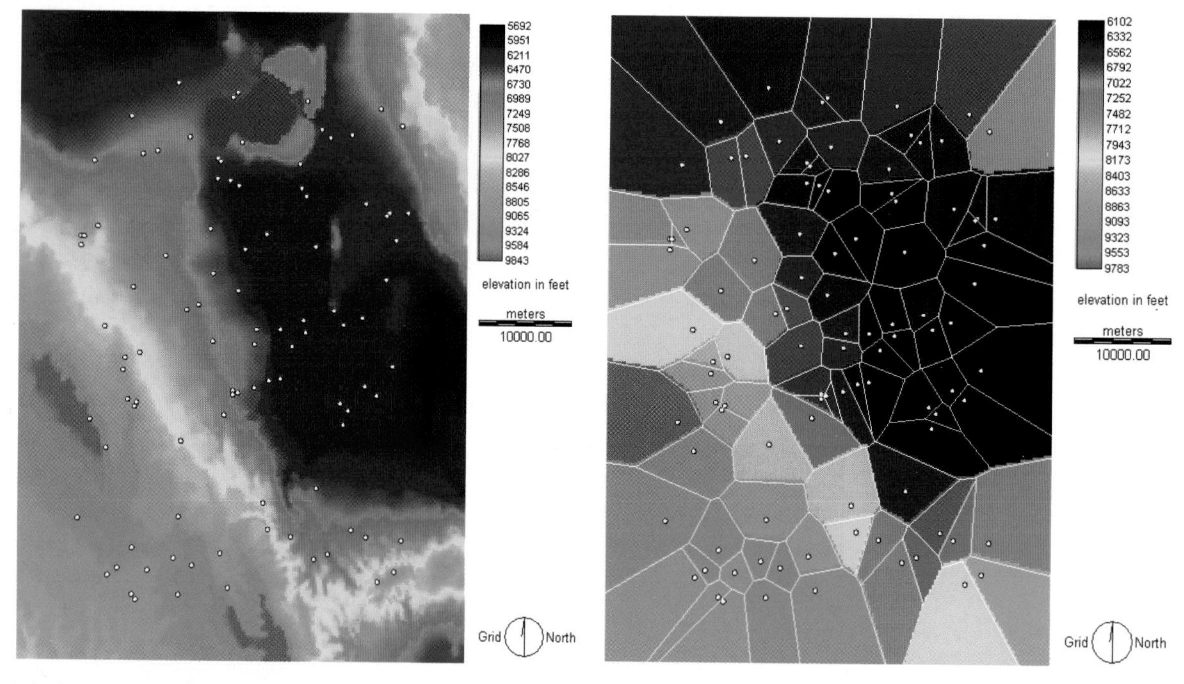

(a) Original elevation surface with sample points

(b) Interpolated elevation – Thiessen polygons

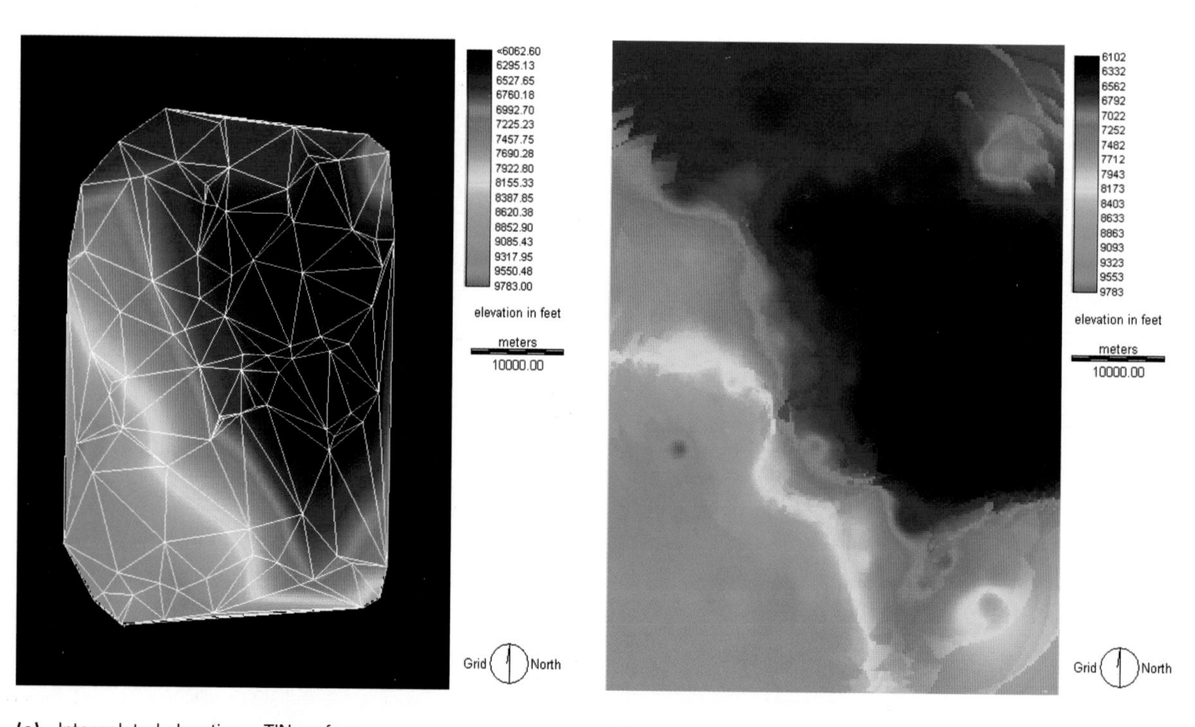

(c) Interpolated elevation – TIN surface

(d) Interpolated elevation – distance weighted average

Plate 12. Three different interpolation methods applied to sampled terrain data.

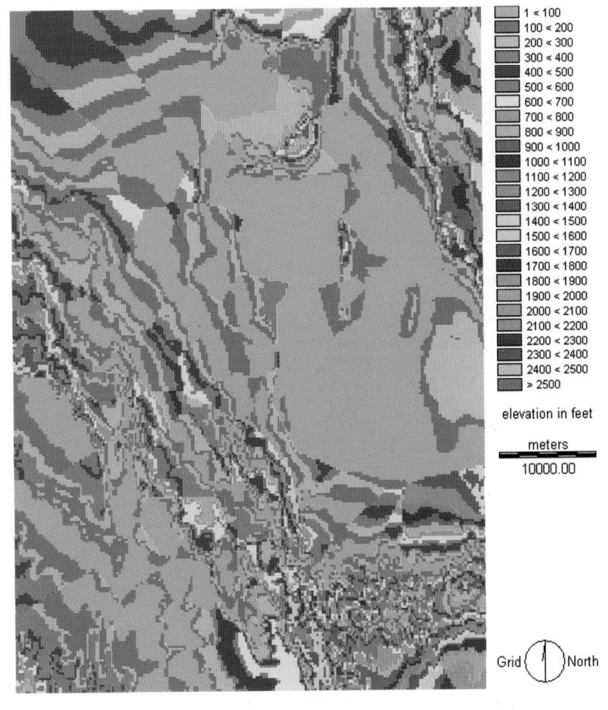

(a) Error map – Thiessen polygon

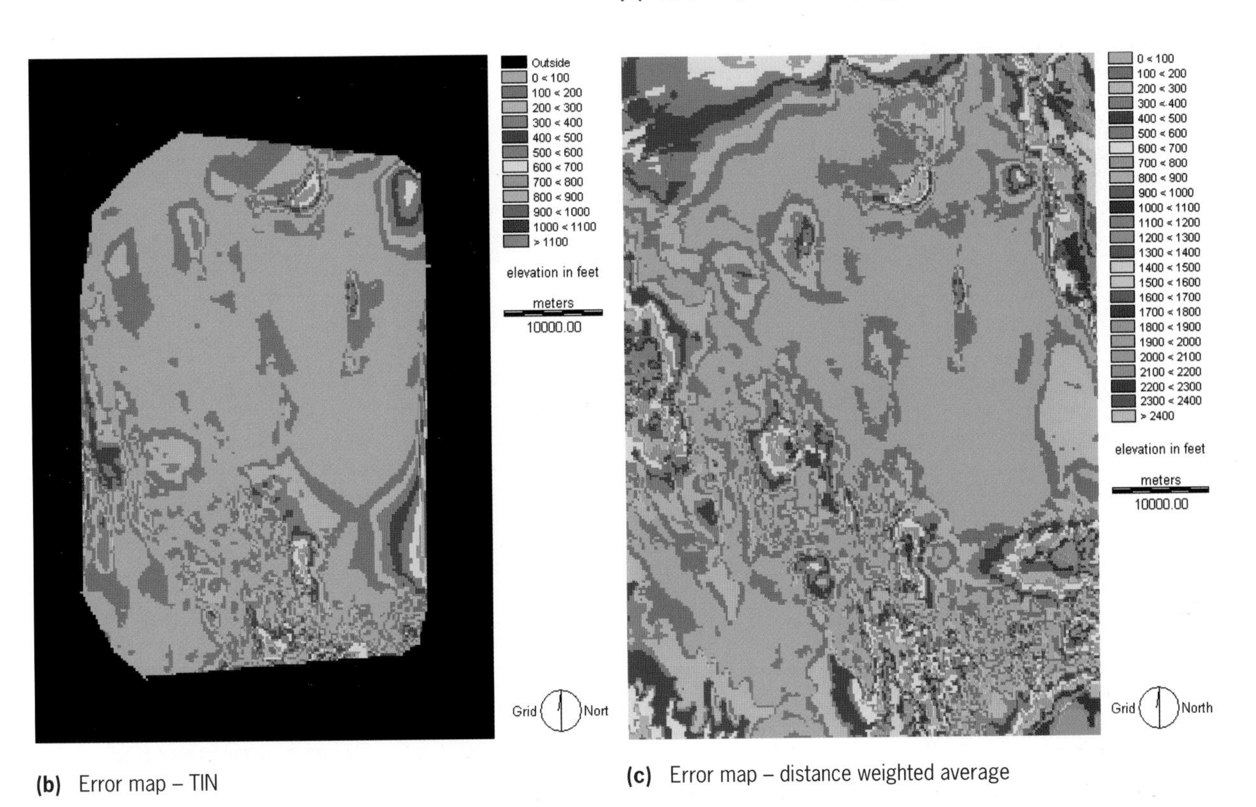

(b) Error map – TIN

(c) Error map – distance weighted average

Plate 13. Error maps showing interpolation errors for the three methods applied to sampled terrain data in Plate 12.

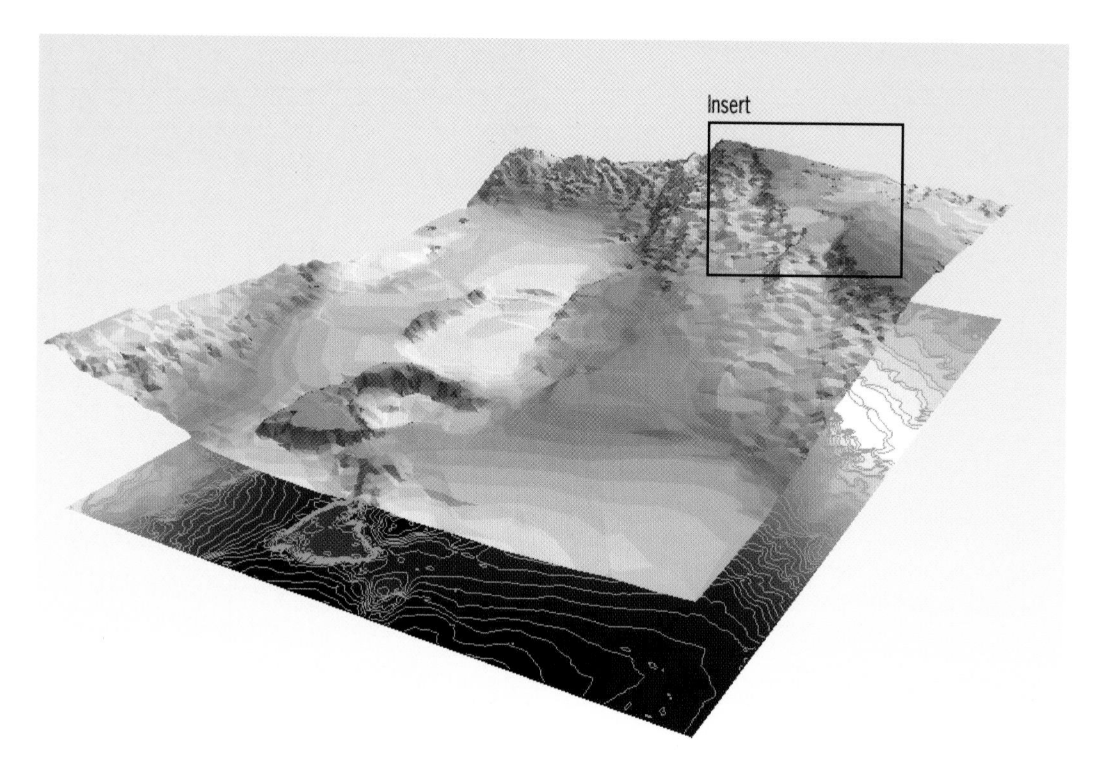

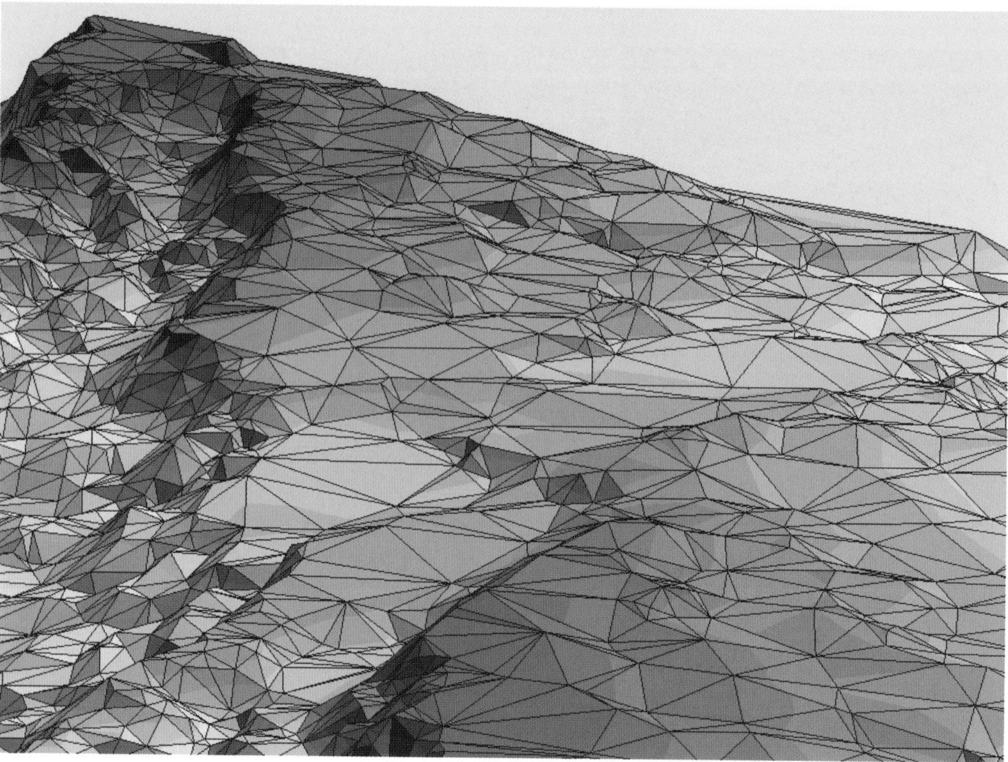

Plate 14. Orthographic perspective of TIN model. Insert shows detail of TIN construction.

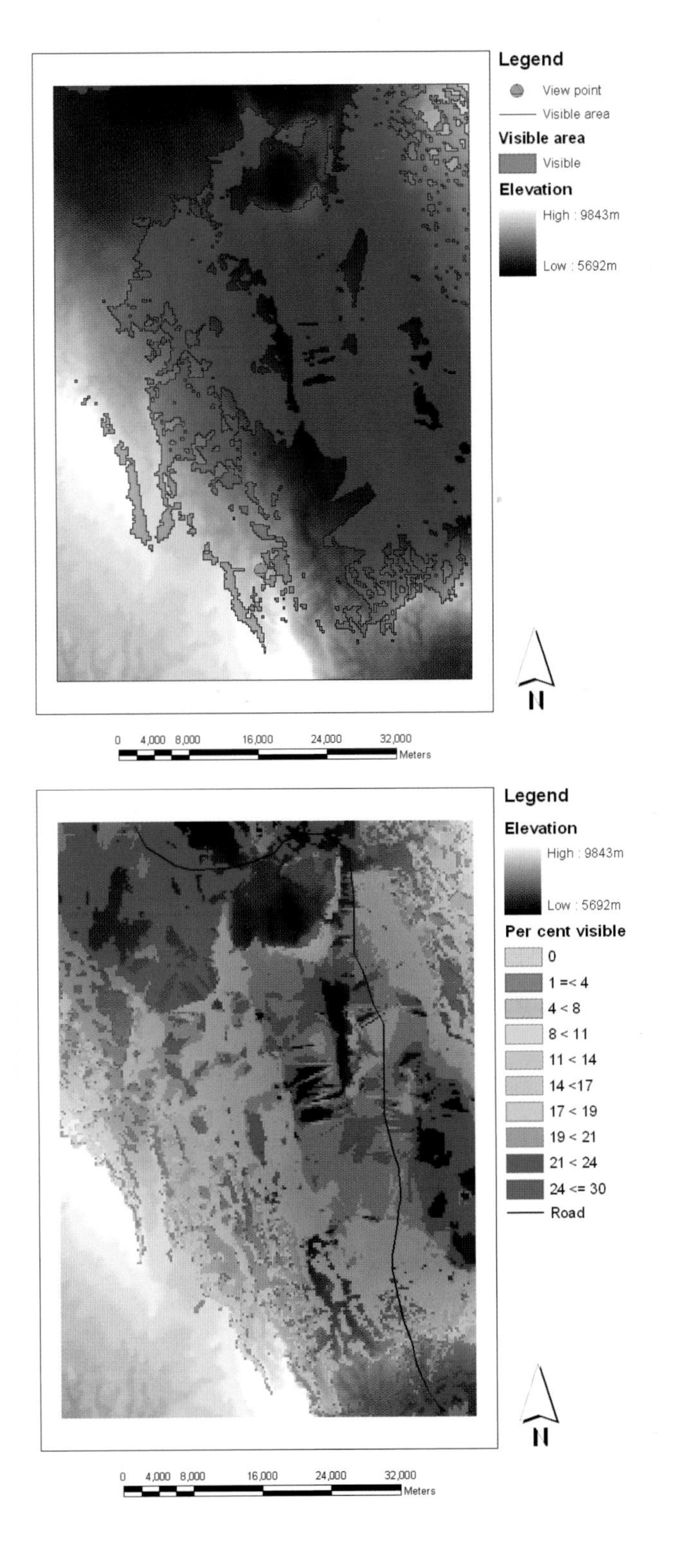

Plate 15. Results of viewshed analyses for single point and linear feature.

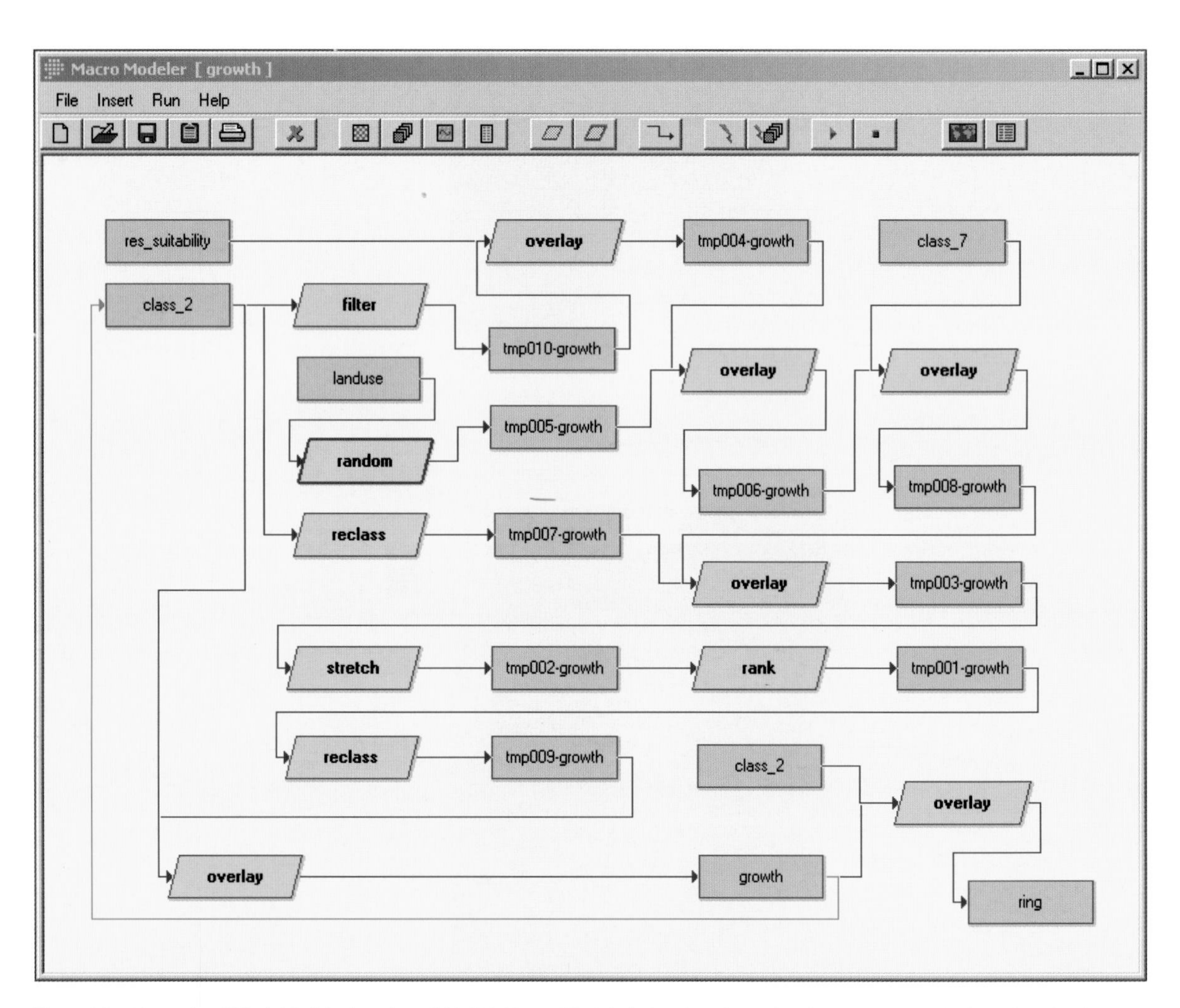

Plate 16. Example of Model Builder interface (Idrisi32 Macro Modeler) showing dynamic urban growth model based on land use and suitability map inputs used to produce map of urban growth areas.

necessary to simulate the complexity of human and environment interactions. As the model becomes more comprehensive, there will be greater data requirements, increased complexity of intra-model interactions, greater opportunity for errors in both model data and model processes, and an increased likelihood that the model outputs will have little relevance to reality. Thus, there is much sense in the saying 'keep it simple'.

For many problems that require a modelling solution the models commonly used by GIS software may be inappropriate. For example, Chapter 3 discussed the limitations of standard GIS data models regarding their handling of time and 3D data. There are certain advanced applications where further problems arise. Global modelling applications are a case in point. Current GIS data models employ a 'flat Earth' approach. Although most GIS packages allow surface modelling of some description (2.5D) and some even permit true solid 3D modelling, rarely do commercially available systems provide facilities for adequately modelling global phenomena. To do this with any degree of realism requires a spherical data model with no edges. Experimental global data models exist. These use a sphere constructed of triangular or hexagonal cells (in a similar fashion to a radar dome or leather football) to approximate the Earth's surface (Goodchild and Yang, 1989). These have found applications in global climate modelling and atmospheric and oceanic circulation models.

Conclusions

Spatial process models are constructed for a variety of different purposes: to help structure ideas, to improve our understanding of a problem or to communicate our ideas to others. Models do not provide answers, but offer a continuous and iterative process from which we gain knowledge about the real world. We can only claim that a model has worked if it has served its purpose. The application of a model allows identification of its weaknesses and allows adjustments to be made. This cyclical nature of model building is often described as a feedback process.

GIS provide a good framework for model building. GIS may need to be coupled with other modelling applications or may be able to provide input data and output capabilities to supplement other modelling software. Since most models have an applications focus, there is a clear need for greater availability of modelling and coupling software within GIS.

The main power of GIS in analytical modelling is that it can provide a spatial database of model variables, and can be used to diffuse model output over geographical space. GIS also offers transformation tools that can be used to integrate disparate data sets required for some models. Data for model calibration can also be handled by GIS.

The problems of using GIS in analytical modelling include the populating of spatial databases (data encoding) at a scale or resolution appropriate for the model. In addition many models are aspatial in character and do not translate easily to a spatial context. The other major limitation of current GIS is its inability to handle temporal data.

If one single characteristic of GIS has to be selected as a key advantage for linking GIS with analytical models, it would probably be the options for the visualization of data inputs and information outputs. The range of visualization options available in GIS are discussed in more detail in the next chapter.

REVISION QUESTIONS

- Outline the differences between models of spatial form and models of spatial process.
- Present, with examples, a classification of process models. Why is it important to develop such a classification scheme?
- Outline the ways in which process models can be linked with GIS.
- Describe how spatial process models can be used to make forecasts of the behaviour ofphysical systems.
- How can GIS be used in the modelling of human processes?
- Describe the MCE approach to modelling the decision-making process.
- Discuss three problems with the implementation of process models in GIS.
- What is PPGIS? When might PPGIS be used?

Further study

There are several good sources of information that can be used to follow up material in this chapter. Goodchild *et al.* (1993), Ripple (1994) and Heit and Shortreid (1991) include details of many environmental GIS applications involving modelling. Specific applications of forecasting and diffusion models are presented in Haines-Young *et al.* (1994) and Price and Heywood (1994). For a review of modelling in human applications Birkin *et al.*, (1996) and Longley and Clarke (1995) provide useful chapters. Batty and Xie (1994a and b) present two interesting papers on embedding urban models within GIS. Fotheringham and O'Kelly (1989) provide a comprehensive review of spatial interaction models.

Several research papers on the integration of GIS and multi-criteria evaluation are available. The papers by Janssen and Rievelt (1990) and Carver (1991b) are essential reading, whilst Thill (1999) and Malczewski (1999) provide comprehensive overviews of the use and application of MCE within GIS.

Batty M, Xie Y (1994a) Modelling inside GIS: Part 1. Model structures, exploratory spatial data analysis and aggregation. *International Journal of Geographical Information Systems* 8 (4): 291–307

Batty M, Xie Y (1994b) Modelling inside GIS: Part 2. Selecting and calibrating urban models using ARC/INFO. *International Journal of Geographical Information Systems* 8 (5): 451–70

Birkin M, Clarke G, Clarke M, Wilson A (1996) *Intelligent GIS: Location Decisions and Strategic Planning*. GeoInformation International, Cambridge, UK

Carver S J (1991b) Integrating multi-criteria evaluation with geographical information systems. *International Journal of Geographical Information Systems* 5 (3): 321–39

Fotheringham A S, O'Kelly M E (1989) *Spatial Interaction Models: Formulations and Applications.* Kluwer Academic, Dordrecht

Goodchild M F, Parks B O, Steyaert L T (eds) (1993) *Environmental Modelling with GIS.* Oxford University Press, Oxford and New York

Haines-Young R, Green R D, Cousins S H (eds) (1994) *Landscape Ecology and GIS.* Taylor and Francis, London

Heit M, Shortreid A (1991) *GIS Applications in Natural Resources.* GIS World Inc., Colorado

Janssen R, Rievelt P (1990) Multicriteria analysis and GIS: an application to agricultural land use in the Netherlands. In: Scholten H J, Stillwell J C H (eds) *Geographical Information Systems for Urban and Regional Planning.* Kluwer Academic, Dordrecht

Longley P, Clarke G (eds) (1995) *GIS for Business and Service Planning.* GeoInformation International, Cambridge, UK

Malczewski J (1999) *GIS and Multi-criteria Decision Analysis.* Wiley, London.

Price M F, Heywood D I (eds) (1994) *Mountain Environments and Geographical Information Systems.* Taylor and Francis, London

Ripple W J (ed.) (1994) *The GIS Applications Book: Examples in Natural Resources: a Compendium.* American Society for Photogrammetry and Remote Sensing, Maryland

Thill J-C (1999) *Spatial Multicriteria Decision Making and Analysis: a Geographic Information Systems Approach.* Ashgate, Aldershot.

8 Output: from new maps to enhanced decisions

KEY QUESTIONS AND ISSUES

- What are the main forms of GIS output?
- What are the basic elements of a map?
- How is three-dimensional output handled by GIS?
- What are the output media used by GIS?
- Why are maps important decision aids?
- What are Spatial Decision Support Systems (SDSS)?
- How has the Internet affected GIS?
- What are VRGIS?

Introduction

Informed decision-making and problem-solving rely on the effective communication and exchange of ideas and information. If GIS is to assist these activities the information it produces must be meaningful to users. Users must be aware of the different forms information from GIS can take (including maps, statistics and tables of numbers). The term 'output' is often used to describe the ways in which information from a GIS can be presented. However, unflattering though the term may appear, it should be remembered that there is a continuity between data, information, decision making and problem solving, and that output – good or bad – will influence the decision-making process.

In the Martian and car analogy, this is where our Martian friend has used the car to reach his destination, in this case Happy Valley. However, now that he has arrived he needs to decide what to do there. He could go cross-country or downhill skiing, ice-skating or tobogganing. Alternatively, he could relax in one of the cafes and do some shopping. In this context the car journey is not the end in itself, rather

the beginning of a new adventure. In a similar way, users should see GIS output not as the final goal of a GIS project but rather as the starting point for informed decision making or problem solving.

The most common form of output from GIS is a map. In many cases the map will be thematic and will illustrate the spatial variation or pattern in a particular variable. For example, a map produced by the Happy Valley fire model will show areas of forest likely to be affected by fire. In other cases, GIS may be used to produce topographic maps such as those used by walkers and skiers. GIS output may also be a single number or list of numbers – for example, a nearest neighbour statistic used to help determine the location of the new Happy Valley supermarket or the correlation between a set of mapped variables (avalanche occurrences and slope aspect). Alternatively, GIS output may be a three-dimensional model of a landscape, produced by draping land use data over a digital terrain surface (see Plates 1 and 2).

As the most common method of visualizing information generated by GIS is the map, this chapter starts by considering the basic principles of map design. An understanding of these is essential for the effective communication of information and ideas. In addition, an understanding of the complexity of the map design process also helps appreciation of the power of maps as a visualization tool. The advantages and disadvantages of alternative forms of cartographic output are then reviewed together with other non-cartographic techniques. The chapter concludes with a brief discussion of the role of GIS output in supporting decision making.

Maps as output

Despite recent advances in computer visualization, the map is still the most elegant and compact method of displaying spatial data. The role of the map is to communicate spatial information to the user. This information may include location, size, shape, pattern, distribution and trends in spatial objects. In designing a map so that it best achieves its objectives, it is necessary to consider a number of key map design elements (Robinson *et al.*, 1995). These include:

- the frame of reference;
- the projection used;
- the features to be mapped;
- the level of generalization;
- annotation used; and
- symbolism employed.

When designing GIS maps we have the element of choice firmly on our side. The choices regarding frame of reference, projection, scale, generalization, content and symbolism are ours. We are in control and can manipulate the individual components of the map to suit our purpose. However, with choice comes responsibility, and the choices made ultimately determine the effectiveness of the map as a communication tool. Despite wide recognition by many that one of the strengths of GIS is its power as a visualization tool (Hearnshaw and Unwin, 1994; Buttenfield and Mackaness, 1991) there is comparatively little literature on producing quality output maps with GIS. This is perhaps understandable as the topic is covered in detail within the cartography literature (see for example Robinson *et al.*, 1995; MacEahren and Taylor, 1994; Monmonier, 1993; Wood and Keller, 1996). We do not go into great detail here on the cartographic principles of map production as that would constitute a book in itself, but summarize the main points and direct you towards appropriate reading for further study.

Since a map can be regarded as a scale model of the real world, it needs some form of spatial referencing so that the user can fix its location in 'real-world space'. A number of graphic and non-graphic devices are used for this purpose. A grid may be used to give a spatial frame of reference with lines representing latitude and longitude or planar (x,y) coordinates. The spacing and labelling of the grid lines provide information on scale, and the orientation of the grid lines indicates which way the map is facing. A north arrow may be used to show orientation without a grid, while a numeric scale or a graphical scale bar is often used to provide information regarding the size of the map relative to the real world. An inset map showing the location of the main map area within its wider geographical setting is a very useful device in indicating approximate geographic location. Figure 8.1 shows how all of these devices may be used, while a quick look at any published map should reveal the use of one or more examples.

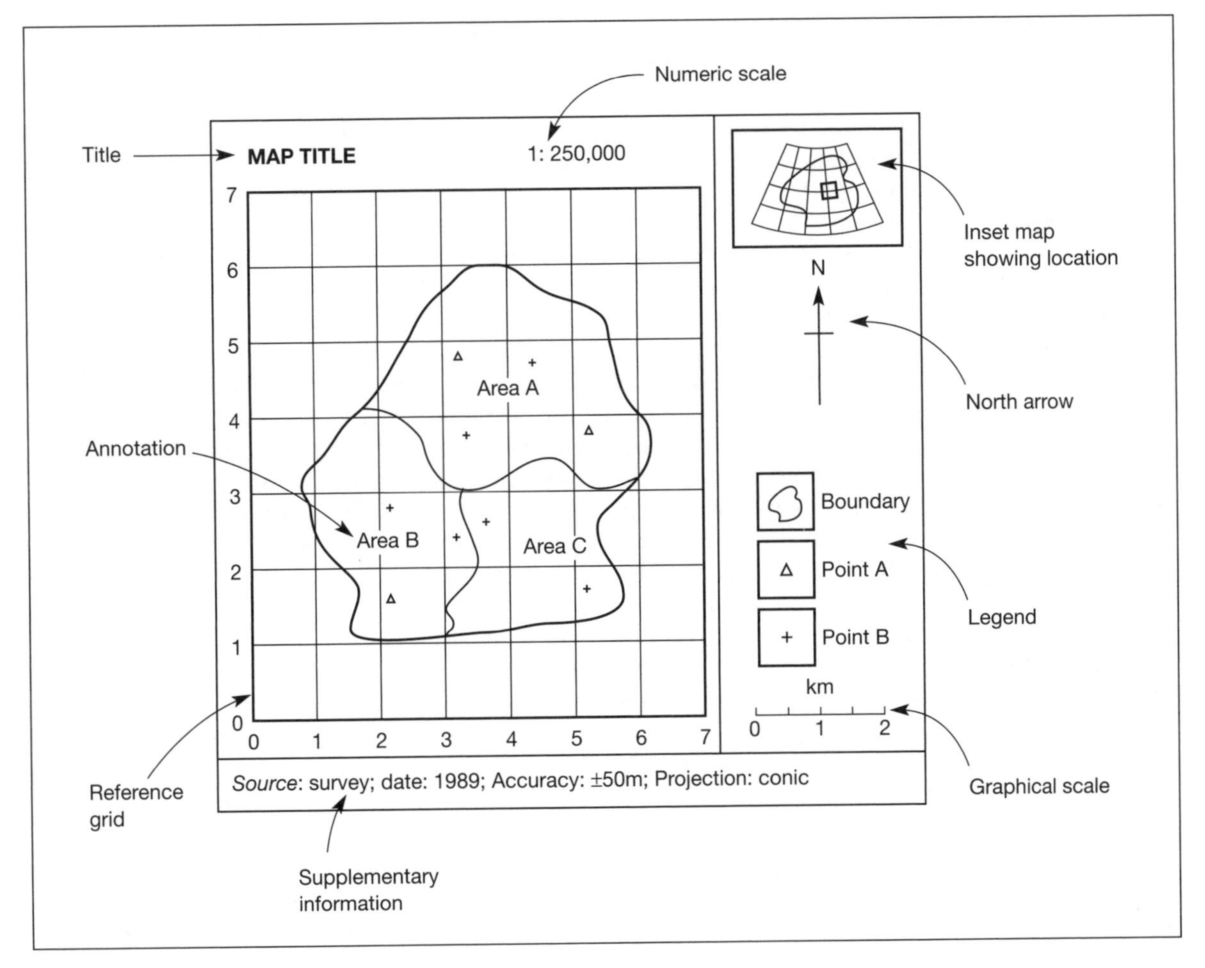

Figure 8.1 Map components

Decisions also need to made about the projection used. Chapter 2 explained how map projections can radically affect basic measurements such as area and distance and how easily GIS can convert from one projection to another. When drawing a map an appropriate map projection must be chosen. While this may not be much of an issue for large-scale maps of small areas, it does become an issue for small-scale maps of continents or the whole world. For large-scale maps of small areas, planar co-ordinates derived from a local projection system are the obvious choice. For maps of much larger areas, hard decisions often need to be made about which aspect of the map is most important: area, distances, directions or lines of true scale.

Returning to the idea that maps are communication devices for spatial information, it is important that the map shows only what is necessary to get across the intended message. A map should not contain anything that may detract from its intended purpose. In other words, 'there should be none of the extraneous objects or fancy flourishes typical of maps showing the location of hidden pirate treasure' (DeMers, 1997: 390). Needless and confusing embellishment is often referred to as 'chart junk' and should be avoided (Tufte, 1983).

The level of detail shown in a map can also determine how well a map communicates information. Too much detail and the map can become difficult to read, too little and essential information is lost. Generalization is important in this context. This is covered in detail in Chapter 2. In creating quality cartographic output, it is necessary to consider both the level of detail used in drawing geographical features (for instance, the number of co-ordinates used to represent a line) and the relative positions in which they are drawn on the map. The high level of precision at which co-ordinates are stored in the GIS data model does not help. Simply plotting these co-ordinates directly onto the map can produce too much detail. For example, in a map showing the road network of the UK that is to be published in a report on nuclear waste disposal, it may be necessary to remove some of the less important minor roads to avoid confusing the overall pattern. In addition, it may be necessary to generalize the detail of main roads by thinning out selected vertices. If the railway network was also to be portrayed on the same map, there is conflict for space where roads and railways run parallel to each other. Plotted at a small scale, the roads and railway lines that follow close and parallel courses will plot directly on top of each other. To avoid confusion between the two features we need to add a slight offset to the different line segments so that they appear to run close to each other but not directly one on top of the other.

As Wood (1993) points out, what you choose to 'leave out' is as important as what you 'leave in'. After all, maps are communication tools and if you choose not to tell someone something there is usually a good reason. In many cases it may be that in omitting detail you can claim to be on the side of the user by improving the clarity of your message. For example, skiers in Happy Valley do not need a trail map that shows the location of every tree. However, the map generalization process is not always as honest as one might expect. Remember the Happy Valley ski area map introduced in Chapter 2 that showed the location of only those restaurants that were owned by the Happy Valley ski company – a generalization that is not really in the best interests of skiers. Monmonier (1995) and MacEahren (1994) provide further insights into how generalizing map information may be done with less than honest intent.

The chosen method of map symbolism can also be important. Symbols on a map are either points, lines or areas. Each symbol can differ in size, shape, density, texture, orientation and colour (Figure 8.2). A poor match between the real world and the symbol used to depict it may confuse the user as to its true nature. It is for this reason that rivers are coloured blue and forests green. If they were coloured red and grey respectively, the user could be forgiven for mistaking the rivers for roads and the forests for urban areas. In our Happy Valley example, it would be a mistake to mix up the colours used to denote the standards of individual ski runs on the resort map. The standard international colour scheme is green (easy), blue (intermediate), red (advanced), black (expert). Reversing this colour scheme could easily result in problems and even legal action from skiers injured on pistes that they thought they would be able to ski safely. Thus, colour is important in influencing the user's understanding of the map. The overall impact of the map can also be affected by colour and symbolism. Black and white maps that use differently shaped symbols and shading patterns can be effective for simple patterns and are certainly easier and cheaper to reproduce than colour maps. However, colour may be necessary to effectively represent complex spatial patterns where detail would be lost in black and white maps.

The shape and pattern of the symbolism used also need to bear some relation to the feature being represented. For example, it would be bad cartographic practice to use an anchor symbol to show the location of an airport and a lighthouse symbol to show the location of a public telephone. Returning to our road and rail example, if the railway lines were plotted as dashed black lines (———) and the main roads as black lines with short perpendicular hatching (++++++) instead of the other way around, the map user may get the wrong impression since the crosshatched line symbol bears a closer resemblance to real railway lines. Careful choice of shape and pattern of symbolism used can influence the user's impression of the map.

Finally, density and texture of shading can affect the impression given by a map. In a monochrome choropleth map, high values may be represented by dense shading patterns (high density and coarse texture), whereas low values may be represented by lighter shading patterns (low density and fine texture). The case of choropleth mapping is discussed further in Box 8.1.

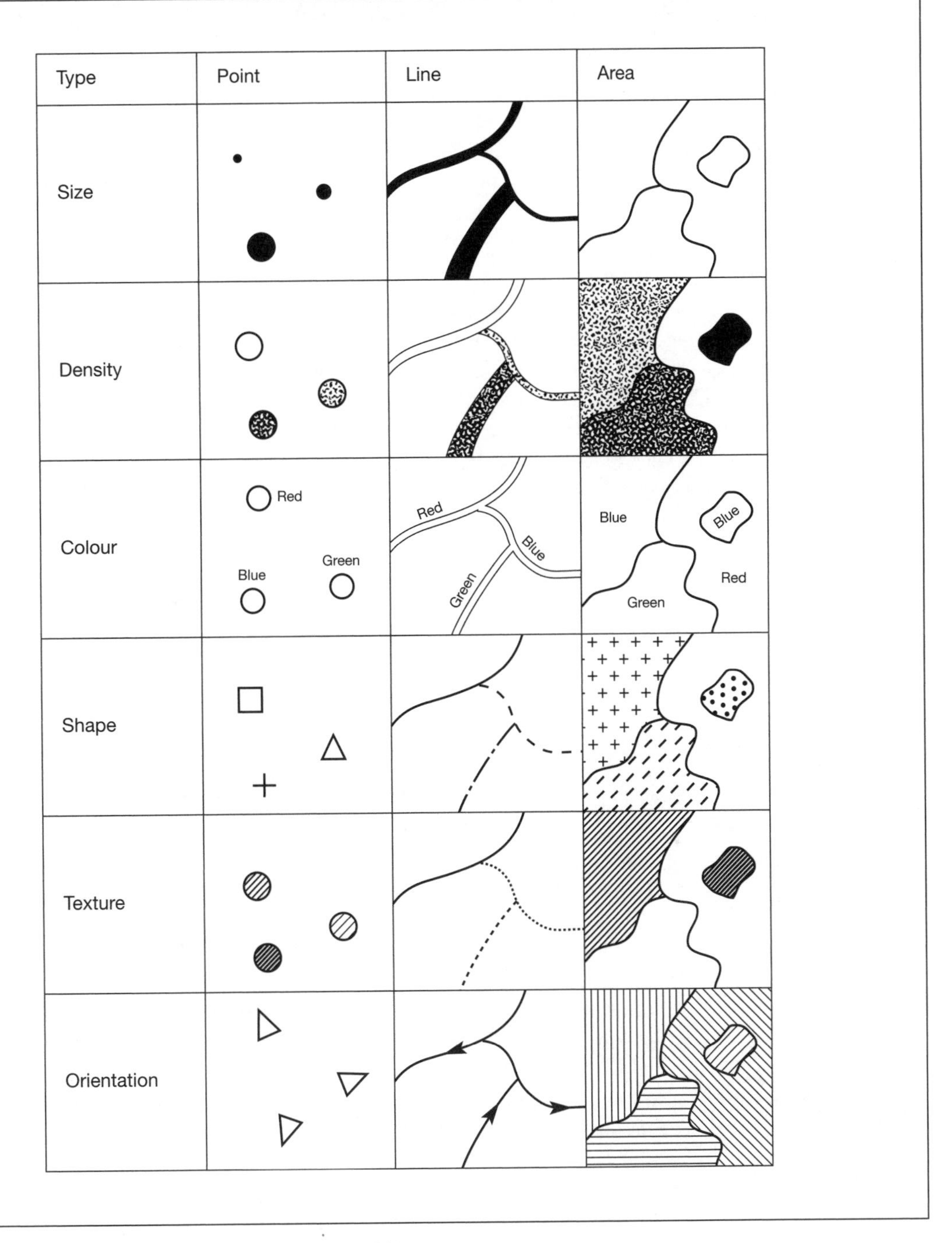

Figure 8.2 Cartographic symbolism. (After Bernhardsen, 1992)

BOX 8.1

Choices in choropleth mapping

A thematic map that displays a quantitative attribute using ordinal classes is called a choropleth map. Each class is assigned a symbolism, and this is applied over an area feature (Chrisman, 1997). Choropleth maps are frequently used to present classified data on socio-economic variables such as population or unemployment where these are mapped by regions, but can be adapted to a wide range of uses. Conventions in choropleth mapping have long been a source of discussion among cartographers and geographers alike (Monmonier, 1972; Evans, 1977; Kostblade, 1981). Three basic problems present themselves in drawing this kind of map. These are:

1 *Choice of shading patterns.* Shading pattern can influence the user's reading of the map. It is normal to use 'the darker the greater' convention when specifying shading patterns. However, confusion may arise between maps where a high value (for example, in unemployment or pollution values) is seen as negative, and those where a high value is seen as positive, such as per capita income or crop yields. The use of colour has similar connotations. Bright colours are often used for and interpreted as high values. There is also a high artistic element in the choice of shading patterns. Mixed use of different shading patterns and textures in the same map can be extremely messy, while unfortunate interactions between adjacent shading patterns can lead to a visual effect known as moiré vibration. Here, shading patterns made up of closely spaced parallel lines give the illusion of movement in the shading pattern. Another convention is to avoid the use of solid white and solid black shading patterns in monochrome maps.
2 *Choice of classification system.* In specifying the classification system of the shading scheme the cartographer needs to specify both the number of classes used and the class intervals allocated. As a

Table 8.1 Example of class interval systems

Interval system	*Description*
Equal interval	Splits data into user-specified number of classes of equal width. Can create classes with unequal distribution of observations and/or classes containing no data
Percentile (quartile)	Data are divided so that an equal number of observations fall into each class. When four classes are used this method is referred to as quartiles. Efficient use is made of each class to produce a visually attractive map. The range of the intervals can be irregular and no indication is given of the frequency distribution of the data
Nested means	Divides and subdivides data on the basis of mean values to give 2, 4, 8, 16, etc., classes. Provides compromise between equal number of observations per class and equal class widths
Natural breaks	Splits data into classes based on natural breaks represented in the data histogram. Provides commonsense method when natural breaks occur but often data are too scattered to provide clear breaks
Box-and-whisker	Splits data using mean, upper and lower quartiles, outlier and extreme values derived from box-and-whisker plot. Accurately describes distribution of the data.

BOX 8.1 CONTINUED

rule, the greater the number of classes used, the more confusing the map. In most cases a maximum of five classes is deemed sufficient to display the variation in the data without creating an unnecessarily complex map. There are many different methods of specifying class intervals (Table 8.1). Some are based on arithmetic rules, while others try to match the statistical patterns in the data being presented.

3 *Choice of spatial unit.* The choice of spatial units used to map the data may be outside the direct control of the cartographer, as it is very difficult to disaggregate data beyond its minimum resolution or translate the data into another set of non-matching units. However, some freedom of choice does exist in making decisions about aggregation of the data into larger spatial units. Amalgamation of data in this manner can lead to a loss of information by masking internal pattern and variation that may be important. The modifiable areal unit problem (MAUP) can also mean that the boundaries of the spatial units themselves act to hide underlying patterns in the data (Chapter 6, Box 6.6).

As a result of these choices, choropleth mapping is a technique for the communication of spatial patterns in attribute data that is open to abuse. Unscrupulous cartographers can draw choropleth maps that have shading patterns, class intervals and spatial units specifically chosen to (mis)represent a particular view of the data. This is common in maps making political capital from dubious statistics (Monmonier, 1991). The maps in Figure 8.3 are based on the same data and spatial units but exhibit different patterns, simply due to the choice of class interval. The user therefore needs to be aware of these issues and study the map and its legend critically before arriving at any conclusions about the data presented.

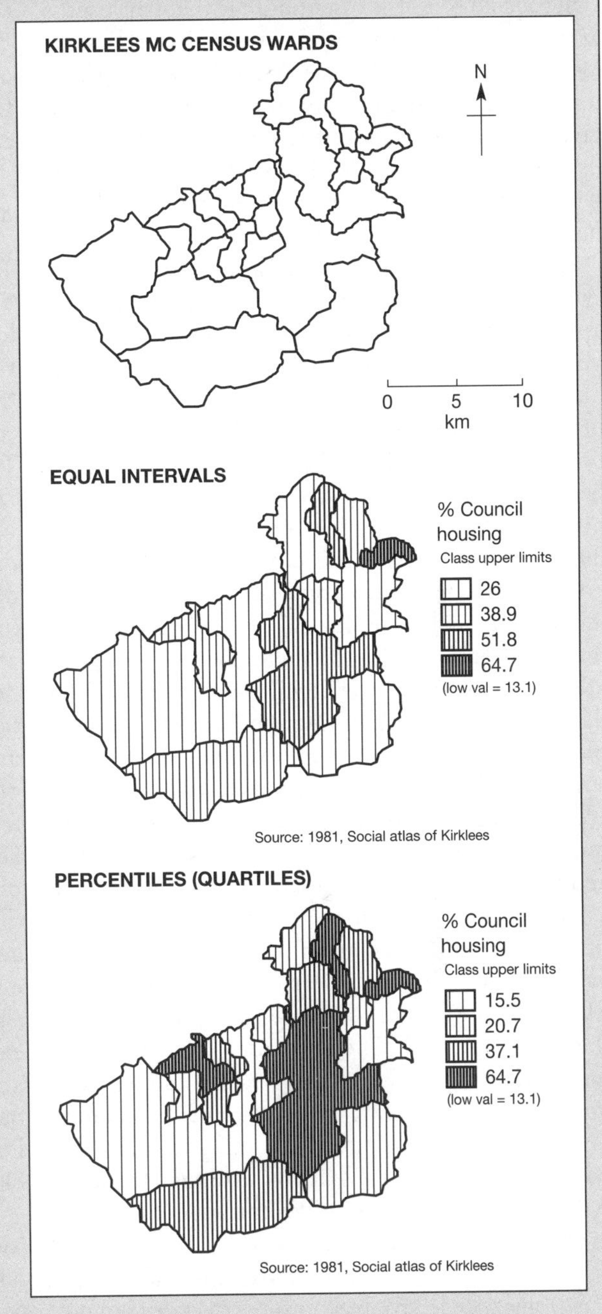

Figure 8.3 Percentage of council housing by census ward

To ensure correct interpretation, whatever the symbolism used, it is essential that the map user is given the necessary information by a key or legend. A key tells the user what the shading patterns, colours, line and point symbols mean and gives information about the class interval system used (see Figures 8.1 and 8.3). Without a key it would be difficult to interpret the mapped information accurately and confidently.

Just as irrelevant detail and inappropriate symbolism can swamp the message in a map, then appropriate and economic use of annotation can help bring a map 'to life' and provide added meaning. Annotation is textual and graphical information that either labels the map and features presented within it, or provides the user with supplementary information. The visual and numeric guides to the frame of reference described above may be considered a form of annotation. More obvious uses of annotation include the map title, legend and labels given to mapped features. Supplementary text may be added to a map to provide information about the data used and the mapping methods. For example, information may be provided about data sources, quality estimates, projection system, reference system, copyright and other issues. 'Parts undiscovered' printed on a map showing the geography of the 'Great Solar Eclipse' of 14 July 1748 leaves the user in no doubt as to the quality of the mapped information in certain geographical regions. This is referred to as metadata. Metadata and quality estimates are discussed further in Chapter 10. The inclusion of metadata on digital GIS maps is important because in their digital form they are more readily adapted for other purposes. Therefore, metadata acts in a similar way to the health and safety warnings found on consumer products – it guides the users to appropriate use of the product.

Map design is as much an art as it is a science. The primary objective is to ensure effective communication of the map's theme(s). This involves many technical decisions about scale, generalization and reference systems, but at the same time artistic decisions about positioning of map elements (title, legend, scale bar and north arrow) and the colour and symbolism used. The artistic element can largely be a matter of personal preference, but certain conventions apply in certain circumstances. A good starting point for developing an appreciation of cartographic design is the book by Wood and Keller (1996), which looks at map design from an applied and theoretical perspective.

Alternative forms of cartographic output

Traditional maps normally assume that the observer is positioned directly overhead and is looking vertically down at the surface of the Earth. Maps are drawn as plan views and at a fixed spatial scale. In recent years new forms of cartographic output have become more common. These include cartograms, three-dimensional views and animation. These are considered below.

At first sight cartograms look like normal maps. However, instead of showing the location of objects or variables using standard Euclidean co-ordinates, the objects depicted are mapped relative to another variable such as population density or relative distance. In this manner, distances, directions and other spatial relationships are relative rather than absolute. A common use of cartograms is for drawing maps of urban mass-transport networks, such as the famous London Underground map. In these examples, routes are plotted as generalized lines showing the sequence of stops and connections rather than showing the actual route taken. In this manner the user's understanding of the map and the spatial relationships between stops is greatly improved at the expense of true scale of distance between stops and the actual route taken. This type of cartogram is known as a *routed line cartogram*. Figure 8.4 shows a routed line cartogram for the Happy Valley ski bus service.

A related form of cartogram shows space distorted from a central point on the basis of time taken or cost incurred to travel to a series of destination areas. These are known as *central point linear cartograms*. Another use of cartograms is to depict the size of areas relative to their importance according to some non-spatial variable such as population or per capita income. These are referred to as *area cartograms*. A good example of these area cartograms can be seen in Dorling (1995).

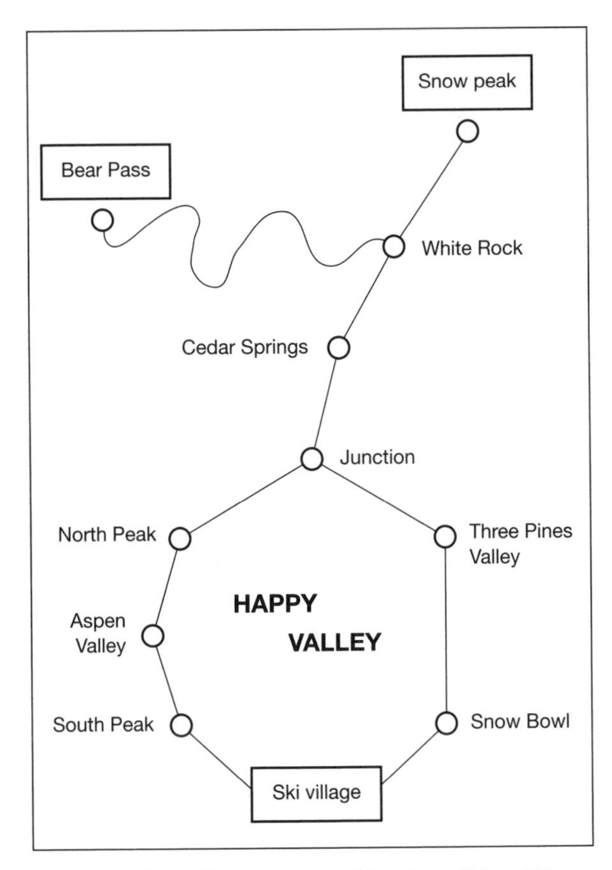

Figure 8.4 Routed line cartogram of the Happy Valley ski bus service

Three-dimensional (3D) views of spatial information are becoming a popular method of visualization. These display (*x,y,z*) co-ordinate information using an orthographic projection based on the angle of view (0–360 degrees around the compass), view azimuth (0–90 degrees from horizontal to vertical) and view distance (distance of viewer from the image space). The resulting image is shown as a fishnet or wire-frame diagram, an example of which is shown in Plate 1. This is an excellent means of displaying topographic data. It is also possible to drape other images over the wire-frame map. Thus, added realism may be created by draping land cover data derived from satellite imagery or aerial photographs over 3D terrain (Plate 2). In the case of our Happy Valley ski example, the 3D topographic maps issued to skiers to help them navigate their way around the many lifts and pistes within the resort are an example of this kind of GIS output. To produce these the Happy Valley GIS team has created several 3D wire-frame images of the ski area upon which it draped digitized aerial photographs together with relevant coded vector data for ski lifts, ski pistes and hotels. An artist and graphic designer produced the final topographic maps complete with legend, annotation and corporate logo.

Animation can be useful when there is a strong temporal element to the data being displayed. Tobler (1970) first used animation as a means of identifying patterns in spatio-temporal data sets. A series of maps is drawn, each showing the spatial pattern of the observed variable at a particular time. When 'played back' in quick succession as a series of frames, the change of the spatial pattern over time can be seen. This technique has proved useful for identifying temporal patterns in point data sets and in visualizing diffusion processes. 'Computer movies' have been used to analyse urban growth (Tobler, 1970) and crime incidents (Dorling and Openshaw, 1992). Animation may be used in a similar way to display all manner of spatio-temporal processes, from the development of forest fires and pollution plumes to changes in hydrological variables and population densities. By combining the two techniques of 3D mapping and animation, modern GIS packages can create realistic 'fly-through' animation using a sequence of three-dimensional images, each viewed from a different perspective. This is a powerful means of terrain visualization, similar to the interactive computer animations seen in flight simulators.

Non-cartographic output

While the map is undoubtedly the main form of GIS output, we must not ignore other non-cartographic forms. Tables and charts containing spatial and non-spatial attribute information derived from the GIS database are important forms of output. Numerical or character information linked to spatial objects (points, lines, areas or surfaces) are often best displayed as tables or in charts. In many cases the output from a GIS analysis may be a single number or character string – for example, a nearest neighbour statistic or the character string returned from a spatial query. This kind of information is best displayed temporarily on the computer screen.

A recent development in the display of non-cartographic information is the *linked display*. This approach makes use of the dynamic linking capabilities provided by window-based operating systems and graphical user interfaces (GUIs) to connect different data display windows. Thus, actions in one window are reflected in displays of linked data in other windows. For example, if a user is simultaneously viewing a choropleth map in one window and a scatterplot of the same data in another, then queries performed on the scatterplot will be echoed in the choropleth map and vice versa. To illustrate the value of this, we return to avalanche data in the Happy Valley case study. The piste wardens are interested in identifying patterns in the occurrence of avalanches that are not immediately apparent when inspecting the raw data. By creating multiple scatterplots of the variables collected for each avalanche recorded in Happy Valley, and dynamically linking them to a map of avalanche events, the piste wardens are able to investigate statistical and spatial patterns simultaneously. Many GIS packages are now exploiting this functionality to dynamically link tables and charts to map displays. The advantage of this approach to data output is that it provides the user with a highly interactive method of simultaneously exploring both the spatial and aspatial aspects of data.

Spatial multimedia

Some GIS packages offer facilities for the display and playback of multimedia to supplement cartographic images. Photographic images, text, graphics, sound and video can be linked to map displays to add value to a traditional map. These media provide the user with additional contextual information about the spatial detail in the map and may act as an aid to understanding. Certainly, being able to look at photographs and watch video of an area is helpful in developing an improved understanding of what a mapped region actually looks like, especially if you have never visited the area in question. For example, the Happy Valley marketing department has developed a multimedia ski map for the ski area that includes video descents of all the ski pistes, which are started by selecting the piste on the trail map. Outside pure marketing applications such spatially referenced multimedia has many benefits. In a utility organization, viewing pipeline condition from photographs or video may be a useful exercise when planning a programme of repairs in a particular area. Heywood *et al.* (1997) consider how multimedia can be used with GIS output in an educational context to inform users about environmental problems. Others have explored how sound can be used to add value to mapped information. Shiffer (1995b) incorporated the sound associated with the take-off and landing of aircraft into a GIS developed to explore the environmental impacts of airports. Whilst multimedia may be exciting, too much can combine to produce a kind of technological overload.

Mechanisms of delivery

In the previous sections of this chapter we have reviewed several forms of GIS output, ranging from maps to animated 'fly-through' visualizations and multimedia presentations. The mechanism, or medium, of delivery of these varies markedly. Two basic types of output medium can be defined: those that are permanent (known as hard copy), and those that are ephemeral (soft copy).

Hard copy may be maps or other GIS output that are printed or plotted on paper or film for the user to take away. A bewildering array of output hardware is available on the market. These devices include dot-matrix printers, bubble-jet and ink-jet printers, laser printers, pen plotters, thermal printers, photographic image setters and electrostatic colour plotters. Each produces a different quality of output using various technologies at varying costs. Table 8.2 summarizes a range of hard-copy output devices and their characteristics.

Ephemeral output is essentially based around the computer screen. It is ephemeral in the sense that the image is digitally reproduced on the computer screen and cannot be taken away, but is

Table 8.2 Hard-copy output devices

Device	*Colour*	*Quality*	*Resolution*	*Cost*
Dot-matrix	mainly mono	low	low	low
Ink-jet	mono/colour	medium	medium	low
Pen plotter	mono/colour	high	high	medium
Laser	mono	high	high	medium
Colour laser	colour	high	high	high
Thermal wax	colour	high	high	high
Electrostatic	colour	high	high	high

viewed by the user for a time and then cleared from the screen. This is typically used by analysts sitting in front of a GIS terminal, or by users accessing images over the World Wide Web using browser software. Increasingly map-based information is being served using Wireless Application Protocol (WAP) technology to users with Personal Digital Assistants (PDAs) and mobile phones. Linked to a GPS, these devices and systems can combine instant local maps and user locations to present almost boundless possibilities for applications from personal navigation systems to real-time monitoring.

There is a wide variety of screen types that can be used for display of output. These include cathode ray tubes (CRT) and liquid crystal displays (LCD). Screens may be monochrome (black-and-white or green-screen) or colour. Most operate by representing the image as a grid of pixels (picture elements) in a similar fashion to the raster data model.

Slide shows are a commonly used ephemeral method of demonstrating a particular GIS application to interested parties. These are simply an ordered selection of screen shots illustrating a sequence of operations and GIS outputs. More advanced demonstrations may be interactive, allowing the user (who may have little or no GIS training) to choose display and/or analysis options and carry out spatial search and query operations on-screen. Multimedia may be included in such systems as appropriate to add contextual information.

Recent research has focused on the development of *virtual reality* (VR) displays of spatial data. Much of this research and development work is taking place on the Web using the virtual reality modelling language (VRML) (Raper, 1997). An interesting development in this field has been the application of VR and the Web in producing so-called 'virtual field courses' for student use (Fisher *et al.*, 1997). The merit of these websites is that they can either serve as preparatory tools for real field studies or be used to allow students to undertake 'field work' in remote or dangerous locations such as Antarctica, where conventional field courses would be impractical and too costly. Box 8.2 describes developments in Virtual Reality GIS.

Serving maps on the web

The use of the Internet to disseminate maps and spatial information has been perhaps the biggest development in GIS over the past few years. Initially the Internet was used to give wider access to GIS images and associated reports in their static form as words and pictures. The industry has been quick to develop the possibilities of the web for more interactive access to GIS software and data. GIS-based websites now allow users to interrogate map data, compose new maps and update and share spatial data. Indeed, one of the most frequent searches on the Internet is for maps and map data. Most GIS packages have Internet mapping additions to complement stand-alone systems. These allow users to serve map-based

BOX 8.2

Virtual Reality GIS

Virtual Reality (VR) produces 'virtual worlds' that are used to imitate reality. Virtual Reality GIS (VRGIS) simulates the real world using a GIS database and computer graphics to produce interactive and navigable 3D visualizations. In practice VRGIS is a standard GIS database with VR as the user interface.

Recent advances in computing power, the availability of low-cost yet powerful desktop GIS packages and dedicated VR programming languages such as VRML (Virtual Reality Modelling Language) have made the combination of GIS and VR possible. VRGIS provides the power and functionality of GIS together with the intuitive and easy-to-use interface of Virtual Reality.

Imagine being able to interrogate a GIS database or retrieve data simply by pointing with your virtual 'finger' or grabbing it with your 'hand'. Or being able to perform a viewshed analysis by 'walking' to the viewpoint and turning through 360 degrees to see what is visible. These are just two examples of what might be possible with VRGIS. Some of the features of VRGIS are (after Faust, 1995):

- Realistic representation of real geographic areas.
- Free movement of the user within and outside the selected geographic terrain.
- Standard GIS capabilities (such as query, select and spatial analysis) in a 3D database.
- A visibility function that is a natural and integral part of the user interface.

However, many applications are not 'true' VRGIS but are 'loosely coupled' systems in which spatial data are transferred from standard GIS packages using VRML for display using VR software (Berger *et al.*, 1996). In this case the 3D visualization capabilities of the VR software tools are used to augment the 2D cartographic tools of the GIS.

The number and type of applications of VRGIS have increased rapidly over the last few years. Examples include exploration of past and future urban landscapes, environmental impact assessment, visualization of abstract data in the context of real world variables, landscape construction and education.

content to members of their organizations and clients without the need for expensive GIS software and data on local computers. For some organizations this is the ideal way of sharing data between various departments and users, particularly if other networking solutions are precluded by large distances between buildings and offices. Box 8.3 provides examples of how web-based GIS might be employed by the Happy Valley Ski Company. Further applications of web-based GIS are described in Chapter 13, and more details of web-based GIS are provided in Chapter 4.

GIS and spatial decision support

At an organizational or strategic level the most important aspect of GIS output is its use in spatial decision support. Central to this is that GIS provides a much-needed framework for approaching,

BOX 8.3

Happy Valley Internet GIS

The Happy Valley management team wish to improve the sharing of data within the company; improve resort monitoring and management; improve the marketing of the ski resort to potential customers; and improve the service they provide to their existing customer base. They have employed a web programmer to work with the GIS and marketing teams to address these broad aims through the features and functions of the website.

Data sharing between departments

Data on the physical and human resources of Happy Valley, together with information on customers and finance, are important resources that need to be continuously available to employees in a range of

BOX 8.3 CONTINUED

departments. Copies of GIS and non-spatial data are placed on the company's Intranet to allow employees to access essential information from their desktops. Changes to the database are password protected to make sure that only genuine changes made by authorized users are permitted. Updates made to the central database are immediately available to all users.

Dynamic piste maps

GIS-based piste maps are available from the company's web page. These allow existing and potential customers to check the condition of ski runs before making a visit. The condition of the ski runs is monitored by ski rangers who log the information on PDAs equipped with GPS receivers. Wireless communications transmit information about piste condition and location from the PDAs to the piste database. The GPS location is used to link the new information on piste condition to the relevant stretch of piste in the GIS database. Information on which lifts are open and the length of any queues is similarly transmitted to the piste database by wireless link or land-based telephone line. Customers using the web to access piste maps see current information, displayed in real time. Clicking on any of the pistes and ski lifts gives access to more detailed information, including condition histories and seasonal summaries.

Web cams

Several web cameras (web cams) are placed around the resort. The location of web cams is marked on the piste maps as 'hot spots'. Customers clicking on the hot spots can pull up a real-time video image of the view from the web cam. They can see the weather, the condition of the pistes, length of the lift queues and whether there are any spaces left in the car parks. This service is also provided to hotels in the resort via cable.

Resort monitoring

It is essential for the smooth running of the resort that managers have instant access to information about conditions in the resort. In addition to the information provided by the dynamic piste maps and web cams, resort managers have access to data from electronic monitoring devices around the resort that are linked via landline or telemetry to the central database.

These data include:

- the volume of people going through ski lift turnstiles;
- the volume of traffic on the resort's road network;
- the location and movement of ski buses;
- the number of spaces left in car parks;
- the location of ski ranger;
- the weight of the snow pack;
- data from strain gauges located on avalanche-prone slopes;
- data on meteorological conditions around the resort.

All data can be displayed dynamically on maps of the resort. Meteorological data are used by the resort's met office to provide short-, medium- and long-term weather forecasts.

Navigation and traffic maps

The Happy Valley website contains a link to a third party route manager and traffic monitoring website. This allows customers to plan the quickest route to the resort and also check current traffic conditions.

Booking service

A booking service is available from the Happy Valley website. Customers can check the availability and cost of hotels, ski school places and ski hire. A map of services in the Happy Valley resort helps customers choose the services they want. Detailed information about a particular hotel or ski school can be viewed simply by clicking on its location. If a service provider has their own website a link to this is provided. For some hotels, virtual tours of facilities are available. The map of services is customizable by the user, so that only those services of interest are displayed. Bookings and payment can be made online. Once an online transaction is made, the information is passed directly to the central database and to the service providers.

Despite the cost of setting up the monitoring systems and Internet and Intranet access tools, the Happy Valley resort management are confident that the benefits of these developments will be better resort management, increased revenue from service commissions and increased visitor numbers.

supporting and making spatial decisions. Decision-making as a scientific discipline has a much longer history than GIS. As such, much of the decision-making theory is well established and has its own specialized literature. Within the wider decision research field, computers have been used to develop decision support systems (DSS). Geoffrion (1983) identifies six distinguishing characteristics of DSS.

1 Decision-support systems are used to tackle ill-structured or semi-structured problems that may occur when the problem and objectives of the decision maker cannot be coherently specified.
2 Decision-support systems are designed to be easy to use, making sophisticated computer technology accessible through a user-friendly graphic user interface (GUI).
3 Decision-support systems enable users to make full use of the data and models that are available, perhaps by interfacing external routines or database management systems.
4 Users of decision-support systems develop a solution procedure using models as decision aids to generate a series of decision alternatives.
5 Decision-support systems are designed for flexibility of use and ease of adaptation to the evolving needs of the user.
6 Decision-support systems are developed interactively and recursively to provide a multiple-pass approach.

GIS has been referred to as a special kind of decision-support system dealing with problems that involve a high degree of spatiality (Densham, 1991). This is perhaps not entirely an accurate view, as GIS *per se* does not meet all the six characteristics of a decision-support system as defined by Geoffrion (1983). However, GIS can provide a framework for the development of 'spatial decision-support systems' (SDSS), particularly when either loosely or tightly coupled with other modelling software (Chapter 7).

SDSS are an extension of DSS as defined by Geoffrion (1983). They share the same characteristics, but in addition provide spatial data models, the means of inputting and displaying spatial data and additional spatial analysis tools. Some of the requirements for an SDSS are met by standard GIS, but others, such as advanced spatial analysis tools and the user-friendly GUI (specified for DSS by Geoffrion, 1983) are not. One of the most important rules governing the use of GIS for SDSS is that GIS themselves do not make decisions – people do. In SDSS the emphasis is on the use of spatial data and in GIS it is on supporting people and the decisions they need to make.

Maps as decision tools

Maps have a long history of being used to support decision-making. Ever since maps were first used as a means of navigation, they were used as a form of decision-support tool. Which is the best route to take between one place and another? What is the best means of transport? What will I need to take with me? Military planners and explorers have all relied on maps. Good maps often meant the difference between success and failure, even life and death. It is not unusual, therefore, to find that maps play a very important role in modern decision-making, be it for retail planning, government policy making, business management, facilities management or, indeed, military planning and exploration. In two of our case study examples, the output from GIS has been used for decision-support in one way or another. This is illustrated in Box 8.4. In many cases digital spatial information is being used through either GIS or custom-designed software to create spatial decision-support systems (SDSS).

Conclusions

It is difficult to beat the map as a method for communicating spatial information. It is an accepted part of our culture and has been refined and developed over a far greater time than any of the new visualization tools being developed by today's computer scientists. However, it is important to recognize that maps come in many guises, each with its own subtle distinctions, and, furthermore, that each map tells a different story.

The application of GIS to spatial problem-solving has meant that many more people are producing maps as part of their day-to-day jobs.

BOX 8.4

Examples of GIS used in a decision-support role

Zdarske Vrchy

In the Zdarske Vrchy GIS output was used to help determine new areas for consideration as nature conservation zones. GIS was used to generate maps showing a range of different scenarios for new protected areas given a range of environmental considerations. These maps were then used to structure discussions about nature conservation plans for the Zdar region. The GIS output acted as a catalyst for informed debate between a range of interested parties, including landowners, conservationists, government officials and international organizations. In this respect the GIS helped bring professionals from disparate backgrounds together to exchange ideas and information. The focus for their discussions was the maps produced from the GIS. The ease and flexibility with which new maps could be produced showing changes to the spatial distribution of protected areas following changes in environmental criteria offered additional benefits to the decision-making process.

Radioactive waste disposal

In the case of radioactive waste disposal in the UK, GIS was used to help decide the best or near-optimum sites for nuclear waste disposal. Carver (1991b) suggests that the combination of GIS and multi-criteria evaluation (MCE) represents a major contribution towards the development of effective SDSS. It is suggested that GIS/MCE-based systems may be used to assist decision makers in site identification and to plan evaluation, since radioactive waste disposal is a typical example of an ill-structured or semi-structured and complex spatial problem of the type referred to by Geoffrion (1983). Site search procedures could significantly benefit from the use of SDSS from the initial survey stage to the final siting decision and public inquiry. It is further envisaged that a stand-alone or networked GIS-based SDSS in a committee room could create significant improvements in the way siting decisions are made. In addition, these types of SDSS may have an important role to play in providing more effective means of public participation and consultation throughout the site search process by allowing for greater public input and feedback to decision makers. Later work suggests that this may best be achieved by SDSS installed on the World Wide Web to ensure as large an audience as possible (Carver *et al.*, 1997).

No longer is a professional cartographer required to produce a map. Maps have become a throw-away commodity generated by the GIS analyst in response to the changing needs of decision makers. The automated nature of much of the GIS map production process – with the software automatically adding default legends, keys and scales – means that maps can be produced in minutes rather than the weeks or months associated with the traditional process. There is often a price to pay for this freedom – and poor-quality output that does not communicate its message effectively to users abounds. Interesting examples are shown in Plates 10 and 11. These are typical examples of GIS output and yet break several cartographic rules on numbers of classes, choice of colour scheme and missing classes. Some of the interactive map composition tools available on the Internet will only serve to compound this problem by making it easier for people without cartographic training to design and produce inappropriate map products. It would be too much to expect the GIS analyst or the decision maker using a desktop GIS to become a professional cartographer overnight – such skills have to be learnt. In this chapter we have tried to provide some basic pointers for map design as well as recommend sources of further study. Maps are powerful tools and users should acknowledge the responsibility they take on when they choose to use them for communicating ideas and information.

Besides maps, there are a number of other output options available to the user of GIS. These include a range of other cartographic presentation techniques such as cartograms, computer movies and animated map sequences. GIS output may also take the form of individual numbers, statistics or tables of values. In addition, GIS output can now be integrated with other media to produce multimedia presentations. For future developments it is suggested that the Web will play an increasingly important role in the dissemination of GIS outputs to a wider audience.

However, it is important to remember that GIS output, be it as a paper map or as part of a multimedia presentation, or a virtual reality experience, is rarely an end in itself. Usually, it is a catalyst that facilitates some decision-making or problem-solving situation.

This chapter concludes Part 1 of the book on the 'fundamentals' of GIS. We have now dealt with definitions of GIS, spatial data, data models and data management, data input, data analysis and spatial modelling, and output. The second part of the book deals with a selection of major GIS 'issues' including the historical development of the technology, data quality, human and organizational issues and project management. To conclude Part 2 we give some thought to where the future of GIS may lie.

REVISION QUESTIONS

- Why is map design an important consideration when creating maps using a GIS?
- What are the main elements of a map?
- What are cartograms and how do they differ from traditional map output?
- When may non-cartographic output be more appropriate than drawing a map?
- What is the difference between permanent and ephemeral output?
- What mechanisms are available for the delivery of GIS output?
- What are the six characteristics of a DSS and an SDSS?
- Describe how the Internet has changed GIS.

Further study

While this chapter has given a summary of the subject of GIS output, it is not possible within the confines of this book to cover all the relevant material in depth. This is especially true for many of the cartographic principles of map design. Therefore to develop your skills as a cartographer you should read further texts. An excellent starting point is Robertson (1988). Wood and Keller (1996) provide a more detailed look at how computer technology can be used to advantage in cartographic design. Taylor (1991a) compares cartography and GIS and charts their convergence. There are several specialist texts that cover some of the developments mentioned in this chapter. These include Peterson (1995), McMaster and Shea (1992) and MacEahren and Taylor (1994). When considering the use and misuse of maps as tools communicating spatial information, there is an excellent series of books by Monmonier (1993, 1995, 1996) entitled *Mapping it Out*, *Drawing the Line*, and *How to Lie with Maps*.

Of the more general GIS texts, DeMers (1997) offers a chapter on GIS output, and Maguire *et al.* (1991) contains two relevant chapters: on visualization (Buttenfield and Mackaness) and on generalization of spatial databases (Muller).

There are several texts available about the Internet and GIS. These include Kraak and Brown (2000) on web cartography and Plewe (1997) *GIS Online*.

Buttenfield B P, Mackaness W A (1991) Visualization. In: Maguire D J, Goodchild M F, Rhind D W (eds) *Geographical Information Systems: Principles and Applications*. Longman, Vol. 1, London, pp. 427–43

DeMers M N (1997) *Fundamentals of Geographic Information Systems*. Wiley, New York

Kraak M-J, Brown A (eds) (2000) *Web Cartography: Developments and Prospects*. Taylor and Francis, London

MacEahren A M, Taylor D R F (1994) *Visualization in Modern Cartography*. Pergamon Press, New York

Maguire D J, Goodchild M F, Rhind D W (eds) (1991) *Geographical Information Systems: Principles and Applications*. Longman, London

McMaster R B, Shea K S (1992) *Generalization in Digital Cartography*. Association of American Geographers, Washington DC

Monmonier M (1993) *Mapping it Out*. University of Chicago Press, Chicago

Monmonier M (1995) *Drawing the Line*. Henry Holt, New York

Monmonier M (1996) *How to Lie with Maps*. University of Chicago Press, Chicago

Muller J-C (1991) Generalization of spatial databases. In: Maguire D J, Goodchild M F, Rhind D W (eds) *Geographical Information Systems: Principles and Applications*. Longman, London, Vol. 1, pp. 457–75
Peterson M P (1995) *Interactive and Animated Cartography*. Prentice-Hall, Englewood Cliffs, New Jersey
Plewe B (1997) *GIS Online: Information Retrieval, Mapping and the Internet*. Onword Press, Santa Fe.
Robertson B (1988) *How to Draw Charts and Diagrams*. North Light, Cincinnati, Ohio
Taylor D R F (1991a) *Geographic Information Systems: the Microcomputer and Modern Cartography*. Pergamon Press, New York
Wood C H, Keller C P (eds) (1996) *Cartographic Design*. Wiley, Chichester

Part 2

Issues in GIS

9 The development of computer methods for handling spatial data

KEY QUESTIONS AND ISSUES

- What were the methods of handling spatial data before GIS?
- Why did computer methods for handling spatial data develop?
- How did computer methods for handling spatial data evolve?
- Which disciplines have contributed to the development of GIS?
- What are the important events in the development of GIS?
- How has GIS spread into different application areas?

Introduction

GIS are considered by Longley and Clarke to have had a 'short but impressive history' (1995: 1). Summarizing the important developments in GIS is difficult as there are no GIS archives and little critical examination of the history of GIS (Pickles, 1995). Many of the developments in GIS have been technology- or application-led, rather than guided by the development of theory (Martin, 1991). Particular personalities and projects have played a major role in the pace and nature of developments in different countries. The influence of policy constraints and different commercial settings have also been important. Developments have often been the result of co-operation and integration, matching the integrating nature of the technology. Many different disciplines have been involved. These include those related to the technology: computer mapping, databases, computer science, geography, remote sensing, data processing, mathematics and statistics; and those concerned with methodological and institutional issues such as systems theory and information management (Coppock and Rhind, 1991; Tomlinson, 1990).

This chapter considers the way GIS have developed to their current state. First, methods of handling spatial data used before computers were commonly available are examined, since these

give an insight into what we require computers to do, and how they can help (or hinder) existing practice. Computer methods for handling spatial data existed before GIS, so these are reviewed, then developments in GIS are discussed together with developments in a selection of complementary disciplines. To conclude we examine reasons for different rates of growth in different countries and the role of policy makers in the development of GIS. The chapter does not attempt to present a comprehensive history of GIS, but aims to give some context for the systems and concepts we work with today.

Handling spatial data manually

Imagine a world before computers were commonly available – maybe back in the 1940s or 1950s. Spatial data were encountered most often as maps. Maps then, as now, could be found in most households, shops and businesses. Maps were used for a plethora of applications: identifying areas for new developments, routing delivery vehicles, planning weekend walks or targeting services.

Maps were important in sieve mapping, a technique that combines several map layers to identify sites meeting a number of criteria (Burrough, 1986). The concepts of sieve mapping were introduced in an important book *Design with Nature* (McHarg, 1969). Each map layer is placed in turn on a light table and areas of interest highlighted. The layers produced are then placed on top of each other to identify areas of overlap. Sometimes the areas of importance on each map are shaded differently to indicate the relative importance of resulting maps. In the early days, McHarg and his colleagues used carefully selected colour pens to represent different values which would be meaningful when overlaid (Anon., 1995). The techniques of sieve mapping were used by planning agencies, nature conservation agencies and many others until relatively recently. The radioactive waste case study introduced in Chapter 1 is a good example of the implementation of sieve mapping in a GIS context. Until the arrival of GIS, the analysis undertaken for NIREX would have been done by hand.

Sieve mapping is inaccurate and can cope with only relatively coarsely mapped data because fine details are obscured in the overlaying process. Problems also arise if there are more than a few map layers involved, since even with transparent media and a good light source, it is difficult to see through layers of different colours.

Another common use of maps was for route finding, for planning deliveries or an emergency evacuation from a disaster zone. Measuring shortest distances on a map by hand is a tedious process involving the use of a piece of string. The string is used to trace the route, then the length of string used is measured and converted to distance using the map scale. The use of a manual or digital map measurer can speed up the process. There are many problems with these methods: they are inaccurate, difficult to repeat, and time-consuming. It is difficult to check alternative routes quickly or take changing local conditions (for example, road repairs, speed limits or one-way streets) into account. Knowledge of local conditions and main routes is necessary.

Calculating areas from maps is even more difficult without help from computers. Even now, those without access to digitizers and GIS or similar technologies sometimes rely on tracing areas onto graph paper and then 'counting squares' to obtain area estimates. An innovative technique is said to have been used by the US Bureau of Census for area estimation. Their method involved cutting out polygons from a paper map and weighing them to gain a measure of relative area (Tomlinson, 1988).

These manual techniques of spatial data handling were in widespread use until computers became generally available and they share many problems. These include the slow speed of update of data, the slow speed of access to data, difficulties in extracting relevant data, difficulties in analysing data, inaccuracies in data and analysis, and the inflexibility of the data storage media.

There are other problems associated with the media used. Paper maps stretch and shrink (due to folding or surrounding climatic changes), are always out of date, require complete redrawing for minor changes, contain inconsistencies in the way features are represented, are easily destroyed, require considerable space for storage and can be difficult to transport.

Other spatial data, for example printed census data, also have features that prevent efficient use. For those researching historical events, population census records contain a wealth of information. However, these need to be combined with birth and death records, marriage certificates and other documents for effective reconstruction of the lives and places of the past. Locating such records may be difficult since data sets may have been dispersed or even destroyed. It may be necessary to wade through files, books or microfiche to locate the exact details required. Moreover, without training, the old-fashioned script may be almost impossible to decipher. Other problems relate to inconsistencies and changes in spellings, and errors in records.

With all these problems associated with spatial data handling and analysis, it was hardly surprising that cartographers and other spatial data users were keen to exploit and explore computer technology.

The development of computer methods for handling spatial data

It is primarily advances in computer technology that have enabled the development of GIS. The decreasing cost of computer power over the last few decades has been one stimulus for wider use of GIS. There are other developments that have had an important impact: improvements in graphics technology, data access and storage methods, digitizing, programming and human–machine interfaces; and developments in systems theory. Martin (1991) considers that current GIS 'have grown out of a number of other technologies and a variety of other application fields, and are thus a meeting point between many different disciplines' (Martin, 1991: 2). Some of the areas of prime importance are databases, remote sensing, computer-aided design and computer graphics. Hardware developments in screens, printers and plotters and input devices such as digitizers and scanners have also had a major impact on GIS developments. Some examples are given in Box 9.1.

BOX 9.1

Developments in geographical computing

Graphics

In the 1960s, graphics facilities were basic, compared with current capabilities. Output from computers in graphics format was obtained using line printers that printed blocks of single characters to produce crudely shaded areas (Mather, 1991; Sinton, 1992). Since GIS relies heavily on graphical output this is an area of important research and development. In the future, graphic displays and output devices will continue to have a big impact on GIS.

Data access and storage

Data access and storage are other areas where important advances have been made. Sinton (1992) cites an example of a 1966 $2 billion mainframe computer that took minutes to access specific data. He contrasts this with a $50,000 workstation of the 1990s, which can access data on disk in milliseconds. Software and data are commonly exchanged on CD-ROM and further developments for data storage and exchange, such as wireless technology and the Internet, will change GIS again.

Data input

Data input has always been a problem for GIS and a time-consuming and error-prone process. In the early days digitizing involved moving a cursor across the map and pressing a button at the point to be recorded. This would cause a card to be punched with the point's co-ordinates (Rhind, 1987). Imagine the size of the card stack required for an outline of the Americas. Then imagine dropping them all on the way to the card reader! Recent developments in data input include scanning and automatic line following (Chapter 5).

The areas of computing developments presented in Box 9.1 are still important to GIS. However, it is the areas of computer cartography and spatial statistics that have produced many of the features of current GIS. Most, but not all, early developments were in North America and the UK (Coppock and Rhind, 1991). In the USA the Public Health Service, the Forestry Service, the Bureau of Census and the Harvard Graphics Laboratory all played a role in the development of GIS (Coppock and Rhind, 1991). In the 1960s there was little commercial development in the field, but groups such as the British Oxford Experimental Cartography Unit, as well as a number of individual researchers, were working with digital technology.

Much of the early research and development work in GIS was carried out in different and quite separate fields, particularly in computer cartography and spatial statistics. Only after advances had been made in both these areas could the systems we now know as GIS emerge. In other areas of IT, such as computer-aided design (CAD) and automated mapping/facilities management (AM/FM), developments have taken place that have led to the development of hybrid systems. These take some elements from CAD or AM/FM and others from GIS. In addition, the development of the systems approach as a theoretical framework for the development of GIS concepts and techniques has had an important influence on GIS. This is considered first below, then developments in computer cartography, spatial statistics, CAD and AM/FM are reviewed.

The systems approach

GIS attempts to produce a computer model of the real world to assist problem-solving and decision-making. General systems theory, one component of the systems approach, suggests that to understand the complexity of the real world, we must attempt to model this complexity. Unlike other scientific methods, which involve breaking down systems into component parts for study, general systems theory attempts to put back together the pieces of reality that science has 'dismembered' for analysis (Huggett, 1980). A macroscopic, rather than a microscopic, view is taken (Odum, 1971). This holistic systems approach to modelling complexity developed from the ideas of writers such as Boulding (1956), Von Bertalanffy (1968), Churchman (1968) and Ackoff (1971).

Systems theory is the conceptual approach to applying general systems theory to problem solving. Using systems theory, all aspects of the real world can be modelled as human activity or physical systems, or as hierarchies of systems. The avalanche prediction model presented in Chapter 8 is a complex system, which could be viewed as an element of a much larger system examining the hydrology and meteorology of the Happy Valley region.

Systems theory considers how a process works. Thus, it is a meta-discipline, taking ideas from other areas and applying to many fields, in a similar manner to mathematics (Reeve, 1997). In the same way GIS is a meta-discipline, linking fields of study such as computer science, remote sensing, cartography, surveying and geography (Goodchild, 1995). There are many parallels between the way in which systems theory and GIS view the world. GIS is all about the integration of data from a wide variety of sources (Aronoff, 1989; Burrough, 1986); these data are frequently the distillation of ideas from different disciplines.

There are many examples of the application of systems analysis, the methodology for implementation of systems theory. These include models of the global hydrological cycle (Ward, 1975) and models of social systems and of the universe (Huggett, 1980). Perhaps more important for GIS is the use of systems analysis as a methodological framework for the development of computer software and systems that can handle users' information requirements. Two approaches, hard and soft systems analysis, have been used for the development of computer systems, although the boundary between them is fuzzy. Both methods may be used to help improve our understanding of a problem or to develop computer systems (Chapter 12).

Computer cartography

Computer cartography (also known as digital mapping, automated cartography or computer-aided cartography) is the generation, storage and editing of maps using a computer (Mather, 1991). The development of computerized GIS is closely linked with the development of computer cartography. Two examples of developments in computer cartography that have links to the development of GIS are outlined in Box 9.2.

The pioneering developments in computer cartography, such as those outlined in Box 9.2, were initiated by map producers, not map users. Their interests were in speeding up and improving the process of map production. Therefore areas such as name placement, plotting, the use of different

BOX 9.2

Developments in computer cartography

The Atlas of Britain

This atlas (Bickmore and Shaw, 1963) is widely recognized as one of the catalysts for the development of computer methods for handling spatial data in the UK. Although published in 1963, work on this traditional atlas had started in the 1950s. Many cartographers and designers were involved and by the time of publication there was both praise and criticism for the work. The cartography was praised, but criticisms were levelled at the out-of-date material included and the unwieldy format of the atlas, and the book made a loss for its publishers (Rhind, 1988). The problems faced during production led Bickmore and his team at the Oxford Experimental Cartography Unit to consider using computers to check, edit and classify data and to experiment with graphics. Experiments were started in the late 1950s into the automatic placement of names on maps and methods of digitizing. These investigations led to the development of the first free-cursor digitizer and programs for line measurement, projection changing and data editing (Rhind, 1988).

SYMAP

SYMAP was probably the first commercially available automated cartography package (Goodchild, 1988). It was first marketed in the USA in about 1965 and could produce crude but useful choropleth maps. Production of maps was quick, but the quality of the output was poor since this was dictated by the output device of the time – the line printer. The line printer achieved area shading by repeated printing of the same character or number. Mather (1991) describes time-consuming early versions of SYMAP. Use involved the production of decks of punched cards held together with elastic bands that were read by a card reader. A deck of cards was submitted to an operator and output collected later. Later, however, file input was possible and other important developments included the implementation of previously subjective cartographic procedures such as interpolation (Goodchild, 1988) and SYMVU, the first three-dimensional mapping package. Both SYMAP and SYMVU were still in widespread use well into the 1980s.

fonts and the production of colour separations for printing were important (Tomlinson, 1988).

Spatial statistics

Spatial statistics and geographical analysis methods were an area of much research and development in the late 1950s and early 1960s as computers became available to help the statistician. Initially, developments were in areas such as measures of spatial distribution, three-dimensional analysis, network analysis and modelling techniques (Tomlinson, 1988). These developments took place remote from any GIS or computer cartography developments primarily because the methods were developed using synthetic data sets and analyses were generally conducted only once. This is in marked contrast to the typical GIS 'analysis', which uses real-world data, and where analysis may be repeated many times.

Many of the methods developed in the middle of the twentieth century by spatial statisticians are still not available in GIS. Despite the 'quantitative revolution' in geography during the 1960s and the huge increase in the application and teaching of spatial statistics, these methods are still not appreciated or required by many users, and are not included in new GIS products and releases. Some of the techniques, notably those used in geodemographics and location planning (gravity models and location-allocation models), appear in separate packages. However, these frequently remain in the academic domain or as tools for independent experts to use for the provision of information to clients.

AM/FM and CAD

Automated mapping/facilities management (AM/FM) and computer-aided design (CAD) are two areas of technology closely related to GIS. Early AM/FM systems were used to store information about utility facilities and link external files to electronic maps. The maps and their related databases contained information about entities such as power lines or transformers (Reina, 1997). With the realization that location was important to the operation of most utility organizations, the use of AM/FM took off in the 1970s. In common with GIS, it took the enthusiasm of individuals to encourage widespread dissemination of AM/FM (Reina, 1997).

CAD is used to enter and manipulate graphics data, whereas GIS is used to store and analyse spatial data (Henry and Pugh, 1997). Traditional users of CAD (in architectural design, engineering and facilities management) used CAD for the production, maintenance and analysis of design graphics. CAD provided no functions for the production or analysis of maps. However, users have become aware of the value of having spatial information attached to their data. Thus, use is now made of GIS/CAD hybrid packages which provide tools for the creation, editing and manipulation of graphical entities (such as site plans or design drawings) along with the creation of spatial relationships between these entities.

BOX 9.3

CGIS: an early GIS

The Canadian Geographic Information System (CGIS) used data collected for the Land Inventory System and was developed as a result of the requirements of the Canadian Agriculture and Development Act. It was designed to produce maps of the crops that areas of land were capable of producing and to map land capability for forestry (based on soil, climate, drainage and physical land characteristics). Land within the survey area was classified according to wildlife potential, taking into account all factors that were favourable or unfavourable for certain species. The land inventory also classified land for recreation. The problem of measuring and comparing all this information necessitated a technological solution. The approach adopted was also influenced by the volume of data required, the huge size of Canada, the need for a consistent spatial referencing system and the need for map linkage methods (Crain, 1985). The search for a computer solution led to the development of CGIS. The date usually quoted for the beginning of CGIS is 1964, but the system was not fully operational until 1971. CGIS was capable of tackling the 1650 'person-year' task of comparing 6000 maps in days, or selecting areas that were suitable for forage crops but not suitable for grain growing or forestry in a few hours (Symington, 1968). CGIS has been modified significantly since then, but it is still a significant component of the Canada Land Data System. Many currently accepted and used GIS concepts, terminology and algorithms are originally from CGIS. It was the first general-purpose GIS to go into operation; the first system to use raster scanning for efficient input of large numbers of maps; the first GIS to employ the data structure of line segments or arcs linked together to form polygons (which has now become a standard); and the first GIS (in 1975) to offer remote interactive graphics retrieval on a national basis (Crain, 1985).

Over the years CGIS has been modified and improved to keep pace with technology and the requirements of users. For example, microcomputers are now used for data input and analysis. However, the overall components of CGIS have remained constant. There is a subsystem for scanning input and raster editing that allows editing and verification of scanned images and some auto-processing (for example line thinning, closing of small gaps and raster to vector conversion). Additional subsystems include the interactive digitizing and editing subsystem and a cartographic output subsystem.

CGIS has been applied to a wide variety of areas and topics. The land inventory data, which took 12 years to gather, has been used in land-monitoring programmes focusing on changes in fruit-growing areas near to urban centres, acid rain monitoring and toxic waste siting. Considerable benefits have been obtained from the system, including cost savings, cost avoidance, improved administrative decisions, wiser land use, better-managed national parks and more effective agriculture and forestry industries (Crain, 1985; Symington 1968).

The Canadians recognized from the beginning that good, accurate and relevant information was vital for the success of the GIS (Symington, 1968). CGIS was still being used in the 1990s to help with land use planning, pollution incidents and resource management (Heywood, 1990).

The development of GIS

GIS and map analysis developments began around the same time as related developments in computer cartography and spatial statistics. They were prompted by the limitations of hard-copy maps, problems with overlaying data sets and the increasing size and number of available data sets (Tomlinson, 1988). The initial developments (in the 1960s) were technical and were aimed at developing a set of spatial data handling and analysis tools that could be used with geographical databases for repeated problem solving (Tomlinson, 1990). The technical issues for early consideration included digitizing, the development of data structures to allow updating and analysis, the compacting of line data to ease storage and the development of computer methods that would operate on large or small sets of data. Unlike the cartographers, initial developments were not focused on the production of high-quality output, but on the analysis of data. One of the first systems called GIS appeared in Canada in 1964 (Box 9.3).

Goodchild (1995) considers that the roots of current GIS lie in the 1960s. He suggests that the main developments of the decade were the adoption of a layer approach to map data handling in CGIS (Box 9.3), and pioneering work by the US Bureau of Census that led to the digital input of the 1970 Census (Box 9.4).

In turn, these developments influenced work at the Harvard Graphics Laboratory, which led directly to the production of some of the first commercial GIS software, including the package ARC/INFO (Table 9.1).

BOX 9.4

Topology for the US Census – the development of DIME

The development of the GBF-DIME (Geographic Base File, Dual Independent Map Encoding) data format by the US Bureau of the Census in the late 1960s was a major step forward in GIS data models. GBF-DIME files are computerized representations of detailed street maps for the major urban areas of the USA that have proved an invaluable aid to census taking. The list below outlines some of the major developments during the development of GBF-DIME:

1965 Census Small Area Data Advisory Committee is established

1966 A Census Use Study (CUS) is planned and begins in New Haven, Connecticut

1967 CUS begins testing procedures for the 1970 Census, and starts computer mapping experiments. Problems are faced with the conversion of analogue maps into digital format. Each street intersection in a normal US rectangular street grid is being digitized 8 times.

Mathematician James Corbett presents the principles of map topology to CUS and these are used to overcome the problem. Street intersections and blocks are numbered. Street segments are encoded by reference to the areas to the left and right of them, as well as to the nodes that they connect (to node and from node). This model gives some redundancy that assists automated error checking.

George Farnsworth of CUS names the new process DIME, and details of the process are presented to URISA and published.

1970s During the 1970s GBF-DIME files were digitized for all US cities

DIME was a practical innovation – it assisted efficient digitizing and error removal, and it supported choropleth mapping of census results. It is at the heart of many vector data models, including that used by ArcInfo. The topological edit is perhaps the most significant contribution of the model allowing the creation of a perfect replica of the elements on the original analogue map (US Bureau of Census, 1990).

DIME files are also a key component of the current TIGER system that is, in turn, a critical part of the National Spatial Data Infrastructure. The DIME files, and the organizations formed to produce them, also became the foundation for the geodemographics industry in the USA.

(Sources: Mark *et al.*, (no date); US Bureau of Census, 1990; Sobel, 1990)

Table 9.1 The development of ESRI

Year	Development
1969	ESRI founded by Jack and Laura Dangermond as a privately held consulting group in Redlands, California
1970s	ESRI is involved in applications such as site selection and urban planning which lead to the development of many of the technical and applied aspects of GIS
1981	ARC/INFO GIS for minicomputers launched. Later the product is shifted to UNIX workstations and PCs The first ESRI user conference attracts 18 participants
1980s	ESRI becomes a company developing commercial software products
1986	PC ARC/INFO is launched
1991	ARCVIEW is launched (desktop mapping and GIS tool)
1992	Between January and July 10,000 copies of ARCVIEW are distributed ArcCAD is launched (a hybrid GIS/CAD package)
1995	SDE (a client-server product for spatial data management) and BusinessMap (a consumer mapping product) are launched
1996	ARC/INFO for Windows NT is launched together with the Data Automation Kit (geodata creation software for Windows) Atlas GIS is aquired by ESRI The annual User Conference attracts nearly 6000 users and business partners from 80 countries A new three-storey research and development centre is opened in Redlands
1997	ESRI employs more than 1100 staff, 830 in Redlands, California. The company is still wholly owned by Jack and Laura Dangermond. ESRI has formal ties with over 500 application developers and value-added resellers using ESRI products
1998	ArcData Online (Internet mapping and data site) launched
1999	ArcInfo 8 released. ArcNews circulation exceeds 200,000
2000	Geography Network for publishing, sharing and using geographic information on the Internet launched
2001	ESRI celebrates 32 years of providing software and services to the GIS industry

Source: adapted from ESRI, 2001

Other technical developments of the 1960s should not be forgotten. These included improvements in the conversion of scanned data into topologically coded data, and the implementation of computer-based methods for efficient point-in-polygon analysis (Tomlinson, 1988).

The 1970s

Developments in GIS and related disciplines in the 1970s were many and varied, assisted by rapid advances in computing technology. New computer cartography products entered the market (for example GIMMS, MAPICS and SURFACE II). In the UK the first multi-colour standard series map to be produced digitally appeared in 1971. Soon after this the Ordnance Survey began the routine digitizing of topographic maps to speed up the production of conventional paper maps (Rhind, 1987).

In the 1970s, computer technology was advancing generally, with memory sizes and processing speeds increasing and costs falling. Interactive graphics became available and blind digitizing (digitizing without a graphical display) could be avoided (Tomlinson, 1990). These technological developments were advantageous to all fields using computers. The decade was one of the 'lateral diffusion' of the technology, with more users and more subject areas being introduced to GIS (Tomlinson, 1988).

The 1970s saw the first conferences and published work on GIS. Conference proceedings and other publications began to appear around 1974 (Coppock and Rhind, 1991). The first UK meeting of academics to discuss GIS was in 1975 (Unwin, 1991). This was part of an increasing awareness of and communication about GIS, which included the first texts on GIS published by the International Geographical Union (Tomlinson, 1990).

During the mid-1970s, developments were still occurring in two separate areas: in GIS, which were now commercially available with limited analysis capabilities; and in computer cartography, where good quality graphics and plotting and easy data editing were still the focus of improvements.

One major contribution of workers in the 1970s was the recognition that the problems facing GIS were not just technical. Tomlinson *et al.* (1976) were among the first to recognize that there are as many problems associated with the management side of implementing an information system as with the technical side, and probably more.

The 1980s

By the 1980s the demand had grown for good graphics and data analysis, real-time querying of databases and true topological overlay. Computers had by now found their way into all aspects of cartographic production (Taylor, 1991a) and computer cartography, and GIS and spatial analysis converged to produce the GIS of the late 1980s and 1990s.

In the UK the term GIS was considered to be in common usage by the mid-1980s (Rhind, 1987). Spatial modelling methods began to be included in software, but GIS developments and demand for this aspect of GIS have lagged far behind the developments in spatial analysis. Developments were, in general, still related to improving technology, but, owing to price decreases, hardware was no longer such a limiting factor (Maguire, 1989). The decade saw a massive increase in interest in the handling of geographical data by computers, and this led to the rapid evolution of systems in response to user needs rather than any clear theoretical structure (Martin, 1991).

One of the major events of the decade, in computing terms, was the production of the first microcomputers using the Intel 80386 chip in 1986. This has paved the way for a new generation of microcomputer-based GIS products (Taylor, 1990). Significant technical developments in the 1980s included the increased use of the raster data model as the cost of computer memory decreased, and the development of the quadtree data structure (Goodchild, 1988). Some of these developments were assisted by new GIS research initiatives. The USA's National Center for Geographic Information and Analysis (NCGIA), spread over several university sites, and the UK's Regional Research Laboratories (RRLs) were perhaps the two largest initiatives to begin in the 1980s (Goodchild and Rhind, 1990; Masser, 1990). Other research programmes began in countries such as Australia, the Netherlands and France. Frequently, the remit for these centres included raising awareness of GIS and training, as well as the development of new techniques and applications. Training was facilitated by the appearance of the first textbooks, including Burrough (1986) and Peuquet and Marble (1990); by the establishment of the *International Journal of GIS* in 1987; and by discussions accompanying the development of GIS curricula and teaching materials such as the AUTOCARTO syllabus (Unwin *et al.*, 1990) and the NCGIA's core curriculum (Goodchild and Kemp, 1990).

Tomlinson (1990) cites an estimate that over 1000 GIS and computer cartography systems were in use in the USA by 1983, and forecasts then were for 4000 systems by the end of the decade. Developments during the 1980s were not necessarily restricted to the players who had been involved thus far. Developing, as well as developed, countries were seeing applications for GIS and international and continental projects were emerging. For the developing nations a steady expansion in applications took place, particularly after 1986 (Taylor, 1991b). The major areas of application were still the natural environment, national and regional government and the utilities. Many new projects relied on inter-organization co-operation and exchange of data (for example, CORINE, Box 9.5). This initiated new debate on data issues such as transfer format, costs and copyright and the general trend towards information being regarded as a commodity (Openshaw and Goddard, 1987).

Besides CORINE (Box 9.5) there are other environmental examples of inter-organizational projects, including WALTER, the terrestrial database for rural Wales (Martin, 1991), and socio-economic examples such as REGIS (the Regional Geographic Information Systems project) in Australia (O'Callaghan and Garner, 1991). The utilities in the UK also investigated GIS, realizing that there were benefits to be gained from the exchange of data for pipes and cables. Savings could be made if repeated 'digging up' of road surfaces for repairs and maintenance could be avoided. The Dudley joint utilities digital mapping trial, initiated in the UK in 1982, aimed to build one database on one computer that could be accessed from the offices of the utilities and local government (Mahoney, 1991). The project highlighted discrepancies in the boundaries of the operating units of each utility, and led to the development of a standard format for the exchange of map data (NJUG, 1986).

BOX 9.5

CORINE: An international multi-agency GIS project

One example of an international, multi-agency GIS project is the European Commission's CORINE (Coordinated Information on the European Environment) programme, initiated in 1985 (Rhind *et al.*, 1986). This GIS was designed as a general environmental database that would encourage the collection and co-ordination of consistent information to aid European Community policy (Rhind *et al.*, 1986; Mounsey, 1991b). It was used initially to address the setting up of an inventory of sites of scientific importance for nature conservation, acid deposition and the protection of the Mediterranean environment. Mounsey (1991b) summarizes the nature of the project and reviews the lessons learnt. She considers that some of the constraints on the development of environmental databases at that time were the availability of data, education of users and need for substantial resources. In addition Mounsey suggests that the establishment of environmental databases should be pro-active (that is, not as a reaction to existing problems), and conducted according to systematic methodologies (such as systems design) with the full support of all interested parties.

The main achievements of CORINE were:

- the standardization of practices and methods for recording environmental data;
- the demonstration of the feasibility of a single database to meet a variety of end-user needs; and
- the development of similar activities at national level.

GIS in the 1990s

The GIS of the early 1990s were commercially available general-purpose packages containing many of the 1960s analysis operations and new developments in technology such as multimedia. Gradually, spatial analysis and decision-making were becoming more important, and those systems that are purely for data display and retrieval have become available separately. In many cases GIS have become part of a wider decision-support or management information system.

The number and range of application areas are still increasing and GIS have infiltrated university and school curricula. One of the problems in the 1990s is that advances in technology have overtaken the average user and software so that the full potential of the hardware is not being exploited. It is now recognized that the technical problems associated with GIS are solvable in most cases. More difficult are the management, implementation and integration of GIS into the work and culture of an organization. Other problems are associated with the nature, quantities of, and access to, data. The range of available spatial data sets is increasing apace. However, many countries still do not have access to an accurate, up-to-date and complete digital topographical base coverage. Solving these management, organizational and social issues will be far more difficult than solving the technical problems of GIS. Some of these

BOX 9.6

GIS and North West Water

GIS activities at North West Water plc (NWW) began with AM/FM, and have developed over the years into the current system, which consists of an open-systems AM/FM/GIS. This is applied in areas such as call centre operations, maintenance management and field communications. The current system has survived deregulation and organizational change, and includes a database of 10 million items which serves over 300 users. The system integrates customer records with technical, environmental and management data. Sixty

Table 9.2 GIS developments at North West Water

Date	*Organization development*	*GIS development*
1989	NWW is privatized after a century as a government-owned utility	
1990		The first suggestions of a new IT project are made
1991		A 'wish list' is produced which became part of 'Vision 2000' strategy
1992		The decision to rebuild the information management systems to help the expansion of NWW is made. A GIS is to be at the heart of the new system
	After a management 'shake-up' an expansion strategy is scrapped	The IT project survives these organizational changes and continues. Software and applications are developed by IT staff, and database creation, involving the conversion of hard-copy tables and schematics, is outsourced. Sewer data are assembled from a variety of sources, including new field data
1995	NWW acquires Norweb (an electricity distributor) to create United Utilities, the UK's first integrated utility company. The two companies continue to operate separately	
1996	A new subsidiary company, Vertex, is created to handle billing, customer service and software and hardware maintenance for United Utilities	
1997		The current system uses a proprietary database linked to GIS software. The open architecture of the system allows spatial data to be integrated into a variety of the organization's activities. • Current users: over 300. • Current database size: 10 million items, 20 gigabytes of data. • Predicted users: 1000. Project completion imminent

Sources: Vertex, 1997; Reina, 1997

BOX 9.6 CONTINUED

thousand maps covering 73,000 km of water mains and sewers in north-west England are in the system, combined with Ordnance Survey digital background data and other company-specific data. Much of these data were digitized by external organizations. Some were easy to produce whilst others required field work to fill in gaps in the databases. Some of the key events in NWW's organizational and GIS history are summarized in Table 9.2.

(Sources: Reina, 1997; Vertex, 1997)

issues will be returned to in later chapters of this book, particularly Chapter 11. Further details of the GIS of the 1990s can be found in Chapter 13.

The diffusion of GIS

The development of GIS within individual organizations is often difficult for outsiders to observe. Two case studies are presented to illustrate the changes occurring in GIS technology and applications over the last three decades. Table 9.1 (adapted from ESRI, 2001) presents the history of ESRI, the commercial company responsible for the software package ARC/INFO, one of the leading GIS products. Box 9.6 presents the development of GIS applications at North West Water plc, a major water utility company in the UK.

In addition to the development of GIS within individual organizations, it is interesting to look at the nature of the diffusion process in different countries and different user groups. Rhind (1987), Kubo (1987) and Shupeng (1987) consider the growth of GIS in the UK, Japan and China, respectively. Masser *et al.* (1996) present research findings on GIS diffusion in local government in nine European countries, including the

BOX 9.7

The development of GIS in the UK

The development of GIS in the UK has been slower than in some other countries, notably the USA. Coppock (1988), Rhind and Mounsey (1989) and Rhind (1987) have suggested a number of reasons for this, including a general lack of resources in the UK, the cost of adopting new techniques, a huge backlog of analogue data, the centralized nature of government decision making, the availability of large-scale printed maps, a lack of awareness of the benefits of GIS and the existence in a small country of a culture that does not readily accept change. The slow pace of development may also have been due to the small size of the UK, and thus the lack of necessity for any major land inventory systems such as in the Canadian CGIS example.

A series of government reports charts the development of GIS in the UK. These include a review of the Ordnance Survey in 1978 (the Serpell Report); the review of remote sensing and digital mapping published in 1981 led by Lord Shackleton; and, perhaps most importantly, the 1987 Report of the Committee of Inquiry into the handling of geographic information, often referred to as the Chorley Report after the committee's chairperson (Department of the Environment, 1987). The Chorley Committee made recommendations relating to digital topographic mapping, data availability, data integration, awareness raising in GIS and the role of government in GIS development (Rhind and Mounsey, 1989; Department of the Environment, 1987). One of the recommendations of the Chorley Report was that a Centre for Geographic Information should be established. The Association for Geographic Information (AGI), launched in January 1989, has gone some way towards fulfilling this role. It is a '...multidisciplinary organization dedicated to the advancement of the use of geographically related information.... It aims to increase awareness of the benefits brought by the new technology, and assist practitioners...' (Shand and Moore, 1989: 9). In addition to the promotion of the technology, the AGI is active in co-ordinating and organizing interest groups, collecting and disseminating information about GIS, developing policy advice on GIS issues, and encouraging GIS research.

UK, Netherlands and Greece. Similarities can be seen in all cases, with the national agencies, utilities and local government playing key roles as early users of the technology. The problems preventing more widespread uptake also show similarities, with the unavailability of digital data and lack of awareness and training in GIS commonly cited. Box 9.7 takes the development of GIS in the UK as an example.

The demand for, interest in and range of applications of GIS grew rapidly in the late 1980s and early 1990s. There are three main factors that are helping to fuel this growth. First, the amount of data available over computer networks, including the Internet, and on disks and CD-ROMs ready for immediate input has increased. This can save many hours manually keying in data. Second, the demands of geographers and other spatial information users for computer power and computer processing methods appropriate to their work has grown and, in some cases, outgrown more traditional analysis and management methods. Along with this, the power of computers has increased further and costs have fallen. Desktop and portable computers are now available that have the same power as those which, not so long ago, would have filled whole rooms. So, there is an increasing demand for more computer power and bigger computers. Applications of GIS have been developed in one field after another, first land management, through utilities management and on into property development, logistics and marketing. Who knows what the next application area will be.

In some cases it may still be that the demands of users are beyond the capabilities of current GIS. For example, Goodchild (1995) has suggested that the ability of GIS to handle only static data is a major impediment to its use in modelling social and economic systems.

Conclusions

Coppock and Rhind (1991) identify three 'ages' of GIS: the pioneering age, the age of experimentation and practice and the age of commercial dominance. These stages in the development of GIS, together with their characteristics, are summarized in Table 9.3. Other information has been included in the table to try to give a summary of the development of GIS over the last three decades.

There has not been space in this chapter to cover all of the disciplines contributing to GIS

Table 9.3 The stages of GIS development

Stage and date	*Description*	*Characteristics*
1 Early 1960s–1975	Pioneering	• Individual personalities important • Mainframe-based systems dominant
2 1973–early 1980s	Experiment and practice	• Local experimentation and action • GIS fostered by national agencies • Much duplication of effort
3 1982–late 1980s	Commercial dominance	• Increasing range of vendors • Workstation and PC systems becoming available • Emergence of GIS consultancies and bureaus • Launch of trade journals such as *GIS World* (USA) and *Mapping Awareness* (UK)
4 1990s	User dominance, vendor competition	• Embryonic standardization • Systems available for all hardware platforms • Increasing use of PC and networked systems • Internet mapping launched

Sources: adapted from Tomlinson, 1990; Coppock and Rhind, 1991

developments or likely to influence the field in the future. However, there are some developments that deserve a brief mention. First, the closer integration of remote sensing and GIS. With the emergence in the 1990s of GIS really capable of handling and analysing both vector and raster data, the use of remotely sensed data in GIS has grown. In addition, developments in remote sensing, including the launch of new satellites, increased data resolution and access to new data (such as those from the former Eastern Bloc) are helping to promote the further integration of the two disciplines. Developments in surveying technology, particularly in global positioning systems, which give users their geographical position, and frequently also their altitude above sea level, at any point on the surface of the Earth, also have a big part to play in the future of the technology. Longley and Clarke consider that 'GIS has now reached a crossroads in terms of potential directions it may take in the future' (1995: 1). Some of the possible directions that GIS may take are discussed in Chapter 13.

REVISION QUESTIONS

- How were map data used before the introduction of computer methods? What were the problems with these methods of data handling?
- Briefly list the major events in the development of GIS during the 1960s, 1970s, 1980s and 1990s.
- How have application areas for GIS changed over the last 40 years? In which disciplines were GIS first applied, and in which areas are new applications being developed today? Can you suggest any reasons for these trends?
- List three individuals who have played a role in the development of GIS, and give brief details of their contributions to the field.
- What have been the major developments in IT over the last three decades? How far do the developments in GIS over the same period reflect these general developments in information processing and handling?
- Describe briefly the nature of GIS in the 1990s. What are the current developments that will affect the nature of GIS in the current decade?

Further study

The history of GIS has, in general, been poorly documented. There are, however, a number of useful sources of information, including some highly personal and very interesting reviews of developments. Coppock and Rhind (1991) offer a good overview of the history of GIS. Tomlinson (1990) outlines the development and applications of CGIS, and considers general GIS developments in the 1970s and 1980s. A special issue of the journal *American Cartographer* in 1988 presented a number of papers by eminent GIS professionals, and gave them the opportunity to reflect on the development of GIS. Coppock (1988), Rhind (1988), Goodchild (1988) and Tomlinson (1988) offer personal views from different sides of the Atlantic on the major developments during the early years of GIS. In a more recent article in *GIS World* magazine (Anon., 2001), McHarg reflected on methods introduced in his book *Design with Nature* (McHarg, 1969), and offered comment on the present and future of GIS. The same magazine included an interview with Roger Tomlinson in 1996.

Foresman (1997) offers interesting 'perspectives from the pioneers' written first-hand by many of those involved in early developments. Major projects, including CGIS and the development of the TIGER data structure, are described and developments in a range of agencies are reviewed. Contributions are mainly from US authors, but there are chapters that outline developments in Canada, Europe, Australia and international applications.

For chronological accounts of GIS development, there are a number of timelines that can be consulted. National Research Council (1997) present a timeline that was constructed by participants at a workshop on the future of spatial data and society. The participants included many famous GIS personalities, and the unedited timeline that they produced can be seen online. Rana *et al.* (2001) are engaged in a project to collate a GIS timeline. Their timeline, viewable online, can be searched and includes some interesting graphics, such as output from SYMAP.

Further details of the ESRI and Vertex developments introduced in this chapter can be found on the Web (ESRI, 2001; Vertex, 1997) and in Reina (1997). Masser *et al.* (1996) is an interesting book for those interested in the diffusion of GIS. Although focused on local government applications, there are many general lessons to be learnt and interesting case studies from a variety of European countries.

Anon. (1995) *GIS World* interview: Ian McHarg reflects on the past, present and future of GIS. *GIS World 8* (10): 46–9

Anon. (1996) *GIS World* interview: Roger Tomlinson the father of GIS. http://www.geoplace.com/gw/1996/0496/0496feat2.asp

Anon. (2001) *GIS World* interview: Ian McHarg reflects on the past, present and future of GIS. http://www.geoplace.com/gw/2001/0601/0601mem.asp

Coppock J T (1988) The analogue to digital revolution: a view from an unreconstructed geographer. *American Cartographer* 15 (3): 263–75

Coppock J T, Rhind D W (1991) The history of GIS. In: Maguire D J, Goodchild M F, Rhind D W (eds) *Geographical Information Systems: Principles and Applications*. Longman, London, Vol. 1, pp. 21–43

ESRI (2001) ESRI Timeline http://www.esri.com/company/about/timeline/flash/index.html

Foresman T (ed.) (1997) *The History of GIS*. Prentice Hall, New Jersey

Goodchild M F (1988) Stepping over the line: the technological constraints and the new cartography. *American Cartographer* 15 (3): 277–89

Masser I, Campbell H, Craglia M (1996) *GIS Diffusion: the Adoption and Use of Geographical Information Systems in Local Government in Europe*. GISDATA 3, Taylor and Francis, London

McHarg I L (1969) *Design with Nature*. Doubleday, New York

National Research Council (1997) *The future of spatial data and society*. National Academic Press, Washington DC

Rana S, Haklay M and Dodge M (2001) GIS Timeline. http://www.casa.ucl.ac.uk/gistimeline

Reina P (1997) At long last, AM/FM/GIS enters the IT mainstream. *Information Technologies for Utilities* (Spring): 29–34

Rhind D (1988) Personality as a factor in the development of a discipline: the example of computer-assisted cartography. *American Cartographer* 15 (3): 277–89

Tomlinson R F (1988) The impact of the transition from analogue to digital cartographic representation. *American Cartographer* 15 (3): 249–63

Tomlinson R F (1990) Geographic Information Systems – a new frontier. In: Peuquet D J, Marble D F (eds) *Introductory Readings in Geographical Information Systems*. Taylor and Francis, London, pp. 18–29

Vertex (1997) http://www.vertex.com

10 Data quality issues

KEY QUESTIONS AND ISSUES

- Error and quality: concepts and terminology
- How can errors in spatial data be described?
- What types of errors arise in GIS?
- What are the sources of error in a typical GIS project?
- How can GIS errors be modelled and traced?
- How can errors in GIS be managed?

Introduction

In many industries, whether they are in the manufacturing or service sectors, quality control is everything. Without a quality product that is properly designed, well made and reliable, customers will go elsewhere for the goods and services they require. GIS users strive for quality products from their systems, and hope to produce good quality results and output.

The computing saying 'garbage in, garbage out' recognizes that if you put poor quality data into your program, you will output poor quality results. The saying applies to GIS since the results of analysis are only as good as the data put into the GIS in the first place. Previous chapters have shown how easy it is to combine maps, create buffer zones, interpolate point data sets and perform dozens of other operations with a GIS to produce complex and attractive maps. However, errors in input data may be compounded during these analyses, and errors in classification or interpretation may create misleading final output for end-users. Maps produced by a computer have an inherent feeling of quality and authority. Naive users have a tendency to believe what they see on the screen, and complexity, clever colours and intrinsic visual appeal may fool users into thinking that GIS output is of a high quality. But what is a high-quality product? What exactly are

good quality data? How can we describe and recognize poor quality output?

Two issues are particularly important in addressing quality and error issues: first, the terminology used for describing problems, and second, the sources, propagation and management of errors. Describing data problems in GIS is difficult since many of the words used are also common in everyday language. Words such as quality, accuracy and error not only mean different things to different people but also have precise technical definitions. The terms used for data errors and quality are introduced at the start of this chapter, since the first step to solving problems is to be able to recognize and describe them. The remainder of the chapter outlines the types and sources of errors in GIS to help you recognize and deal with problems at the appropriate stage of a GIS project. Techniques for modelling and managing errors are also considered. These techniques are important, as few data sets are error-free. GIS users must learn to manage and live with the errors their data and systems produce. Perhaps more importantly, GIS users should document the limitations of their source data and the output generated. Such documentation will ensure that output will be of value to others and will permit appropriate future use of data. For example, our Martian friend might be less likely to buy an old vehicle from a car auction (which might have high mileage, several owners and no service history) than a smart little car with quality assurance documentation from an established dealership (with guaranteed low mileage, one careful owner and a full service history). The same principles apply to GIS data. Quality control should be impeccable because confidence in GIS analysis and results relies heavily on having access to reliable, good quality data.

Earlier in the book we compared the fuel in the Martian's car to data in a GIS. In the same way that a poor quality fuel may cause problems with the running of the car, poor quality data will introduce errors into your GIS.

Describing data quality and errors

GIS users must describe data problems before they can solve them. So, before discussing the sources of data problems and their implications and management, this chapter considers the terms used to describe data problems in a GIS context. These terms will be used later in discussions of the sources of error in GIS.

Quality is a word in everyday use that is very difficult to define. It has different meanings for different people and in different contexts. A dictionary definition of quality is 'degree of excellence' (Collins, 1981). In GIS, *data quality* is used to give an indication of how good data are. It describes the overall fitness or suitability of data for a specific purpose or is used to indicate data free from errors and other problems. Examining issues such as *error*, *accuracy*, *precision* and *bias* can help to assess the quality of individual data sets. In addition, the *resolution* and *generalization* of source data, and the data model used, may influence the portrayal of features of interest. Data sets used for analysis need to be *complete*, *compatible* and *consistent*, and *applicable* for the analysis being performed. These concepts are explained below.

Flaws in data are usually referred to as errors. Error is the physical difference between the real world and the GIS facsimile. Errors may be single, definable departures from reality, or may be persistent, widespread deviations throughout a whole database. In the Happy Valley database a single error in a point data set would occur if a co-ordinate pair representing a ski lift station were entered incorrectly, perhaps with two digits in the wrong order. A more systematic error would have occurred if the co-ordinates for all the ski lift stations in the data set had been entered in (y,x) order instead of (x,y).

Accuracy is the extent to which an estimated data value approaches its true value (Aronoff, 1989). If a GIS database is accurate, it is a true representation of reality. It is impossible for a GIS database to be 100 per cent accurate, though it is possible to have data that are accurate to within specified tolerances. For example, a ski lift station co-ordinate may be accurate to within plus or minus 10 metres.

Precision is the recorded level of detail of your data. A co-ordinate in metres to the nearest 12 decimal places is more precise than one specified to the nearest three decimal places. Computers store data with a high level of precision, though a high level of precision does not imply a high level of accuracy. The difference between accuracy and precision is important and is explained in Box 10.1.

Bias in GIS data is the systematic variation of data from reality. Bias is a consistent error throughout a data set. A consistent overshoot in digitized data caused by a badly calibrated digitizer, or the consistent truncation of the decimal points from data values by a software program, are possible examples. These examples have a technical source. Human sources of bias also exist. An aerial photograph interpreter may have a consistent tendency to ignore all features below a certain size. Although such consistent errors should be easy to rectify, they are often very difficult to spot.

Bias, error, precision and accuracy are problems that affect the quality of individual data sets. Data quality is also affected by some of the inherent characteristics of source data and the data models used to represent data in GIS. Resolution and generalization are two important issues that may affect the representation of features in a GIS database.

Resolution is the term used to describe the smallest feature in a data set that can be displayed or mapped. In raster GIS, resolution is determined by cell size. For example, for a raster data set with a 20-metre cell size, only those features that are 20 × 20 metres or larger can be distinguished. At this resolution it is possible to map large features such as fields, lakes and urban areas but not individual trees or telegraph poles because they measure less than 20 × 20 metres. Vector data can also have resolution, although this is described in different terms. Resolution is dependent on the scale of the original map, the point size and line width of the features represented thereon and the precision of digitizing.

Generalization is the process of simplifying the complexities of the real world to produce scale models and maps. Cartographic generalization is a subject in itself and is the cause of many errors in GIS data derived from maps. It is the subjective process by which the cartographer selectively removes the enormous detail of the real world in order to make it understandable and attractive in map form. Generalization was introduced in Chapter 2 and discussed further in Chapter 8. Further examples of how cartographic generaliza-

BOX 10.1

Accuracy and precision

It is perfectly feasible to have a GIS data set that is highly accurate but not very precise, and vice versa. The difference between accuracy and precision is best illustrated using diagrams. Imagine the Happy Valley biathlon has just taken place. Four contestants in the shooting have produced the results illustrated in Figure 10.1. Contestant A was accurate but not precise; contestant B was inaccurate but precise; contestant C was neither accurate nor precise; and contestant D, the winner, was both accurate and precise.

Extending this illustration to GIS, imagine the target is geographical space. The 'bull's eye', or central cross on the target, is the true geographical location of a feature we are trying to locate and the athlete's 'hits' on the target represent points collected by surveying or positioning techniques.

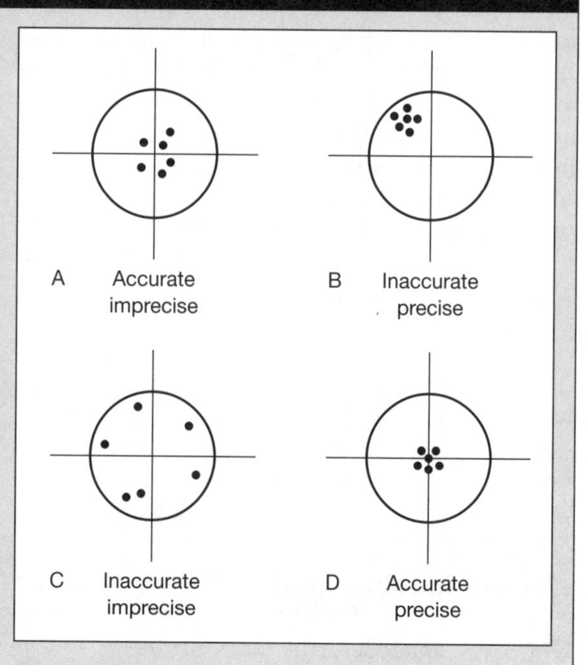

Figure 10.1 Accuracy versus precision

tion may influence data quality are given in Box 10.2. These cartographic practices, whilst understandably necessary for clear map production, cause problems when using paper maps as data sources for GIS. Positional inaccuracies are introduced through generalization of real-world detail, representation of area features as points, exaggerated line thickness and lateral dislocation of adjacent features.

There are further data quality considerations if data sets are to be used in a GIS. *Completeness*, *compatibility*, *consistency* and *applicability* are introduced here.

A complete data set will cover the study area and the time period of interest in its entirety. The data are complete spatially and temporally, and should have a complete set of attribute information. Completeness of polygon data is relatively easy to determine. Errors in attribute data might be seen as polygons lacking attribute information, and errors in spatial data may be seen as polygons

BOX 10.2

The influence of cartographic generalization on data quality

1. A ski piste has two distinct edges and follows a sinuous course. On a map, there is limited space to represent this. The cartographer may opt to draw the piste as a single blue line without many of the smaller bends, while still retaining the general trend of its course, thus producing a generalized representation.
2. Area features on a map may be generalized to points. This extreme form of generalization is common in small-scale maps, where important features such as towns and villages can be too small to be drawn as area features. They are effectively reduced to a single annotated point to communicate their name and location.
3. The cartographer may draw features larger than they actually are in order to emphasize their importance. This is useful to highlight important communication routes. For instance, a road is drawn on a 1:50,000 map as a broad red line approximately 0.8 mm wide. If we multiply this by the map scale, it suggests the road is 40 metres wide, whereas in reality it is in the order of 10 metres wide.
4. Linear features that run close together, such as a road running beside a river, may have the distance separating them on the map considerably exaggerated to avoid drawing them on top of each other.
5. Scale is very important in determining the level of generalization present on a map. This can be illustrated by considering the cartographic portrayal of line and area features at various map scales. In Figure 10.2 the generalization of a village is shown at three different scales. At 1:10,000, the village is drawn as a series of points, lines and polygons representing individual features such as trees, fences and buildings. At 1:50,000, the village may be reduced to a few lines and polygons representing the roads and main groups of buildings. At 1:500,000, the village has become a single point at the intersection of a few main roads.

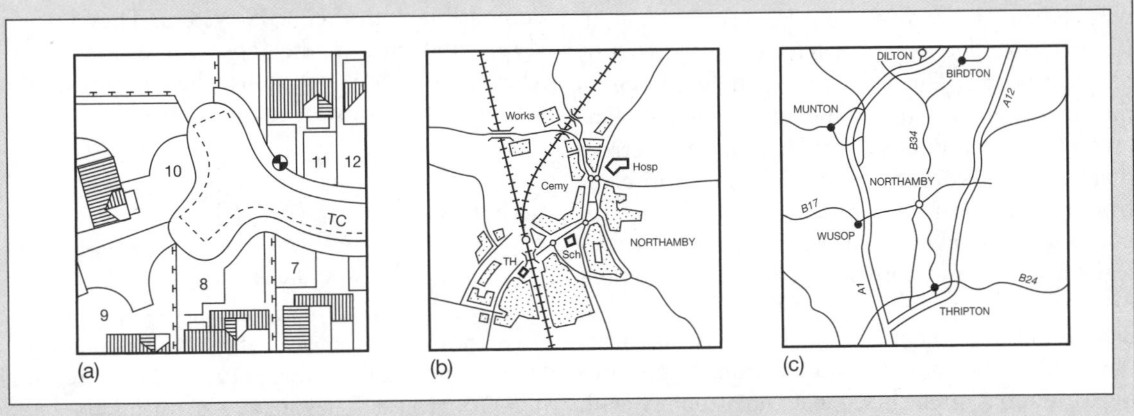

Figure 10.2 Scale-related generalization: (a) 1:10,000; (b) 1:50,000; (c) 1:500,000

with two sets of attribute data. Completeness of line and point data is less obvious as features may be missing without showing up in the database. The only way to check is by comparison with another source. Sample data, whether taken via a random, regular or systematic survey, are by definition incomplete. In this instance, completeness is a matter of having a sufficiently large sample size to represent adequately the variation in the real world. In the case of time series data, completeness is very difficult to define, let alone assess. Almost all time series data in GIS are referenced to discrete moments in time. However, time is continuous. A set of maps relating to a series of discrete times is therefore incomplete, and requires broad assumptions to be made regarding change in the intervening periods.

Compatible data sets can be used together sensibly. With GIS it is possible to overlay two maps, one originally mapped at a scale of 1:500,000 and the other at 1:25,000. The result, however, is largely worthless because of incompatibility between the scales of the source documents. Maps containing data measured in different scales of measurement cannot be combined easily. Overlaying a map of snow depths in Happy Valley (in metres – ratio scale) with a map of piste and non-piste areas (binary classification – nominal scale) using a union operation would produce a meaningless output map.

To ensure compatibility, ideally data sets should be developed using similar methods of data capture, storage, manipulation and editing. Consistency applies not only to separate data sets but also within individual data sets. Inconsistencies can occur within data sets where sections have come from different source documents or have been digitized by different people. This will cause spatial variation in the error characteristics of the final data layer. One area of the final data set may contain more errors than another. Problems of inconsistency also come from the manner in which the data were collected. For example, in Happy Valley there are meteorological stations that record snowfall. The equipment used to measure snowfall is of different ages and designs and thus the accuracy of the measurements obtained varies from station to station. If meteorological stations are included in a GIS as points, with snowfall records as their attributes, inconsistencies will be introduced to the database. Another example is administrative boundaries. These are subject to constant revision as a result of international disputes and internal reorganization, thus positional inconsistencies may occur where boundary revisions have taken place.

Applicability is a term used to describe the appropriateness or suitability of data for a set of commands, operations or analyses. For example, interpolating height data using Thiessen polygons (Chapter 6) may be an unsuitable match between data and technique, as height data vary continuously (not abruptly as assumed by the Thiessen polygon method). Alternatively, applicability can be used to describe the suitability of data to solve a particular problem. It is inappropriate to use height data to estimate some socio-economic variables, such as car ownership, but height data may be applicable to estimating the number of days in a year with snow cover.

The criteria introduced above provide some pointers for gauging the overall quality of GIS databases. However, in the same way that different terms can be interpreted in different ways, the quality criteria regarded as important by data creators and users are likely to be different. Suppliers of digital topographic maps to the Happy Valley GIS team may regard their data as of high quality if they are complete and up to date at a range of scales, and if they deliver a known and measurable degree of accuracy. Their reputation depends on their ability to supply a product of consistent standard. However, different criteria may be important for the data users. The GIS team may require precise data on particular themes, and data that are compatible with their GIS system. They want all slopes and cliffs to be represented consistently. They may be unwilling to use the data for decision support if they know it contains errors.

Sources of error in GIS

Having the terminology to describe flaws in data is a good first step to providing quality GIS. Awareness of the sources of problems comes next. Spatial and attribute errors can occur at any stage in a GIS project. They may arise during the definition of spatial entities, from the representation of these entities in the computer, or from the use of data in analysis. In addition, they may be present in source data, arise during conversion of data to digital format, occur during data manipulations and processing, or even be produced during the presentation of results. The best way to explore how and where errors may arise is to view them within the context of a typical GIS project.

Errors arising from our understanding and modelling of reality

Some of the errors arising in GIS have a source far away from the computer. Errors can originate from the ways in which we perceive, study and model reality. These errors can be termed conceptual errors, since they are associated with the representation of the real world for study and communication.

The different ways in which people perceive reality can have effects on how they model the world using GIS. Mental mapping (Box 10.3) provides an illustration of how varying perceptions can manifest themselves in a geographical context.

Our perception of reality influences our definition of reality, and in turn our use of spatial data. This can create real errors and often gives rise to inconsistencies between data collected by different

BOX 10.3

Perceptions of reality – mental mapping

Your mental map of your home town or local area is your personal, internalized representation that you use to help you navigate on a day-to-day basis and describe locations to other people. You could commit this to paper as a sketch map if required. Figure 10.3 presents two sketch maps describing the location of Northallerton in the UK. They would be flawed as cartographic products, but nevertheless are a valid model of reality. Keates (1982) considers these maps to be personal, fragmentary, incomplete and frequently erroneous. Consider the mental maps that a group of residents and visitors in Happy Valley might draw. The residents might stress their homes, place of work and community facilities. Visitors would include their hotel, ski school, ski runs and the après ski facilities. In all cases, distance and shape would be distorted in inconsistent ways and the final product would be influenced by the personality and experience of the 'cartographer'.

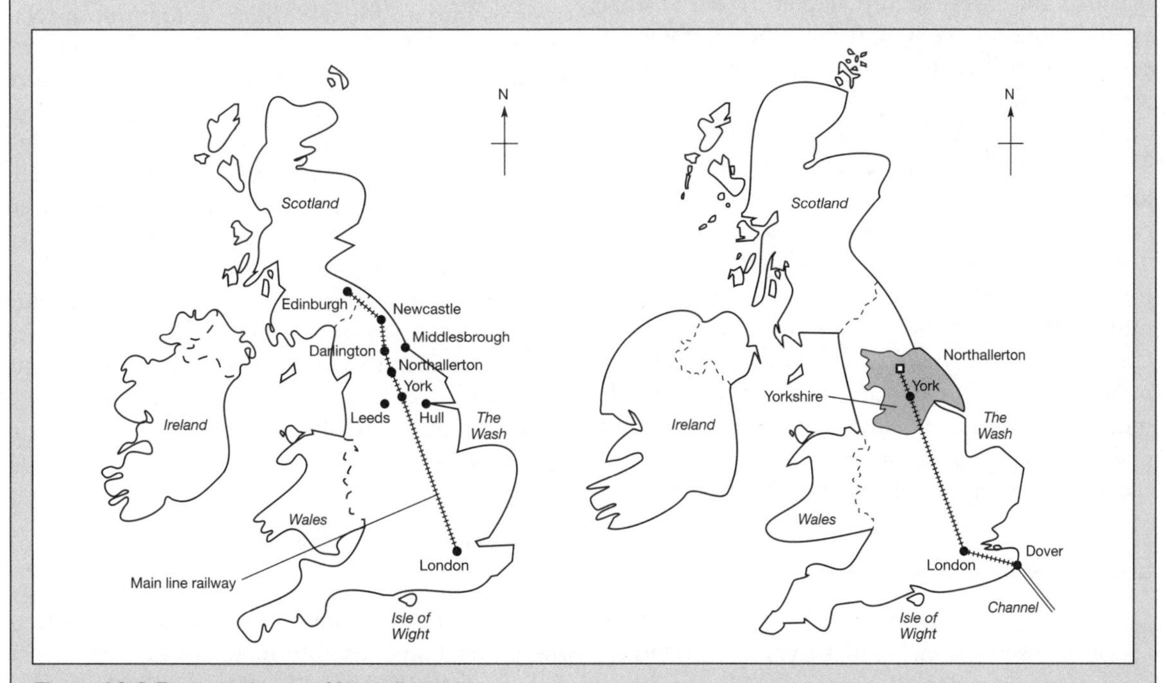

Figure 10.3 Two mental maps of Northallerton

surveyors, maps drawn by different cartographers and databases created by different GIS users.

Different scientific disciplines have developed their own particular ways of explaining and studying reality. Compare the reductionist approach of the physical sciences, such as physics and chemistry, with the wider, more holistic approaches of the environmental sciences such as geography and ecology. The reductionist view attempts to explain and model parts of the world system by isolating individual physical, chemical and biological processes at a micro scale, whereas the holistic view attempts an explanation on the basis of related and interdependent macro- and meso-scale systems. For example, a biologist is more likely to be interested in photosynthesis as a biochemical process, taking place at the level of an individual plant cell or leaf. An ecologist is no less interested in the process of photosynthesis but takes a much wider view of its role. Given such different approaches to science it is understandable that difficulties (such as bias and compatibility) arise when different people's models of space and process are compared.

In GIS we use spatial models to reflect reality. The main models in use are raster, vector, object-oriented and layer-based (Chapter 3). All of these spatial models have limitations when it comes to portraying reality. For instance, the raster model assumes that all real-world features can be represented as individual cells or collections of cells. This is clearly not the case. The vector model assumes that all features can be given a single co-ordinate or a collection of Cartesian co-ordinates. The world is actually made up of physical and biological material, which is, in turn, made up of molecular and sub-molecular matter grouped into complex systems linked by flows of energy and materials (solids, liquids and gases). Whatever GIS model we adopt, it is a simplification of this reality, and any simplification of reality will include errors of generalization, completeness and consistency.

Errors in source data for GIS

Our models of reality in GIS are built from a variety of data sources including survey data, remotely sensed and map data (Chapter 2). All sources of spatial and attribute data for GIS are likely to include errors. Three examples of data sources are examined below to identify possible errors.

Survey data can contain errors due to mistakes made by people operating equipment or recording observations, or due to technical problems with equipment. For example, if sites of damage to ski pistes were being recorded using a GPS, errors in spatial data might occur if there were technical problems with the receiver. Errors in attribute data could occur if features were recorded incorrectly by the operator. A visitor survey of holidaymakers in Happy Valley could include spatial errors if postal codes for the visitors' home addresses were wrongly remembered or recorded. Attribute errors would occur if the characteristics of the respondents were wrongly allocated, or incorrectly noted.

Remotely sensed and aerial photography data could have spatial errors if they were spatially referenced wrongly, and mistakes in classification and interpretation would create attribute errors. It should also be remembered that these data sources are 'snapshots' of reality, showing the location and nature of features at a particular moment in time. Thus, for any changeable features the date of the image used is extremely important.

Maps are probably the most frequently used sources of data for GIS. Maps contain both relatively straightforward spatial and attribute errors, caused by human or equipment failings, and more subtle errors, introduced as a result of the cartographic techniques employed in the map-making process. Generalization is one cartographic technique that may introduce errors and this has been discussed earlier in this chapter. There are other cartographic difficulties, some of which affect other types of data as well as maps. These include the representation of continuous data and features with indistinct (or fuzzy) boundaries. Problems may also be introduced because of sampling and measurement difficulties. These are explained in Box 10.4.

Errors in data encoding

Data encoding is the process by which data are transferred from some non-GIS source, such as the paper map, satellite image or survey, into a GIS format (Chapter 5). The method of data encoding, and the conditions under which it is

BOX 10.4

Examples of errors from cartographic data sources

Continuous data

Continuous data require special cartographic representation. For example, height is usually represented by contours on paper maps. Annotated lines indicate land of a particular height above a datum (usually sea level). Limited additional information is provided by spot heights. Apart from this, no information is given about the height of the land between contours, except that it lies somewhere in the range denoted by the bounding contours. Since height varies continually over space, the vector-based contour and spot height model is always a compromise for the representation of height data.

Fuzzy boundaries

Where changes between entities are gradual in nature, such as between soil types, vegetation zones or rural and urban areas of settlement, indistinct boundaries exist. These 'fuzzy' boundaries are difficult to represent cartographically. As a result, features are often made distinct by drawing sharp boundaries around entities. This will ultimately lead to positional errors and attribute uncertainty in digital map data.

Map scale

Whether the 'fuzziness' of natural boundaries is important depends largely on the scale at which the data are mapped and viewed. For instance, if a soil boundary is mapped at a large scale (for example 1:1250) then the representation of a fuzzy boundary can result in a significant amount of error. However, if the boundary is mapped at a much smaller scale (such as 1:25,000) then the error resulting from the representation, although still there, is no longer so significant. The thickness of the line used to draw the boundary on the map at this scale may well account for much of the fuzziness between the soil types on the ground.

Map measurements

Measurements on maps and other data sources are rarely 100 per cent accurate. Generalization aside, measurements such as contour heights, feature locations and distances have limited accuracy because of human and mechanical error. Human measurement error can result from personal bias (for example, in the interpretation of aerial photographs), typographical errors, equipment reading errors and mistakes in the use of equipment. Examples of machine error may include limitations of surveying equipment, limited accuracy of GPS receivers and equipment calibration errors.

carried out, are perhaps the greatest source of error in most GIS. Digitizing, both manual and automatic, is an important method of data entry, so this section will focus on errors that can arise during digitizing.

Despite the availability of hardware for automatic conversion of paper maps into digital form, much of the digitizing of paper maps is still done using a manual digitizing table (Chapter 5). Manual digitizing is recognized by researchers as one of the main sources of error in GIS (Otawa, 1987; Keefer *et al.*, 1988; Dunn *et al.*, 1990); however, digitizing error is often largely ignored. Beyond simple checks in the editing process, practical means of handling digitizing errors do not exist in proprietary GIS software.

Sources of error within the digitizing process are many, but may be broken down into two main types: source map error and operational error. Source map errors include those discussed in the previous section and in Box 10.4. Operational errors are those introduced and propagated during the digitizing process. Human operators can compound errors present in an original map (Poiker, 1982) and add their own distinctive error signature. Some of the operational errors that can be introduced during manual digitizing are outlined in Box 10.5.

The errors outlined in Box 10.5 occur during the digitizing process. However, errors can arise earlier. For example, before starting to digitize, the map needs to be placed on the digitizing table and registered to enable later conversion from digitizer to geographic co-ordinates. If this is done incorrectly or carelessly then the whole set of co-ordinates digitized during that session

BOX 10.5

Operational errors introduced during manual digitizing

Following the exact course of a line on a map with a digitizer cursor is a difficult task, requiring skill, patience and concentration. Hand wobble, a tendency to under- or overshoot lines, strain and lack of hand–eye co-ordination are all sources of human error in the digitizing process. Jenks (1981) categorizes human digitizing error into two types: psychological and physiological. To these can be added errors resulting from problems with line thickness and the method of digitizing which is used.

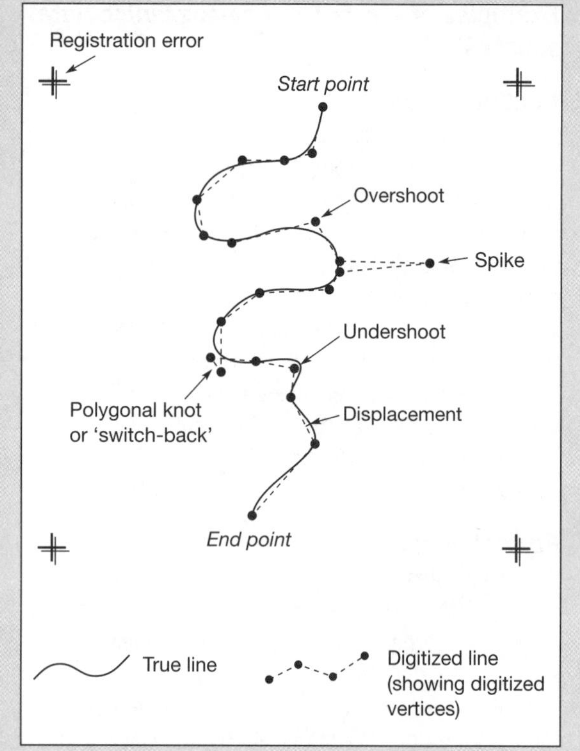

Figure 10.4 Digitizing errors

1. *Psychological errors.* Psychological errors include difficulties in perceiving the true centre of the line being digitized and the inability to move the cursor cross-hairs accurately along it. This type of error results in lateral offset between the true and digitized line and under- or overshoots at corners. Under- and overshoots and lateral displacement of features can be difficult to spot and correct.
2. *Physiological errors.* Physiological errors result from involuntary muscle spasms (twitches and jerks) that give rise to random displacements or 'spikes' and polygonal knots or 'switchbacks' (Figure 10.4). Some of these errors, such as spikes and knots, are easy to spot and correct.
3. *Line thickness.* The thickness of lines on a map is determined by the cartographic generalization employed. All lines are drawn so as to be visible to the user and thickness may in some instances reflect the perceived importance of the feature being depicted rather than its actual size on the ground. A major road may be shown on a map by a line wider than the cross-hairs on the digitizer cursor. Common sense suggests that the true course of the line is along its mid-point, but it is a difficult task for the operator to follow the centre of the line exactly. Some displacement of the cursor either side of the centre line is inevitable, leading to positional errors in the resulting data.
4. *Method of digitizing.* There are two basic methods of manual digitizing: point mode and stream mode (Chapter 5). Point mode digitizing allows careful selection of sample points to create a faithful representation of a line. The method involves a certain amount of generalization by the operator, who decides both the number and location of sample points. The greater the sample size, the greater the detail, but the positional error associated with each point remains much the same. In stream mode digitizing, sampling is determined by the frequency set by the user at the start of the digitizing session. The number of points sampled is controlled more by line complexity, since the operator can move relatively quickly along straight sections but needs to slow down significantly to navigate the cursor along more convoluted sections.

will contain the same error or bias. Care when setting up the map document on the digitizer is therefore essential.

Automatic digitizing, like manual digitizing, requires correct registration of the map document before digitizing commences, but there the similarity ends. By far the most common method of automatic digitizing is the use of a raster scanner. This input device suffers from the same problems regarding resolution as the raster data model. Cell size is important. The smaller the cell size, the greater the resolution of the resulting image and the greater the quantity of the data produced. Cell size is determined by the resolution of the machine being used; a fine resolution may produce data with less positional error than manual digitizer output. Some of the other practical problems faced when using an automatic digitizer are discussed in Chapter 5.

Data input by either manual or automatic digitizing will almost always require editing and cleaning. This is the next source of errors in GIS.

Errors in data editing and conversion

After data encoding is complete, cleaning and editing are almost always required. These procedures are the last line of defence against errors before the data are used for analysis. Of course, it is impossible to spot and remove all the errors, but many problems can be eliminated by careful scrutiny of the data. As a potential source of error, cleaning and editing do not rank high; they are positive processes. Some GIS provide a suite of automated procedures, which can lead to unexpected results. For example, it is common for vector GIS software to contain routines to check and build topology. The closing of 'open' polygons and the removal of 'dangling' lines (overshoots) are controlled by tolerances set by the user. If the tolerances are set too low, then the gaps in polygons will not be closed and overshoots will not be removed. Alternatively, if the tolerances are set too high, small polygons may disappear altogether and large dangling lines, which may be required on networks, may also disappear (Figure 10.5a).

A different problem occurs when automated techniques are used to clean raster data. The main problem requiring attention is 'noise' – the misclassification of cells. Noise can be easy to spot where it produces a regular pattern, such as striping. At other times it may be more difficult to identify as it occurs as randomly scattered cells. These noise errors can be rectified by filtering the raster data to reclassify single cells or small groups of cells by matching them with general trends in the data. The 'noisy' cells are given the same value as their neighbouring cells. Filtering methods for this process are discussed in Chapter 6. Choosing an appropriate method is important as the wrong method may remove genuine variation in the data or retain too much of the noise.

After cleaning and editing data it may be necessary to convert the data from vector to raster or vice versa. During vector to raster conversion both the size of the raster and the method of rasterization used have important implications for positional error and, in some cases, attribute uncertainty. The smaller the cell size, the greater the precision of the resulting data. Finer raster sizes can trace the path of a line more precisely and therefore help to reduce classification error – a form of attribute error. Positional and attribute errors as a result of generalization are seen as classification error in cells along the vector polygon boundary. This is seen visually as the 'stepped' appearance of the raster version when compared with the vector original (Figure 10.6). Box 10.6 provides further examples of errors created during the vector to raster conversion process. The results of this process are often more generalized data with more uncertainty attached. The conversion of data from raster to vector format is largely a question of geometric conversion; however, certain topological ambiguities can occur – such as where differently coded raster cells join at corners as in Figure 10.5b. In this case it is impossible to say, without returning to the original source data, whether the vector polygons should join. Where vector maps have been derived from raster data, conversion may result in a stepped appearance on the output map. This can be reduced, to some extent, by line-smoothing algorithms (Chapter 5), but these make certain assumptions about topological relationships and detail that may not be present in the raster source data.

Most GIS packages provide data conversion software so that their files can be read by other GIS. For example, a GIS program may not be able to read data from a CAD package, so a conversion program is supplied to convert to another format,

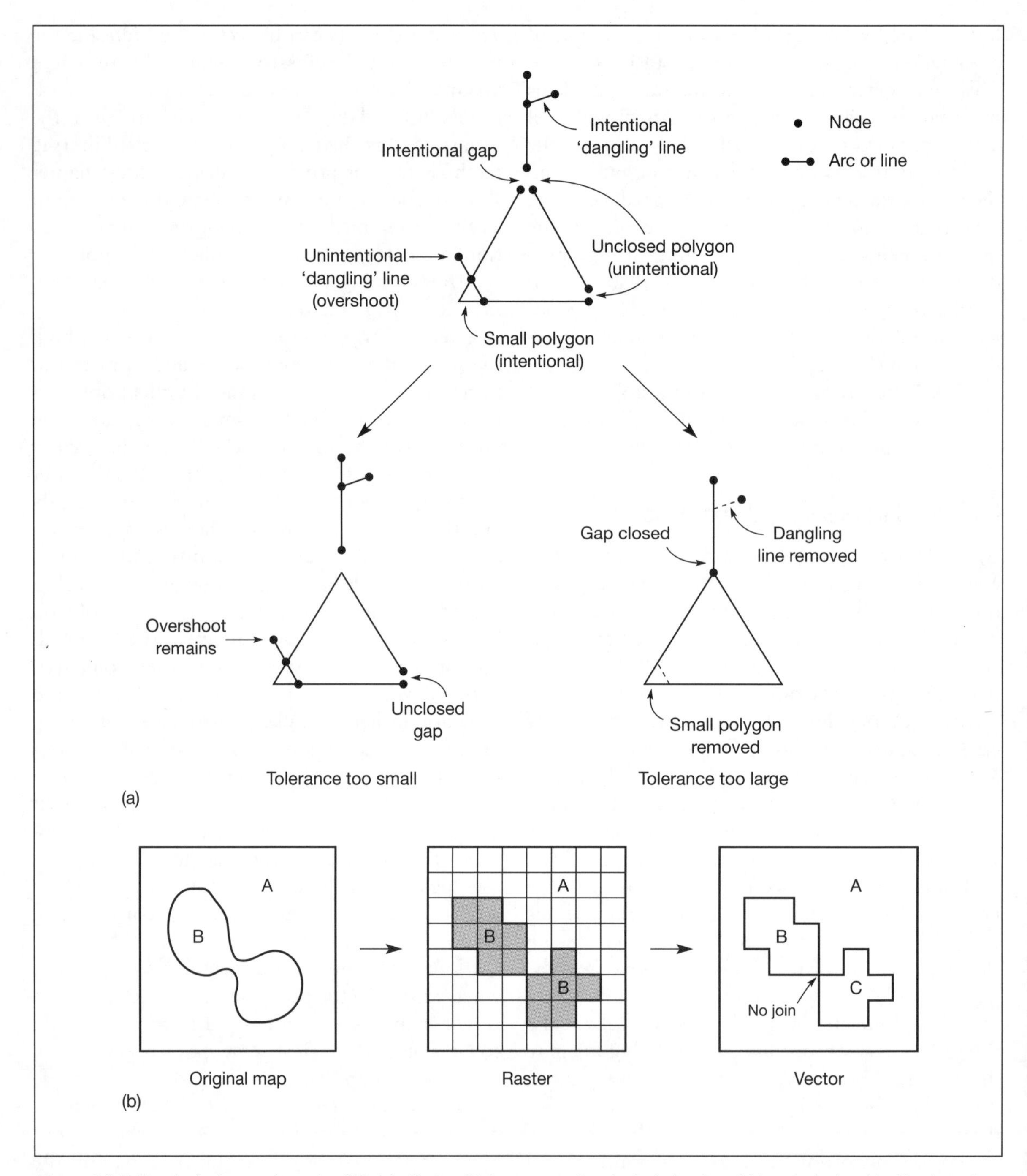

Figure 10.5 Topological errors in vector GIS: (a) effects of tolerances on topological cleaning; (b) topological ambiguities in raster to vector conversion

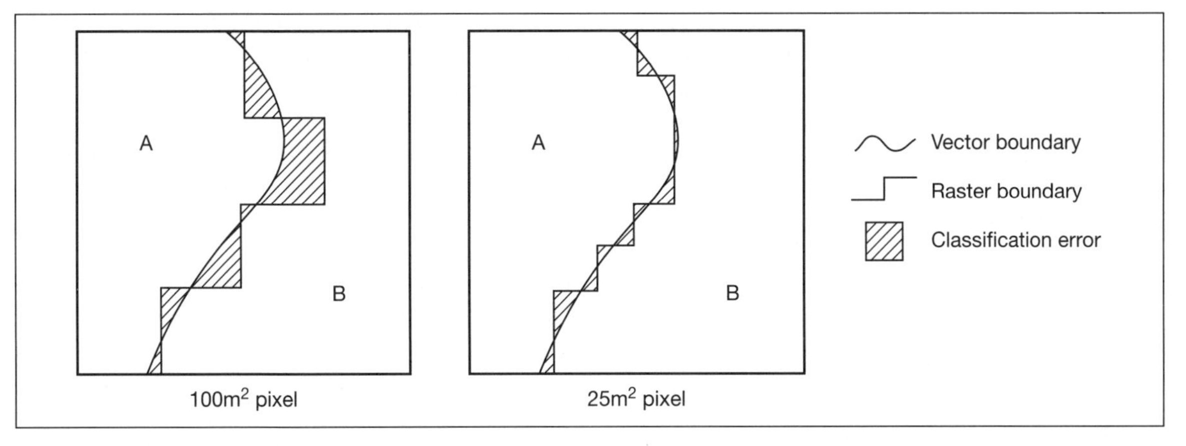

Figure 10.6 Vector to raster classification error

BOX 10.6

Rasterization errors

Vector to raster conversion can cause an interesting assortment of errors in the resulting data. For example:

1. *Topological errors.* These include loss of connectivity, creation of false connectivity (Figure 10.7a), loss of information, where individual vector polygons are smaller than the chosen raster cell size, and ambiguities due to changes in raster orientation and datum. Connectivity problems occur when rasterizing vector polygons separated by narrow gaps or connected via long narrow corridors. Significant features such as these are easily lost if they are narrower than an individual raster cell.
2. *Loss of small polygons.* When individual vector polygons are less than half the area of the raster cell or small enough to be missed by adjacent cell centroids, they may be lost in the conversion process.
3. *Effects of grid orientation.* If two identical grids are placed over the same vector map but at different angles (such that the difference between the angles is not a multiple of 90), the resulting raster maps will be different from each other. It is usual when rasterizing vector data for the grid to be oriented parallel to the co-ordinate system of the source data. However, certain circumstances may dictate that a different orientation is required (such as when integrating rasterized vector data with satellite imagery where the satellite track is not parallel to the vector co-ordinate system or when using data digitized on a different projection system).
4. *Variations in grid origin and datum.* Using the same grid but with slightly different origins (where the difference is not a multiple of raster size) will result in two different raster maps in much the same fashion as with grid orientation.

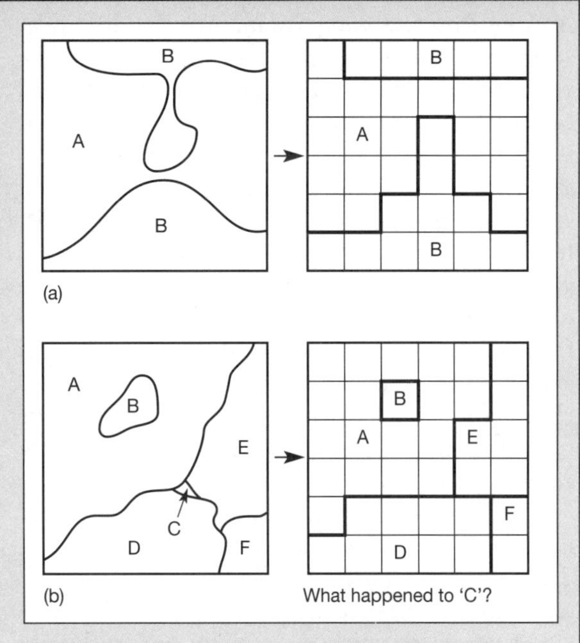

Figure 10.7 Topological errors in vector GIS: (a) loss of connectivity and creation of false connectivity, (b) loss of information

such as DXF, which both can handle. Users naturally assume that these conversion routines work without introducing new errors into the data. However, the transfer of a GIS database from one software package to another can lead to errors through changes in format and definitions. These database transfer errors are a kind of 'mechanical' error that is often overlooked. To overcome this problem standard formats have been developed.

Errors in data processing and analysis

Errors may be introduced during the manipulation and analysis of the GIS database. GIS users must ask themselves questions before initiating a GIS analysis. For example: Are the data suitable for this analysis? Are they in a suitable format? Are the data sets compatible? Are the data relevant? Will the output mean anything? Is the proposed technique appropriate to the desired output? These questions may seem obvious but there are many examples of inappropriate analysis. These include the inappropriate phrasing of spatial queries, overlaying maps which have different co-ordinate systems, combining maps which have attributes measured in incompatible units, using maps together that have been derived from source data of widely different map scales and using an exact and abrupt method of interpolation (such as Thiessen polygons) to interpolate approximate and gradual point data (height data collected using GPS).

GIS operations that can introduce errors include the classification of data, aggregation or disaggregation of area data and the integration of data using overlay techniques.

The way in which we classify and reclassify data affects what we see and therefore how we interpret the data. Choropleth mapping is an example (Chapter 8). Much has been written in the cartographic literature on how the definition of class intervals in this sort of map can radically affect the pattern produced (Monmonier, 1991). Take, for example, a choropleth map of unemployment by region drawn from census data. How the class intervals are defined can affect perception of unemployment across the country. Similarly, how the areas are shaded also influences what we see in terms of pattern and magnitude. Dark colours in this instance can be used for the top two classes to create a more depressing feel.

Classification errors also affect raster data. Classified satellite images provide a reflectance value for each pixel within a specific wavelength range or spectral band (for example, red, near infrared or microwave). Raster maps of environmental variables, such as surface cover type, are derived by classifying each pixel in the image according to typical reflectance values for the range of individual cover types present in the image. Error can occur where different land cover types have similar reflectance values (for example, asphalt car parks may have similar reflectance to mature spruce forest) and where shadows cast by terrain, trees or buildings reduce the reflectance value of the surface. Careful choice of classification method can help to reduce this type of error.

Where a certain level of spatial resolution or a certain set of polygon boundaries are required, data sets that are not mapped with these may need to be *aggregated* or *disaggregated* to the required level. This is not a problem if the data need to be aggregated from smaller areas into larger areas, provided that the smaller areas nest hierarchically into the larger areas. Problems with error do occur, however, if we wish to disaggregate our data into smaller areas or aggregate into larger non-hierarchical units. The information required to decide how the attribute data associated with the available units aggregates into the larger but non-nested units or disaggregates into a set of smaller units, rarely exists. This problem is normally approached by assuming that the data in the source units are evenly distributed by area (for example, population within an enumeration district is evenly distributed across its total area). The data for the new set of units or areas are in this case simply calculated as a function of the area of overlap with the source units. However, such a homogeneous distribution of data within the source units cannot always be correctly assumed. In the case of population, distribution will depend on urban morphology, which can vary over small areas. For example, even population distribution may be assumed across an area of similar housing type, but in an area of mixed land use (for example, housing, parkland and industrial estates) it cannot. Aggregation, disaggregation and classification errors are all combined in the modifiable areal unit problem (MAUP) (Chapter 6).

Error arising from map overlay in GIS is a major concern and has correspondingly received much attention in the GIS literature (Chrisman, 1989; Goodchild and Gopal, 1989; Openshaw, 1989). This is primarily because much of the analysis performed using GIS consists of the overlay of categorical maps (where the data are presented in a a series of categories). GIS allows the quantitative treatment of these data (for example, surface interpolation or spatial autocorrelation), which may be inappropriate. Map overlay in GIS uses positional information to construct new zones from input map layers using Boolean logic or 'mapematics' (Chapter 6). Consequently, positional and attribute errors present in the input map layers will be transferred to the output map, together with additional error introduced by multiplicatory effects and other internal sources. Data output from a map overlay procedure are only as good as the worst data input to the process.

Perhaps the most visual effect of positional error in vector map overlay is the generation of sliver polygons. Slivers (or 'weird' polygons) occur when two maps containing common boundaries are overlaid. If the common boundaries in the two separate maps have been digitized separately, the co-ordinates defining the boundaries may be slightly different as a result of digitizing error. When the maps are overlaid a series of small, thin polygons will be formed where the common boundaries overlap (figure 10.8). Slivers may also be produced when maps from two different scales are overlaid. Of course, sliver polygons can and do occur by chance, but genuine sliver polygons are relatively easy to spot by their location along common boundaries and their physical arrangement as long thin polygonal chains.

Also shown in Figure 10.8 is the attribute error that can result from positional error. If one of the two maps that are overlaid contains an error (such as the missing 'hole' feature in map A that is present as an island in map B), then a classification error will result in the composite map (polygon BA). Chrisman (1989) identifies these two cases as extremes and points out that one type of error (positional or attribute) will inevitably result in the other type of error in the output map and vice versa. Classification errors occur along the common boundary where the slivers are produced as a result of positional error and attribute error.

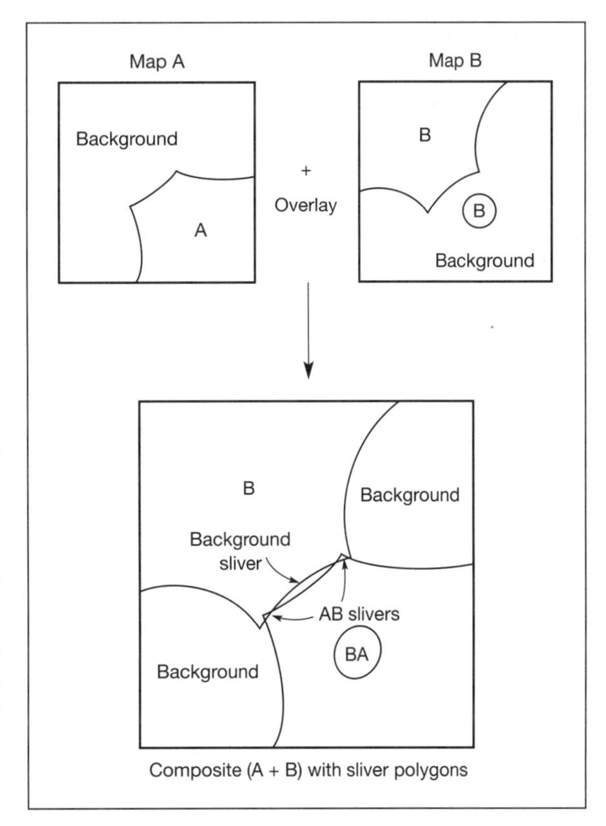

Figure 10.8 Generation of sliver polygons

Errors in data output

From the preceding discussion it should be clear that all GIS databases will contain error. In addition, further errors will be introduced during manipulation and analysis of the data. Therefore, it is inevitable that all GIS output, whether in the form of a paper map or a digital database, will contain inaccuracies. The extent of these inaccuracies will depend on the care and attention paid during the construction, manipulation and analysis of the databases. It is also possible that errors can be introduced when preparing GIS output. For example, if we choose the map as a means of communicating the results of our analysis, it may be necessary to generalize data for the sake of clarity. This will add errors and assumptions into the database.

Finding and modelling errors in GIS

Much research effort has focused on identifying, tracking and reducing error in GIS. One American GIS research organization, the National Center for Geographic Information and Analysis, even placed the problem of error in GIS first on its list of research priorities (NCGIA, 1989). Despite this research effort, little has been done to incorporate error identification, modelling and handling techniques within proprietary GIS packages. This has had the effect of keeping awareness of the general error problem at a low level amongst non-specialist GIS users. However, there are methods for detecting and modelling errors which users can employ to improve confidence in their data and results. Methods for checking data include visual inspection, double digitizing, the examination of error signatures and statistical analysis.

Checking for errors

Probably the simplest means of checking for data errors is by *visual inspection*. This should be included as a matter of course in the data cleaning and editing process. Once in a GIS format, data can be plotted on paper or on the graphics screen for visual comparison with the original source document. Obvious oversights in the digitizing process and missing data will be revealed and can be amended. The attributes of spatial features can be checked by adding annotations, line colours and patterns. Again, comparison with the original map should reveal major errors.

Double digitizing is an error-checking method used by large digitizing operations (for example map producers such as the UK Ordnance Survey). The same map is digitized twice (possibly by different people) and the resulting copies compared to identify inconsistencies. Polygon areas, length of lines, polygon perimeters and the number of sliver polygons produced are calculated and compared. If the two versions are deemed significantly dissimilar, then the map may be re-digitized and the comparison repeated. However, this is a costly and time-consuming method of error checking.

Work on the psychological and physiological aspects of digitizing (Box 10.5) suggests that each operator has a different error signature that describes how they digitize. An operator may, for example, consistently over- or undershoot changes in line direction or show lateral displacement relative to position on the digitizing table. If different digitizer operators do have recognizable and repeatable error signatures, these could be used to effectively cancel out the errors introduced in digitizing.

Various statistical methods can be employed to help pinpoint potential errors. Attribute errors in GIS data sets may stand out as extreme data points in the distribution of data values; or outliers in scatter plots if plotted against a correlated variable (for example, total annual rainfall against number of rain days). A statistical measure of the error attributed to generalization and line thickness can be gained by estimating the total area of the map covered by lines. An example given by Burrough (1986) shows that a 1:25,000 scale soil map measuring 400 × 600 mm may have as much as 24,000 mm of drawn lines, covering 10 per cent of the map's total area.

Error modelling

Two approaches to the modelling of errors in spatial data are epsilon modelling and Monte Carlo simulation. Epsilon modelling is based on an old method of line generalization developed by Perkal (1956). Blakemore (1984) adapted this idea by using the model to define an error band about a digitized line that described the probable distribution of digitizing error about the true line. The shape of a graph of the error distribution is a matter of debate, but the graph is probably a bell-shaped curve. A true normal distribution is unlikely since very large errors represented by the tails of the distribution are rare and easily spotted and removed in the data cleaning and editing process. Some authors believe a 'piecewise quartic distribution' to be most realistic. This has a flatter central peak and tails that quickly run out to zero (Brunsdon *et al.*, 1990). Others have suggested a bimodal distribution on the grounds that digitized points are very unlikely to fall exactly on the line.

In an empirical example based on overlaying employment offices (points) with employment office areas (polygons), Blakemore (1984) found that he could unambiguously assign only 60 per cent of the points to any polygon. The remaining 40 per cent were found to lie within the specified epsilon band width for the employment office area polygons. From this work Blakemore defined the five different categories of containment illustrated in Figure 10.9.

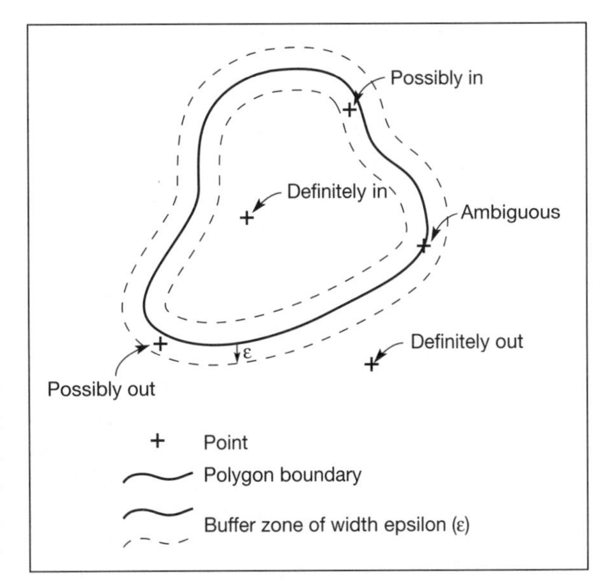

Figure 10.9 Point-in-polygon categories of containment. (After Blakemore, 1984)

The epsilon model has been developed in various ways and with various error distributions to produce a seemingly robust means of error modelling. Given information on the distribution of error in the input data layers, this technique can be used to handle error in GIS operations by allowing for the epsilon band in GIS analyses to improve the confidence of the user in the results (Carver, 1991a).

A Monte Carlo simulation approach has also been used to model the effects of error in GIS overlays (Openshaw *et al.*, 1991; Fisher, 1991). This method simulates the effect of input data error by the addition of random 'noise' to the line co-ordinates in map data. Each input data source is assumed to be characterized by an error model that represents reasonable estimates of positional error thought to be present in a similar manner to the epsilon approach. The random 'noise' added to the map co-ordinates simulates the combined effects of source map and digitizing errors by slightly varying the course of the original shape of the line. The randomized data are then processed through the required series of GIS operations and the results saved. This entire process is repeated many times. If the output is numeric, then the distribution of the results gives some indication of the effects of input data error. If the output is in the form of a map, then the total set of saved maps can be used to draw confidence intervals. These can be overlaid on the normal, unvaried results as a visual means of communicating the likely effects of input data error. The main drawback with this method is the large number of simulations required to achieve a statistically valid result. For GIS processes where the effects of error need to be estimated, up to 100 extra sets of processing may be required. Monte Carlo simulation is only likely to be used for very important analyses requiring a high level of confidence in the results. Although initial research on Monte Carlo simulation focused on its use for modelling positional error in vector map overlay (Openshaw *et al.*, 1991), the approach is equally applicable to modelling attribute or positional error in raster or vector GIS operations.

Managing GIS error

We must accept that no matter how careful we are in the preparation of our data and how cautious we are in our choice of analysis, errors will find their way into the GIS database. To manage these errors we must track and document them. The data quality parameters discussed earlier in this chapter – accuracy, precision, resolution, generalization, applicability, bias, compatibility, completeness and consistency – provide a useful checklist of quality indicators. Documenting these for each data layer in a GIS, and for any new layers produced by analysis, will help subsequent users of the GIS to keep track of data quality. Data quality information is frequently used to help construct a data *lineage*.

Lineage is a record of data history that presents essential information about the development of data from their source to their present format. Lineage information should provide the data user with details of their source, method of data capture, data model, stages of transformation, editing and manipulation, known errors and software and hardware used. In a way, lineage is like the service history of the car in our Martian example. A good knowledge of lineage gives the user a better feel for the data and is indicative of its overall quality. Above all (as in the car's service history) it lets the user know where problems with the data are likely to occur. Unfortunately, just as with our car example, full service histories are something of a rarity in GIS. Depending on the context, documentary evidence may be available. Some digital data suppliers provide good lineage information in their accompanying materials. Users are less likely to maintain good lineage records. They are not helped by the fact that few GIS packages provide any means of storing lineage information.

Together, lineage and consistency give an indication as to the *repeatability* of the methods being used. Could the same set of results from a GIS analysis be obtained if the data were reconstructed from scratch using the same methods and processes as described in the lineage? If consistency in the data sources, methods of capture and data processing is low it is unlikely that the results will be the same. Inevitably this leads to a lack of confidence in GIS data and analysis. Thus, it is essential that all possible means are used to ensure as high a level of consistency and overall data quality as is practical. This is especially important in those cases where the results of a GIS analysis are to be used for real-life decision making where lives, property and money are at stake.

A number of organizations involved in the creation, processing and use of digital map data have attempted formalization of lineage requirements. Both the Association for Geographical Information (AGI) in the UK and the National Committee on Digital Cartographic Data Standards (NCDCDS) in the USA define basic lineage requirements (see Box 10.7).

Lineage information can be recorded manually or automatically. The user may manually create the lineage by recording the information needed directly on paper or in a computer text file. A paper-based scheme involves the user keeping a record in a log book or card system as operations are carried out on the data. Wholly computerized systems of manually recording lineage are perhaps the best option. This is possible if a database system into which the user records lineage information runs concurrently with the GIS. Automatic systems remove the need for user input altogether. These are rare in GIS, but some GIS do produce

BOX 10.7

Basic lineage requirements

1. *Description of source data.* This should contain information relevant to the data source, including its name, date and method of production, date of last modification, producer, reference, map scale and projection. It should also include the necessary backward links so that the user can see how the source data were generated.
2. *Transformation documentation.* Transformation information contains details of the operations and actions that have led to the production of the data in their present form. This includes the commands, parameters and context of use.
3. *Input/output specifications.* This records details of the data files used and the products generated. This information should include descriptions of file formats, transfer formats, input/output procedures, media specifications and any interaction the user makes with the process.
4. *Application-dependent information.* This category has been proposed to record information about the purpose for which a particular data set was generated. This can help the user determine for which applications the data are useful.

log files automatically which can then be converted to lineage information when combined with data-specific information. The benefits of lineage in quality control are summarized in Table 10.1.

Conclusions

Error in GIS is a 'fact of life' but not an intractable problem. Adopting good practice in terms of data capture and analysis should, in most instances, be sufficient to ensure errors are kept to a minimum. Conscientious documentation and the incorporation of lineage information and quality statements on GIS output will help to ensure that end-users of GIS products are aware of their limitations. Box 10.8 outlines some of the known errors in the nuclear waste case study by way of an example of the types of errors that can be documented.

Despite the existance and documentation of errors such as those presented in Box 10.8, some GIS users, like car drivers, continue despite the GIS equivalent of the flashing of the oil warning light, the clouds of steam coming from the radiator and the clanking noise from the rear axle. Like (petrol engine) motor cars, GIS require the correct data (petrol not diesel), data which are fit for the purpose and system (unleaded or leaded petrol), appropriate choice of analysis method (careful driving) and maintenance of correct lineage (good service history).

As GIS become widely used in everyday decision making, the issue of liability raises its head. If a GIS is used to support a decision, which subsequently results in loss or injury where the GIS analysis has been proven to be at fault, who or what is to blame? Is it the decision maker who decided to use GIS? Is it the GIS analyst who directed the GIS work? Is it the GIS technician who pushed the keys? Is it the GIS or the vendor who supplied the software? Is it the data or the data supplier? Chances are that the 'buck' stops with the GIS analyst, for it is they who will have made the crucial decisions regarding what data and methods of analysis to use. Precedents exist in the USA, where people have been successfully sued over faulty GIS analyses. Openshaw (1991) gives an interesting account of these and other issues in his editorial 'GIS crime and criminality'.

With expertise in GIS comes responsibility. Responsibility to ensure that the data and the chosen method of analysis are appropriate to the problem in hand. Responsibility to ensure that possible alternative solutions have been considered and are presented to the decision maker. Responsibility to ensure that the best available data are used and any potential errors are accounted for and communicated clearly to the end-user. As suggested in Chapter 8, GIS output never fails to impress management and decision makers; it is colourful, it has been produced using a computer, it is often complex. Is it correct? Any doubt in the mind of the GIS analyst, whether due to potential data error or to uncertainty about the analysis, should be clearly stated in the output. Be careful in your choice of data. Be careful in your chosen analysis procedure. Communicate any doubts clearly in the results and output.

Table 10.1 Benefits of lineage

Error detection	Lineage helps recreate analysis processes in which previous data sets containing errors have been used
Management accountability	Lineage provides information from which accounting can be undertaken (Aronoff, 1989). It allows the assessment of workloads and efficiency
External accountability	Lineage records the work of each GIS user, which allows others to assess the validity of the work undertaken
Quality reporting	The AGI in the UK and the NCDCDS in the USA define lineage as a requirement for GIS data quality reports. These should include information on data history, positional accuracy, attribute accuracy, completeness and logical consistency

BOX 10.8

Errors in the nuclear waste case study

Source data error and the effects of generalization

Just as with any other GIS application, the source data used in this study are subject to problems of data error. The four siting factors used present different problems:

1. *Geology.* These data were derived from 1:625,000 scale maps of deep hydrogeological environments produced for NIREX by the British Geological Survey (BGS). The requirement of any suitable geological environments is that water flow through the rock strata is either non-existent or so slow that it would take thousands of years for it to return radionuclides leached out of the waste to the surface. Although the surface geology of the UK is accurately mapped, very little is known about geological structures 500–1000 metres underground. The maps showing potentially suitable geological environments were interpolated and extrapolated from a very limited number of deep boreholes, seismic records and 'intuitive guesswork'. Given the complexity of geological structures and the smoothness of the areas identified by the BGS, these data are subject to a high degree of generalization and therefore positional and attribute uncertainty. Given the element of guesswork involved it is very difficult to quantify the level of error involved, but it can only be assumed to be large.
2. *Population.* The population data were 1-kilometre-square gridded population counts derived from the 1981 OPCS Census of Population. The census data themselves can be assumed to be reasonably accurate for the night on which the census was taken. The derivation of gridded population counts from ED-level data for studies of this kind makes certain assumptions about the data and the distribution of people within the enumeration districts. This is referred to as the modifiable areal unit problem (Chapter 8).
3. *Accessibility.* Accessibility data were derived by buffering the road and rail networks at predetermined widths. The road and rail data were taken from Ordnance Survey 1:625,000 digital data. The relatively small scale of this data source means that a significant amount of scale-related cartographic generalization is present. Also, if digitizing error is assumed to be in the order of 0.8 mm either side of the mapped line (Dunn *et al.*, 1990), then at this scale this would account for possible lateral displacement of the digitized line by up to 500 metres.
4. *Conservation.* The location of conservation areas was derived from 1:250,000 scale maps provided by the Nature Conservancy Council (NCC). Again, the effects of scale-related cartographic generalization need to be recognized as well as errors introduced during the digitizing process. At this scale these could account for up to 200 metres lateral displacement.

Choice of criteria

The choice of the criteria used in performing categorical map overlay is critical to the outcome of the site search. Guidelines produced by the IAEA (1983) suggested that geology, population distribution, accessibility and conservation be taken into account. However, no specific criteria specifying how these were to be mapped were included. The criteria used in this case study to produce the maps shown in Plate 5 mimic those chosen by NIREX using the best available digital data sets at the time.

1. *Geology.* Suitable hydrogeological environments were mapped by the BGS and no further processing of these maps was required after digitizing.
2. *Population.* Areas of low population density were defined by NIREX by adapting the Nuclear Installations Inspectorate's (NII) relaxed nuclear power station siting guidelines. These state that there should be a

BOX 10.8 CONTINUED

maximum population of 100,000 people within five miles of the reactor site (Charlesworth and Gronow, 1967). Beale (1987) and NIREX converted this into a population density threshold of 490 persons per square kilometre. Here the same criterion is used, but is mapped using the 1-kilometre gridded population data.

3 *Accessibility*. Local accessibility to the national network of roads and railways is considered more important than 'strategic' accessibility to waste-producing sites based on Weberian location analysis. Good local accessibility is defined as a maximum straight-line distance to a motorway or railway line of 3 km, or 1.5 km to a 'primary route'. Proximity to local transport networks would keep the costs associated with building new access roads and rail sidings to a minimum. These criteria are not based on any formula, but are deemed to be 'reasonable' figures by the analyst.
4 *Conservation*. As with geology, the location of existing conservation areas was provided on maps by the NCC and, after digitizing, no further processing of these was required.

The criteria that are used to map population and accessibility clearly affect the results shown in Plate 5. 'What if?' analyses can be carried out to investigate the effects of specifying different criteria. However, the question of the appropriateness of the criteria still remains. For example, why is a population density of 490 persons per square kilometre considered to be the most appropriate threshold? What makes a site with a population density of 490 better than a site with a population density of 491? Why not use a density threshold of 49 persons? Why not use 4.9? The answer is probably that a lower population density threshold would have excluded too much of the UK land area from consideration, making the task of finding an acceptable site far harder for NIREX. Choosing appropriate threshold criteria and being able to justify them is critical when defending an analysis of this kind (particularly if the siting decisions made are likely to provoke a hostile reaction from local residents).

Analytical error

Any analytical error affecting this case study is likely to come from within the GIS software via the analyst. The opportunity for gross errors in the choice of appropriate analytical tools is minimal since the analysis is limited to categorical mapping (reclassification functions), buffering and map overlay. All the source data are at comparable scales and resolutions. It is conceivable that errors may occur in the processing of the data due to hardware or software faults.

Conceptual error

The above discussion of data error and choice of criteria reveals the difficulties of conceptualizing what at first appears to be a map overlay problem. In the first instance, it is not safe to assume that the IAEA's suggested siting factors (geology, population, accessibility and conservation) are correct for every situation. From a personal point of view you may suggest 'as far away from me as possible!' as an additional factor. In terms of the data used and the criteria chosen we can see that the site search pursued by NIREX was driven to a certain extent by economic and political considerations. By specifying a relatively high population density threshold they were able to keep their options open. Sites they already owned were identified. By defining good accessibility they could keep transport costs to a minimum. NIREX is currently investigating a site near the Sellafield reprocessing plant in Cumbria (the main source of the UK's nuclear waste). This is a choice driven more by economic factors such as low transport costs and political pragmatism than by other geographical considerations.

REVISION QUESTIONS

- What is the difference between accuracy and precision? Give examples of both in a GIS context.
- Write a brief explanation of the following terms: applicability, bias, compatibility, completeness, consistency.
- Conceptual error is related to how we perceive reality. How does conceptual error affect GIS?
- What are the main sources of error in GIS data input, database creation and data processing?
- The modifiable areal unit problem (MAUP) is commonly cited as a difficulty faced when attempting to overlay data layers in a GIS. What is the MAUP, and why does it cause problems for data integration?
- What measures can be taken to provide information about error in GIS data?
- Discuss some of the ways in which errors in GIS data can be avoided.

Further study

The problems of data quality and error propagation in GIS have seen a flush of research papers in recent years. The NCGIA research initiative on the accuracy of spatial data started much work in the USA and produced one of the few books dedicated to the subject (Goodchild and Gopal, 1989). Similarly, in the UK the NERC/ESRC-funded Geographical Information Handling project did much to stimulate research on error propagation and generalization in GIS (see Mather, 1993). Prior to these initiatives, Burrough (1986) was one of the first authors to recognize the problem of error in GIS and dedicated a whole chapter in his book to the issue. A good section on data quality can be found in Longley *et al.* (1999).

Many studies have been carried out into digitizing error (for example Jenks, 1981; Otawa, 1987; Keefer *et al.*, 1988; Bolstad *et al.*, 1990; Dunn *et al.*, 1990). Errors associated with map overlay operations have been addressed by several authors. Sliver polygons are addressed by McAlpine and Cook (1971) and Goodchild (1978); whilst Newcomer and Szajgin (1984), Chrisman (1989) and Walsh *et al.* (1987) consider the cumulative effects of map overlay and map complexity on output errors. The paper by Walsh *et al.* (1987) addresses the problem of errors propagated through vector to raster conversion; this work has been followed up by Carver and Brunsdon (1994). Key papers in the development of techniques of error handling include Brunsdon *et al.* (1990), Openshaw *et al.* (1991), Fisher (1991) and Heuvelink *et al.* (1990).

Bolstad P V, Gessler P, Lillesand T M (1990) Positional uncertainty in manually digitized map data. *International Journal of Geographical Information Systems* 4 (4): 399–412

Brunsdon C, Carver S, Charlton M, Openshaw S (1990) A review of methods for handling error propagation in GIS. *Proceedings of 1st European Conference on Geographical Information Systems*, Amsterdam, April, pp. 106–16

Burrough P (1986) *Principles of Geographical Information Systems for Land Resources Assessment*. Clarendon Press, Oxford

Carver S, Brunsdon C (1994) Vector to raster conversion error and feature complexity: an empirical study using simulated data. *International Journal of Geographical Information Systems* 8 (3): 261–72

Chrisman N R (1989) Modelling error in overlaid categorical maps. In: Goodchild M F, Gopal S (eds) *The Accuracy of Spatial Databases*. Taylor and Francis, London, pp. 21–34

Dunn R, Harrison A R, White J C (1990) Positional accuracy and measurement error in digital databases of land use: an empirical study. *International Journal of Geographical Information Systems* 4 (4): 385–98

Fisher P (1991) Modelling soil map-unit inclusions by Monte Carlo simulation. *International Journal of Geographical Information Systems* 5 (2): 193–208

Goodchild M (1978) Statistical aspects of the polygon overlay problem. In: Dutton G H (ed.) *Harvard Papers on Geographical Information Systems*, Vol. 6, Addison-Wesley, Reading, Mass., pp. 1–21

Goodchild M F, Gopal S (eds) (1989) *The Accuracy of Spatial Databases*. Taylor and Francis, London

Heuvelink G B, Burrough P A, Stein A (1990) Propogation of errors in spatial modelling with GIS. *International Journal of Geographical Information Systems* 3: 303–22

Jenks G F (1981) Lines, computer and human frailties. *Annals of the Association of American Geographers* 71 (1): 142–7

Keefer B K, Smith J L, Gregoire T G (1988) Simulating manual digitizing error with statistical models. *Proceedings of GIS/LIS '88*. ACSM, ASPRS, AAG and URISA, San Antonio, Texas, pp. 475–83

Longley P A, Goodchild M F, Maguire D J, Rhind D W (eds) (1999) *Geographical Information Systems: Principal, Technique, Applications and Management*. Wiley, Chichester

Mather P M (ed.) (1993) *Geographical Information Handling – Research and Applications*. Wiley, Chichester

McAlpine J R, Cook B G (1971) Data reliability from map overlay. *Proceedings of the Australian and New Zealand Association for the Advancement of Science 43rd Congress, Section 21 – Geographical Sciences*. Australian and New Zealand Association for the Advancement of Science, Brisbane, May

Newcomer J A, Szajgin J (1984) Accumulation of thematic map errors in digital overlay analysis. *American Cartographer* 11 (1): 58–62

Openshaw S, Charlton M, Carver S (1991) Error propagation: a Monte Carlo simulation. In: Masser I, Blakemore M (eds) *Handling Geographic Information: Methodology and Potential Applications*. Longman, London, pp. 78–101

Otawa T (1987) Accuracy of digitizing: overlooked factor in GIS operations. *Proceedings of GIS '87*. ACSM, ASPRS, AAG and URISA, San Francisco, pp. 295–9

Walsh S J, Lightfoot D R, Butler D R (1987) Recognition and assessment of error in Geographic Information Systems. *Photogrammetric Engineering and Remote Sensing* 53 (10): 1423–30

11 Human and organizational issues

KEY QUESTIONS AND ISSUES

- What types of GIS applications exist in commercial organizations?
- How can investment in GIS be justified?
- Who are the users of GIS?
- How can an organization choose an appropriate GIS?
- How can successful implementation of GIS be ensured?
- What organizational changes may result from GIS adoption?

Introduction

This chapter moves away from the technical and conceptual issues that underpin GIS and takes a look at some of the human and organizational issues surrounding commercial and business applications. In utilities, local government, commerce and consultancy, GIS must serve the needs of a wide range of users and should fit seamlessly and effectively into the information technology strategy and decision-making culture of the organization. The technical issues for such systems will be the same as those discussed in earlier chapters. For instance, the most appropriate data model must be chosen, the most effective data input method selected and suitable analysis procedures designed. However, successful application of large-scale GIS projects depends not only on technical aspects, but also on human and organizational factors relating to the implementation and management of systems. This has been recognized since the 1970s (Tomlinson *et al.*, 1976), and there has been considerable research into the organizational and human factors that may impede the implementation of GIS (Department of the Environment, 1987; Onsrud and Pinto, 1991; Medyckyj-Scott and Hearnshaw, 1994; Campbell and Masser, 1995; Hernandez *et al.*, 1999). This chapter examines some of the human and organizational issues associated with the choice and implementation of GIS for commercial organizations.

The issues discussed in this chapter are, in many cases, not unique to GIS, but are also faced by organizations seeking to use any information technology for the first time. No doubt our Martian friend would be faced with similar issues when deciding whether or not to use a car as his main means of transport. He has to decide which car to purchase. As he will be the user of the car he must carefully assess his needs and identify his requirements. This will help him set parameters to allow comparison of different makes and models. Good fuel consumption, the reputation of the vendor for after-sales service, the insurance cost, the colour or the dimensions may be important. He can test some of the models, perhaps by seeing demonstrations or taking a test drive. If he goes ahead with his purchase he will find that the car brings major changes to his lifestyle. His relationships with friends may change as he can visit those further away more often, use out-of-town shopping facilities and work further away from his home. He may also find himself with less disposable income as the costs of insurance, servicing and fuel eat into his budget! In summary, the Martian needs to justify investment in a vehicle, review the range and functions of models on offer, formulate a list of his own requirements,

take advice on purchase options and be prepared for the changes that will occur once he acquires his car. Similar questions are faced by organizations considering adopting GIS:

- *What applications of GIS exist?* What GIS are available? What are the options for hardware and software?
- *What are the needs of potential users?* Who are the users of a GIS? How can users be involved in setting up a new GIS? How will GIS impact on their work tasks and jobs? How can their commitment to a GIS project be ensured?
- *Can investment in GIS be justified?* Is GIS a proven technology in the organization's sphere of activity? Will GIS bring benefits to the organization? How can potential benefits be assessed?
- *Which system is appropriate, and how should it be implemented?* How can an appropriate system be chosen? How can a system failure be avoided? Can resistance to a new system be prevented?
- *What changes will GIS cause in the organization?* How can implementation of a new GIS be undertaken so that the new methods of working have minimum impact on the work culture of the organization?

The importance of each of these issues in the process of choosing and implementing a GIS varies depending on the nature of the organization and its function. Awareness of the characteristics of GIS applications, and hardware and software choices that have to be made, are important first steps in appreciating the significance of these issues.

GIS applications

From the early days of custom-designed and specially built systems such as CGIS (Chapter 9), the GIS market has grown to accommodate general-purpose and tailored packages for a wide range of application areas. According to *GIS World* (Anon. 1994), the most frequent users of GIS are in government (over 20 per cent) and education (over 15 per cent), but there are also users in organizations as diverse as pharmaceuticals, oceanography and restaurants. With such a diverse user group it is not surprising that GIS applications vary enormously in their scale and nature. They can serve the whole organization, several departments or just one project (Box 11.1).

GIS applications tend to fall into one of three categories: *pioneering*, *opportunistic* or *routine*. Pioneering applications are found in organizations that either are at the cutting edge of their field, or have sufficient financial reserves to allow them to explore new opportunities. Pioneering applications are characterized by the use of unproven methods. As such they are considered high-risk and often have high development costs (Figure 11.1). However, successful pioneering GIS applications can provide high paybacks by giving considerable advantages over competitors.

Opportunist applications are found in organizations that kept a careful eye on the pioneers and

BOX 11.1

Corporate, multi-department and independent GIS applications

Corporate GIS is developed across an entire organization. GIS is usually implemented using a 'top down' approach to promote data sharing, reduce data duplication and create a more informed decision-making environment. Corporate GIS is appropriate for a utility company where all departments – customer support, maintenance, research and development, logistics, sales and marketing – could benefit from the sharing of data and access to GIS. Local government agencies also benefit from corporate GIS since all departments work within the same geographical area.

Multi-department GIS involves collaboration between different parts of an organization. GIS is implemented in a number of related departments who recognize the need to share resources such as data, and the benefits of working together to secure investment for a system. Multi-department GIS may be appropriate in a retail organization. Marketing, customer profiling, store location planning, logistics management and competitor analysis may all make use of GIS. Other activities, such as product design and planning, brand management or financial services, may be completed without the use of GIS.

Independent GIS exists in a single department and the GIS will serve the host department's needs. These systems can be adopted quickly and are generally task specific. GIS is usually adopted in response to a specific clearly identified need within the department. In a telecommunications company an independent GIS may be used to assist the siting of masts for a mobile phone network.

Corporate and multi-department GIS share many benefits, although the benefits will be at a corporate level where a corporate system is implemented, and only found in cooperating departments in the multi-department case. These benefits include integration of data sets, generation of shared resources, more informed decision-making, increased data sharing and improved access to information. Improved control over priorities, information and staffing can be benefits for independent systems.

Corporate and multi-department systems can suffer where individual departments have different priorities, and where there are different levels of skills and awareness of GIS and spatial data. Independent systems may be hampered by lack of support, isolation and limited lobbying power.
(*Source*: Hernandez *et al.*,1999)

quickly adopted the technology once they saw the benefits. Following this approach, opportunist organizations let the pioneers bear more of the cost and risk of application development. Those who wait too long risk becoming users of routine GIS applications. Routine users adopt a tried, tested and refined product with lower risk and cost.

While there are now plenty of examples of routine and opportunistic GIS applications, truly pioneering applications are rare. This is not surprising, since few organizations are prepared to be the first to apply GIS to new areas because of the high risks and costs that may be involved. For example, in our nuclear waste disposal example, NIREX was unlikely to risk any adverse reaction or publicity to such a sensitive problem by application of what was, in the mid-1980s, a new technology. Despite being, by today's standards, a relatively straightforward GIS application, the use of GIS to perform site search and evaluation analysis was, at the time, the realm of academic research (Openshaw *et al.*, 1989; Carver, 1991b).

Pioneering, opportunistic and routine applications of GIS have permeated the entire range of

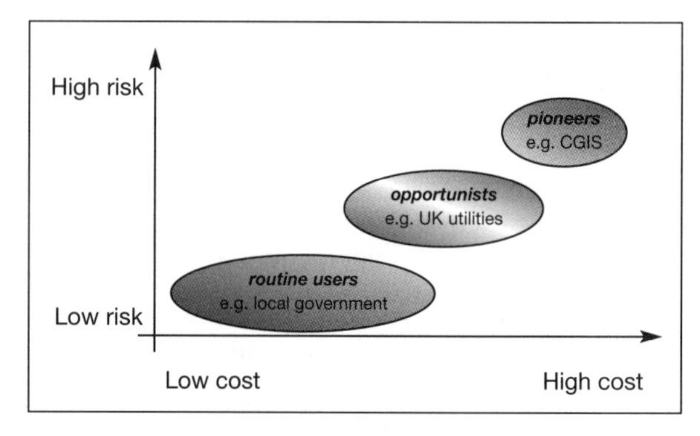

Figure 11.1 Development of GIS applications

BOX 11.2

Business applications of GIS

Operational applications

Operational GIS applications are concerned with managing facilities and assets. For example:

- a utility company may use GIS to identify assets such as pylons and masts in need of routine maintenance;
- a waste management company may use GIS to route waste collection vehicles; or
- a property management company may maintain hotel buildings and grounds with the help of maps and plans from its GIS.

Management/tactical applications

Management (or tactical) GIS applications are concerned with distributing resources to gain competitive advantage. For example:

- a ski holiday company may use GIS to identify appropriate potential customers to receive direct mailings;
- an education department might use GIS to produce options for school closures when deciding how to distribute limited resources; or
- a telecommunications company might use GIS to identify and evaluate possible communications mast sites to serve the maximum possible population.

Strategic applications

Strategic GIS applications are concerned with the creation and implementation of an organization's strategic business plan. For example:

- a ski equipment retailer may decide which geographical areas to target over the next five years after geodemographic analysis using GIS;
- a local government organization may decide on budget reallocations after analysis of population growth or decline in certain areas using modelling and GIS; or
- a catering business may decide to expand a restaurant chain to 100 outlets after analysis of the location of its competitors with GIS.

business activities from operational, through management to strategic (Box 11.2).

In many organizations GIS applications cross the boundaries between the three types of activity introduced in Box 11.2. For example, in a utility organization GIS may be used to help operational maintenance of pipelines, improve the management of marketing services to customers and plan the strategic development of the organization. In many organizations a GIS will initially provide information in support of one activity and, if it is successful, develop so that information can be provided to support other activities. The evolution of CGIS, the Canadian GIS, illustrates this development.

CGIS underwent three major stages of development (Crain, 1985). The system was used first in an operational role for data inventory, with data collection and querying the dominant functions. Then CGIS became more of a management tool with new data being produced through the use of spatial analysis techniques, complex queries and data manipulation. Now the system is used in a strategic role as a decision-support tool for modelling and simulation. The evolutionary path followed by CGIS is typical of many GIS applications where the need to integrate, collate and manage data drives users to become more sophisticated and explore, through analysis, spatio-temporal patterns within the database. In turn, users begin to develop decision support applications.

Therefore, when evaluating whether or not to develop a GIS for a particular application much can be learnt from the experiences of other organizations. An appreciation of the nature of the

Table 11.1 *GIS software types*

Professional	High-end GIS products that offer the full range of modelling and analysis functions – these are the 'true' GIS that fit the definitions of GIS offered in Chapter 1. These are the most expensive software products and are likely to have the least users.
Desktop	Desktop mapping systems have become popular GIS products. These do not offer the full range of GIS analysis or data input functions, but are easy to use and geared towards users who need to display and query data. Less expensive than professional products, these offer a wide range of functions.
Hand-held	GIS software specifically for use in the field on portable and hand-held hardware devices has emerged in the last few years. These systems are used for data capture, navigation and the provision of information to field staff. Many of these products work together with Internet products. Hand-held GIS are likely to be cheaper than desktop systems and relevant to more users. They provide a more limited range of functions.
Component	Tool kits of functions (or components) can be used by an experienced programmer to 'build their own' GIS. Component GIS are particularly relevant to highly specialized applications requiring only a subset of the full range of GIS functions. Some of the resulting systems are embedded within other packages so end users may be unaware that GIS has been employed.
GIS Viewers	A GIS Viewer provides functions that allow the display of files in common formats. Simple query and mapping functions may also be available, but sophisticated analysis will not. These are inexpensive, entry level products designed to encourage the widespread use of a particular vendor's products or specific standards for data exchange.
Internet GIS	Longley *et al.* (2001) consider that Internet GIS now have the highest number of users. These products offer a very limited range of functions at low cost to the widest range of users. These are 'end user' products that are used to develop particular applications – such as online location finding, or data exchange.

(adapted from Longley *et al.*, 2001)

application (pioneering, opportunistic or routine) and the type of business activity being supported (operational, management or strategic) is important. However, in the business world issues of secrecy, power and commercial confidence often prevent the sharing of information and experiences.

GIS hardware and software

The existing information technology (IT) infrastructure of an organization will affect the nature of any GIS implemented. There may be organizational standards for hardware to which new purchases must adhere. Stand-alone PCs may be favoured, or any new installation may have to fit in with the existing IT networks. Only specific 'company-approved' software products may be permitted, and there may be restrictions on the use of e-mail or Internet services. The physical location of personnel may be important in determining the nature of a GIS installation. The GIS and analysts may all be located in a single office or department, or spread widely across an organization.

Given the pace of developments in IT, evaluating and classifying GIS installations according to the hardware and software used may not be the best place to start for the development of future applications. However, it is useful to review the nature of current technology to give a clear picture of the wide range of technology used in GIS installations.

In 1996, *GIS World* published the results from a survey that suggested that the type of computer hardware used in an organization was related to the size of the user community (Anon.,

Table 11.2 Hardware and software questions for organizations considering GIS

Hardware questions	*Software questions*
• What hardware is necessary for the application? • How many computers are needed? • Which computers are needed? PCs, workstations or mainframe? • What storage capacity, memory, speed, etc., must the computers have? • What graphic display devices are needed? • Which peripherals are needed – printers, plotters, digitizers, scanners, data loggers? • Will the GIS be run over a network? Will additional hardware be required for this? • Will any new hardware be compatible with existing hardware in the organization? • What about the reliability and maintenance of the hardware? • How much will it cost? • Is it possible to expand or upgrade the hardware once it is in place? • How long can the hardware be expected to last? • Will new staff be required to install/run/maintain the hardware? • Will GIS be run over the Internet?	• What functions does the software offer? • Will the functions available meet the requirements of the organization? • Are all the functions of a full GIS really necessary, or would a more easily used desktop mapping package be more suitable? • Is the package user-friendly? • Can the package be customized? Does the organization have to do the customizing? • Are there extra modules/functions for more experienced users – for example, is it possible to write macros? • What standards does the software use – for instance, database query language, data transfer formats, handling of address standards? • Can the software exchange data with other software in use in the organization? • What documentation is provided? • What training is available? • Are upgrades to the software going to be provided? • How much will it cost? • Will the software continue to be developed and supported by the vendor? • Which operating system will be required?

1996). In local and state or provincial governments, where there were small numbers of GIS users, the PC platform was prevalent. In similar organizations with more GIS users, more powerful, larger and more costly workstations were commonly used. It was suggested that larger organizations regarded high-speed networking, sharing of databases and improved processing power as important to their GIS applications. This trend in hardware use may be reflected in other GIS user sectors.

The range of GIS and related software can be reviewed by examining annual surveys and directories produced by the trade press. Longley *et al.* (2001) classify the main GIS software packages into six groups as summarized in Table 11.1.

The distinctions between packages are ofen difficult to discern, and are becoming more and more blurred. In all cases it is 'fitness for purpose' that is most important – the package which is chosen must meet the needs of the users it will serve. Table 11.2 offers a list of topics that it would be useful for any prospective GIS user to consider when evaluating hardware and software.

In addition to the proprietary software systems outlined above, many organizations have opted for specially designed customized systems. At the simplest level customization may be adaptation of proprietary software by changing the interface and including a few special functions. On the other hand, customization may require the writing of completely new software to perform specific tasks, or to allow the integration of GIS with other databases and software in use in the organization. Cresswell (1995) offers examples of customization of GIS in the retail sector, but also considers that whilst customized systems would seem to offer an ideal solution for an organization investing in GIS, there are a number of reasons for initially purchasing a proprietary system. Proprietary systems are likely to be cheaper, able to serve a large number of users, easy to use and well supported, and flexible in dealing with different data and problems.

Grimshaw (1994) summarizes the choices available to an organization considering GIS as:

- to develop its own software in-house;
- to purchase a standard software package;
- to purchase a ready developed software system from a third party user; or
- to employ consultants to write new software.

Awareness of the range of GIS applications, and the options for hardware and software, is an essential first step in the GIS implementation process. Comparison with similar organizations can give information about whether the organization is a GIS pioneer or a routine user following an accepted trend for the business sector. Understanding and learning from successful applications of the same nature is one way of attempting to justify investment in GIS, but more formal methods are often required. However, first in any project come the users.

GIS users

Before GIS adoption and implementation it is important to consider just who the users of a GIS will be. Consider, for example, a large sports equipment retailer. It has many stores throughout Europe (including one in Happy Valley), and offers a range of services to customers, including mail order, financial services and a store card. This company, SkiSupplies Inc., has a GIS that it uses for a variety of applications, including:

- location analysis for the siting of new stores;
- target marketing of store card customers;
- managing distribution of stock from warehouses to stores; and
- producing maps of individual stores to help customers find their way around.

The SkiSupplies GIS, therefore, has a number of different users. At one end of the spectrum are the company directors and managers. These decision makers require strategic information from the GIS, but are unlikely ever to use the system in a 'hands-on' way. They are more interested in hard-copy output, such as maps of stores that are performing well, or maps that justify the decision about where to locate a new store. They have a good knowledge of the context of the applications – the retail environment – but little GIS knowledge or experience. They rely on a group of GIS analysts to perform appropriate GIS tasks to provide them with the information they require. These analysts have an understanding of retailing and GIS. They are able to translate the managers' requirements into real GIS analysis. They also have a role in designing and maintaining the GIS, including the development of new applications for the system, so that it will continue to meet the needs of other users and justify the amount of money that has been spent on it. A third group of users is the computer technicians. These users have other responsibilities in addition to the GIS. They are responsible for the wages and personnel computer system, which has no GIS element. They are not GIS experts, but computer specialists who assist with data formatting and input, hardware maintenance and system upgrading. Finally, the customers are also users of the system. For example, on-screen maps are available in larger stores to help customers navigate their way around the store, and these provide additional information on products and services. Although these are not 'true' GIS they are providers of spatial information. Customers also receive target mailing produced as a result of the GIS analysts' manipulations of the customer databases. Such users do not need to know that it is a GIS they are interacting with. They are unlikely to have any retailing, GIS or computer expertise. They are end-users of the products and services offered by the company.

Eason (1988, reported in Grimshaw, 1994) has a similar view of users of information systems. He classified the four groups of users discussed above as 'occasional professional' (with high task expertise and low system expertise); 'application specialists' (with high task and high system expertise); 'computer specialists' (with low task expertise and high system expertise); and 'the public' (with low expertise in both tasks and system). These ideas are shown in Figure 11.2. For most GIS applications a similar classification of users can be achieved or even expanded. Brown (1989, reported in Huxhold, 1991) suggested 10 roles that might be required in a GIS application. These have been included in Figure 11.2.

Understanding the range of potential users is important in the 'user-centred approach' advocated by many researchers for successful GIS implementation (Eason, 1988, 1994; Medyckyj-Scott, 1989). Different types of users have different requirements from GIS and will need different functions. For instance, the managers may be interested in the data output functions and the ability to reproduce the company logo, whereas the analysts may be interested in the analysis functions and the ability to link the GIS with appropriate retail modelling packages. Different users have different skill levels with computers and GIS, and different levels of awareness about what GIS can achieve. Thus, the questions they will ask the GIS will differ. Different users also have varying time available to work with the GIS. For example, the analysts may interact with the system every day, whereas an interested manager, who likes to find out the answers to her own queries from time to time, will need a system that is easy to use, and quick to relearn after a period without using it. To ensure that the GIS will meet the needs of all its users a range of issues must be considered:

- education, awareness raising and training;
- the adoption of standards; and
- GIS usability.

Educating users not only in how to use the new system, but also in the reasons why it has been necessary to develop it in the first place, is an important, but often overlooked, element of introducing a new GIS application into an organization. This education may take many different forms, from the awareness raising generated by articles in a corporate newsletter, to formal training in software development and use.

Adopting standards can help new users to learn and accept a new innovation more quickly. There are many standards associated with GIS for software, hardware and data. Some of the organizations responsible for these are shown in Table 11.3. Adoption of standards with which users are already familiar will reduce the learning curve associated with a new system, and the implementation of standards that are required by legislation or convention will help ensure commitment to applications. Further information on GIS standards can be found in Chapters 5 and 10.

	High application skills	***Low application skills***
High GIS skills	**GIS analysts** *'applications specialists'* e.g. system manager analyst cartographer	**Computer technicians** *'computer specialists'* e.g. programmer data processor database administrator digitizing technicians
Low GIS skills	**Managers** *'occasional professionals'* e.g. end-users decision makers	**Customers** *'the public'* e.g. customers

Figure 11.2 GIS users. (Adapted from Brown, 1989; Grimshaw, 1994; and Eason, 1994)

Table 11.3 *A selection of national and international standards development organizations*

AFNOR	Association Française de Normalisation
AGI	Association for Geographic Information (UK)
ANSI	American National Standards Institute
BSI	British Standards Institute
CEN	European Standardization Organization
DGIWG	Digital Geographic Information Working Group
EUROGI	European Umbrella Organization for Geographical Information
EUROSTAT	Organisation responsible for developing standards for the dissemination of geostatistical data within the European Commission
ISO	International Standards Organization
NMA	Norwegian Mapping Authority
NNI	Netherlands Normalisation Institute
OGC	Open GIS Consortium

(*Source*: adapted from Harding and Wilkinson, 1997)

System usability is affected by many factors: ergonomic aspects such as hardware design and siting; training; support factors such as documentation; and software characteristics such as the interface design. The human–computer interface can be considered to be more than just the design of the graphics on the screen; it is the way in which the users' requirements are translated into actions in the computer, and the way in which the user interacts with the software. The effects of a selection of interface characteristics are reviewed in Box 11.3.

Davies and Medyckyj-Scott (1996) conducted a study of GIS usability. While they found that most users were happy with the usability of their system, they were critical of error messages, documentation and feedback, and infrequent users of GIS were found to encounter more problems. From their observations of GIS users, Davies and Medyckyj-Scott estimate that around 10 per cent of the time spent in front of a GIS screen is spent on unproductive tasks, such as sorting out 'snags' and waiting for the system to process. Clearly there is a long way to go to produce user-friendly software to suit all users.

Standards and interfaces are just two issues that can have a huge effect on users. If users are not kept happy throughout the implementation of a new GIS project then signs of resistance to a project may emerge. Stern and Stern (1993) suggest that some of the ways in which employees could resist the implementation of a new information technology system include inserting or deleting data, copying or destroying files or installing 'bugs'. Some examples of resistance to GIS were identified by a survey, conducted in 1990, of GIS users in the UK (Cornelius and Medyckyj-Scott, 1991). Unwillingness to use the systems, reluctance to change to new methods, and finding faults with improved output were uncovered as problems from unhappy users. Unhappy users may be one reason for under-use of systems, but there are others. In particular, a poorly researched and constructed business case may lead to misplaced investment in GIS.

Justifying the investment in GIS

A business case must be made for almost all GIS projects in commercial organizations, to persuade sponsors to support the project. Clear objectives

BOX 11.3

GIS interfaces

Some of the characteristics of the human–computer interface assist users with the adoption of a new product. Two examples are the visual characteristics of the software, and help and error messages.

Software appearance

Different screen interfaces may be appropriate for different types of user. For occasional users of the SkiSupplies Inc. GIS, a Windows-type GIS interface may be appropriate if other software in the company is also Windows-based. This would enable users to adapt to the GIS quickly, and menus would prompt them as to which functions to use if they had forgotten. For the general public it may be appropriate to restrict access to all of the GIS functions, or to customize the GIS interface to a simple screen with point-and-click buttons. For the most frequent GIS users, point-and-click buttons would be inappropriate and limited, and menu-based interfaces can sometimes be frustrating and time-consuming. It can take longer to find the menu option you require than to type in the command you need. Many typists, for example, have found the switch to Windows-based word-processing programs frustrating, as they have to move their hands to operate the mouse. This is slower than typing frequently used key combinations to achieve the same result. The same might be true for the computer experts and even the GIS analysts. Knowing what they want to achieve, and with some typing skills, a command line interface such as that found in DOS may be quicker. So, it may be necessary to select a GIS, or a range of GIS, which will allow different interfaces to be used by different users.

Help and error messages

Windows-based interfaces have adopted a common format for help and error messages. Information about commands is usually displayed at the bottom of the screen, and context-sensitive help can be obtained using the function keys. Error messages normally appear in 'pop-up' windows that can be closed once remedial action has been taken. These features are helpful for users, and if they are already familiar with the way in which Windows software offers help and advice, a new system can be learnt quickly. However, not all advice from software is helpful. A message to 'ring technical support for help' whilst working with GIS at a field site, away from telephones and electricity, is not very helpful. For a single user, the only GIS user in an organization working on a stand-alone system, the message to 'contact the system administrator' is also unhelpful. A single user may be the system administrator, as well as analyst, data entry technician and cartographer. So, problems can remained unsolved.

must be set for the GIS, and the advantages and disadvantages of introducing a system need to be considered and evaluated. Questions that need to be addressed include:

- What will be the benefits of introducing a GIS?
- How will GIS help to improve the organization's effectiveness?
- What will be the costs of the new GIS?
- Will the expected benefits outweigh the anticipated costs?

One of the most frequently used methodologies at this stage of a GIS implementation is cost–benefit analysis.

Cost–benefit analysis

Cost–benefit analysis is a methodology that is not unique to GIS as it has applications in many other areas. The method involves a thorough assessment of all the costs and benefits expected in association with a new project. Each cost or benefit is given a monetary value, and a graph, sometimes known as a baseline comparison chart, may be drawn up that predicts how costs and benefits will vary in relation to each other over a period of a few years (Huxhold, 1991). An example of such a graph is illustrated in Figure 11.3 (after Bernhardsen, 1992). The point on the cost–benefit graph where the cost and benefit lines cross is known as the break-even point, and this gives an indication of when a new project may start to become profitable – when the initial investment will start to produce returns.

The results of cost–benefit analysis can also be expressed as a payback period (in years) where:

$$\text{Payback period} = \frac{\text{total cost of investment}}{\text{estimated annual revenue}}$$

In GIS projects cost–benefit analysis can be used to justify a project in its entirety from the outset, or to compare the costs and benefits associated with differing GIS solutions or products. For example, Huxhold (1991) compared the costs of implementing and operating a GIS with the costs of not implementing and operating a GIS between 1984 and 1993 for an American local authority. Without a GIS, costs were set to increase fairly steadily over the period under consideration, whereas analysis showed that with a GIS considerable savings would be made. Smith and Tomlinson (1992) offer a methodology for cost–benefit analysis that they applied to the city of Ottawa in Canada. They consider that cost–benefit analysis is the best way of assessing the value of GIS, but admit that there are problems with the technique.

In GIS the main problem with cost–benefit analysis is the difficulty of identifying and quantifying the costs and benefits associated with a new project. Table 11.4 suggests some of the costs and benefits that may be associated with a GIS project. Direct benefits, such as staff savings or cost savings, are easy to quantify. However, other benefits (intangible or indirect benefits) are hard to quantify. These may include, for instance, improved information flow, better decision making and improved access to information (see, for example, Openshaw *et al.*, 1989). Another problem with applying cost–benefit analysis to GIS is the difficulty in identifying initial costs. Particularly important is the high cost of the creation of a digital map base. Burrough (1986) estimated that 46 per cent of the cost of a GIS project was in base map conversion, and Huxhold (1991) estimated that these costs could rise as high as 80 per cent. The situation may be different with respect to more recent business applications, where data may be obtained from secondary sources rather than digitized in-house. The expected lives of hardware, software and data compound this problem. Bernhardsen (1992) considers that whilst hardware and software can be expected to be replaced after two to five and three to six years, respectively, the data for a GIS may last for 15 or 20 years. This means that cost–benefit analysis should be extended for at least this period of time.

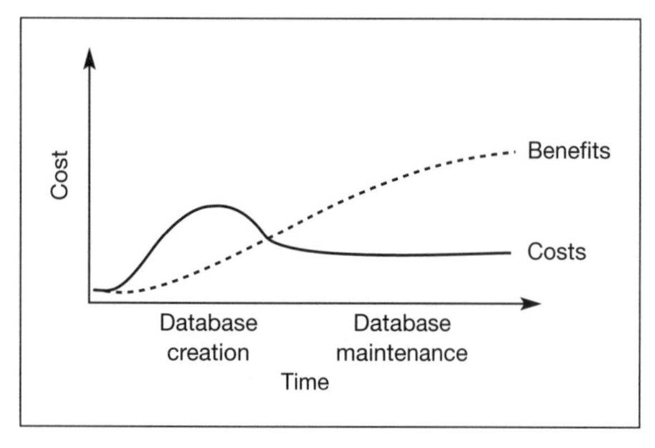

Figure 11.3 Cost-benefit graph. (adapted from Bernhardsen, 1992)

A further difficulty with cost–benefit analysis is that there is, to date, no standard methodology for GIS. Mounsey (1995) even considers that the traditional methods of cost–benefit analysis are inappropriate for some business applications of GIS, where timescales are short, implementations are small and there is unpredictability and change in potential applications, people, business processes, data and technological requirements. She suggests an alternative 'short cut' method for justifying GIS investment in such organizations that involves four stages:

1 Determine the business drives for GIS. This will help to clarify the *strategic benefits* of a new system and interest potential sponsors.
2 Identify the range of applications to assess the scope and practical application of the GIS. This should consider the users of the system, provide a basis for comparing the systems being considered and identify the *operational benefits* of a GIS.
3 Identify the technological framework within which the GIS will fit. Consideration of software, data, hardware and implementation

Table 11.4 Costs and benefits of GIS

Direct costs		*Direct benefits*	
1 Hardware/software	• Hardware • Fewer staff required • Software • Software development • Customization • Consumables (printer supplies, etc.) • Software upgrades • Maintenance and support contracts • Communications networks	1 Savings	• Reduced cost of information production and provision • Fewer staff required • Less space required for, e.g., map storage • Time savings for routine and repetitive tasks
2 Data	• Database creation • Data conversion • Database maintenance • Data updating	2 Increased effectiveness	• Faster provision of information • Greater range of information services provided • Information more readily available • Up-to-date information available
3 Human/admin.	• Insurance • Administration • Security • Training • Rent	3 New products	• New range of output – maps, tables, etc. • Better-quality output
4 Method for choosing system	• Pilot project • Benchmarking • Cost–benefit analysis • Consultancy		
Indirect costs		*Indirect benefits*	
• Increased reliance on computers – vulnerability to failures, changes in software/hardware, etc. • Poorer working environment – noise, heat, tedious tasks for users • Higher-skilled workforce required		• Improved information sharing and flows • Better-informed decision making • Stronger competitive ability • Better-motivated workforce – more career options, less tedious tasks • Greater analysis and understanding of problems • Justification for decisions made • Improved visualization of data	

issues will help to identify potential costs.

4 Develop the plan for implementation.

Finally, having performed a comprehensive cost–benefit analysis there is no guarantee that the benefits identified will actually be realized, so the predicted break-even point and savings may not be achieved. However, if little or no attempt is made to establish the business case for GIS using a methodology such as cost–benefit analysis, there is a greater chance the application will be poorly received within the organization.

Choosing and implementing a GIS

Once the users of a GIS application have been identified and a business case for development established, the organization must choose and implement a system. There are many formal models of information system design and implementation that can be used as a starting point. A variation on just one of these models is presented in Chapter 12 as a framework for the design of a GIS project, and others are introduced in Box 11.4.

However, formal methods such as those introduced in Box 11.4 are not always implemented, and use of these methods is not in itself a guarantor of a successful project. Therefore, this section considers how organizations choose and implement their GIS by reviewing the research findings of surveys (Anon., 1993; Medyckyj-Scott and Cornelius, 1991; Campbell and Masser, 1995) and evidence from case studies. First, the way organizations choose their GIS systems is examined, then how users can be involved in GIS selection and implementation by adoption of a 'user-centred approach' is considered. Some of the methods available for testing and supporting purchase decisions (benchmarking and pilot studies) are also introduced.

Involving users

There have been a number of surveys undertaken to try to find out how organizations choose their GIS. For example, Woodcock *et al.* (1990) considered factors which users in environmental management should take into account when choosing a system. Top of their list was cost, second functionality (the range of operations offered by the software). They considered that users should opt for the widest range of functionality offered. In practice this may not be a sensible decision, since the organization may pay for functions that will never be used. A *GIS World* survey in 1993 put functionality at the top of a list of users' considerations when buying a GIS (Anon, 1993). Other factors were compatibility with existing hardware and software within the organi-

BOX 11.4

Methods for information system design

Four of the most commonly used types of method for information system design are (after Avison and Fitzgerald, 1988; Grimshaw, 1994; and Reeve, 1997):

1 *Structured systems methods.* Structured systems analysis and design (SSADM) is an example of a structured systems design methodology. In SSADM, an understanding of the current working system is developed in order to create a logical model. A second logical model of the required system is developed after consultation with users, and, following implementation of this, users develop a specification of the system they require. A sequence of design phases follows. LBMS (Learmouth Burchett Management Systems) is another widely used example of a structured systems method.

2 *Soft systems methods.* One of the major soft systems methodologies, soft systems analysis (SSA), was developed as a general-purpose methodology for investigating unstructured management problems (Checkland, 1981). These approaches place emphasis on the human aspects of problems, and have been used to investigate organizational aspects of information systems development. An adapted soft systems method is presented in Chapter 12.

3 *Participative methods.* ETHICS (effective technical and human implementation of computer-based systems) is a participative information systems design methodology. The method aims to develop systems that will improve the job satisfaction of those who use them, and thus improve system use and acceptance. All those affected by the system – users, managers, customers and suppliers – are involved in the design process.

4 *Hybrid methods.* MULTIVIEW is a hybrid design methodology, incorporating ideas from soft systems analysis and data analysis techniques. A number of different perspectives of system development are taken to allow the incorporation of broader human issues as well as technical issues into the system design process.

zation, technical support, the reputation of the vendor and cost. The issue of compatibility with other systems also came top of the list of criteria that were important to utilities (Battista, 1995).

In almost all cases the key to success is considered to be the adoption of a *user-centred* approach to GIS purchase decisions (Burrough, 1986; Eason, 1988; Medyckyj-Scott and Cornelius, 1991). Details of one user-centred approach are given in Box 11.5.

However, a user-centred approach can be difficult to implement, since, as Campbell and Masser (1995) found, users do not always know what they want. They stress that defining user needs is one of the hardest, yet most critical aspects of the GIS implementation process. Evidence of other implementation problems is found in reports of failed applications (one example is presented in Box 11.6) and in the reasons why organizations do not implement GIS.

Apart from the financial constraints associated with adopting GIS, there are other reasons why an organization may decide not to implement GIS:

- There may be a lack of senior management support.
- There may be concerns about the technology and data conversion problems.
- There may be a lack of knowledge and awareness about GIS.
- There may be internal political disputes or constraints.
- There may be concerns about the time needed to implement a system.

This list clearly shows that the issues that prevent the uptake of GIS are predominantly human and

BOX 11.5

User-centred GIS implementation

A user-centred approach to GIS implementation involves all users from the outset in any decision-making related to the GIS. This helps to overcome the problems faced where managers and designers make assumptions about users' requirements. In addition, involving users helps to:

- specify functionality requirements;
- ensure user commitment and co-operation;
- identify areas where organizational change may be required; and
- limit resistance from individuals to any new system.

Medyckyj-Scott and Cornelius (1991) suggest a four-stage approach to achieving a user-centred implementation:

1. Establish an appropriate management structure, with time for meetings to discuss human and organizational factors as well as technical issues.
2. Explain the benefits and rewards that the new GIS will offer. Employees who are aware of the benefits are less likely to offer resistance.
3. Allow time for human and organizational changes to occur. Training and extra time to adjust to new working practices may be necessary.
4. Ensure that there is a carefully planned implementation process, backed up by proper resources, appropriate staffing and a positive attitude to the changes expected.

There are various ways in which a user-centred approach can be implemented (Eason, 1988, 1994). First, users can be directly involved in the choice of system, using methods such as user-needs surveys and data audits. Then, users can be involved in benchmarking exercises and pilot projects. They must be aware of the implications and changes that will result from the implementation of the system. They should also have a direct input into the strategy for implementation that is adopted. In all cases, the users must be involved and care must be taken to evaluate their needs, raise their GIS awareness, train them and provide appropriate support (Medyckyj-Scott, 1989).

BOX 11.6

A problematic implementation!

In 1992, the London Ambulance Survey updated its computerized command and control system by including among other things some level of GIS functionality. According to newspaper reports at the time, the new computer system cost well over £1 million and involved 40 networked 486 PCs together with a number of servers. It incorporated vehicle tracking, GIS and other off-the-shelf software (Hobby, 1992; Mullin, 1992). It was considered to be a 'pioneering' application.

As soon as the system was implemented, problems started to occur. Calls from those needing an ambulance were lost in the system, and the response time for ambulances to reach accidents and emergencies increased. Several deaths were initially blamed on the failure of the new information system. The staff had no back-up system to turn to and problems continued until the old system could be reinstated. A number of causes for these problems were identified. It was suggested that there had been insufficient testing of the system and insufficient training of the staff (some new users had limited IT experience), and that the system was overcomplicated (Arthur, 1992). The method of selecting the system also came in for criticism. It was suggested that the Ambulance Service had not examined applications in other similar organizations before making its choice, and that no competitive tendering had been carried out (Mullin, 1992).

Page (reported in Arthur, 1992) implied at the time that the failure of the system might have been due to incomplete specification of requirements, and that the system had not been designed from the 'people outwards' – a user-centred approach had not been adopted. He also suggested that the failure was due not to the computer system (as there was little evidence of widespread technical problems), but rather to the method by which the system had been integrated into the working practices of the organization. Commenting on the official report on the incident in 1993, Arthur (1993) reported the findings that management had made mistakes in connection with procuring, installing and operating the system. It was estimated that it would take five years to properly introduce the system, including testing and training.

organizational in nature. The reasons why GIS implementations fail are not dissimilar. Eason (1994) identifies three reasons for GIS failure:

1. Organizational mismatch (where the system introduced does not meet the needs of the organization).
2. User acceptability (where the system has negative implications for user groups, which may result in resistance to the system).
3. Non-usability (where users may face technical or other difficulties when trying to use the system).

Again, Eason advocates a user-centred approach to help avoid these problems, and suggests that users should be involved in user-needs surveys, benchmarking and pilot studies to help select an appropriate system (Eason, 1994).

User-needs surveys

Clear articulation of user needs is essential to assist the development of an appropriate list of requirements for a GIS. This requires talking and listening to all potential users, from management to technicians. Goodwin (1995) reports the process used by the retailers 'Boots the Chemist' for its choice of GIS. The first phase of its GIS development project was to talk. Talking went on within the information systems department and with all those who might have business applications for the proposed GIS. These discussions resulted in a list of all of the potential users and applications for the GIS. In turn, this allowed the writing of a technical specification for the system that included details of the functionality required.

Evaluation of users' needs may be difficult if users are new to GIS. As stated previously, users do not always know what they want. Awareness-raising activities, such as demonstrations or seminars, may be helpful, although it is probably best to start with an evaluation of potential users' current roles and information requirements. Clarke (1990) considers that user-requirements analysis can be conducted through interviews, documentation reviews and workshops. Topics that need to be considered are workflows, the characteristics of spatial data used in these workflows and the information products required (hard copy, on-screen displays, reports and data in various output formats). Assessment of these topics should result in an understanding of what data are being used, by whom, and how they are being collected, processed, stored and maintained. As well as giving a basis against which to evaluate the acquisition of GIS, this will help to develop data requirements for issues such as accuracy, update frequency and classification. In addition, issues such as the number of users, response times required and data volumes will help with the identification of hardware requirements. Performance indicators developed during user-requirements analysis can be evaluated during benchmarking or pilot studies.

Benchmarking

Benchmarking is a technique often employed to help with the decision about which software package to select. The prospective buyer organizes a benchmark test. Three or four system vendors will be invited for a day or more to the buyer's site and will be provided with a list of tasks that the buyer would like their GIS to be able to perform. The buyers may also supply data, and possibly some expertise from within the organization to assist and advise. The vendors will then attempt to demonstrate that their system will meet the buyer's requirements. Outputs from the benchmarking process may include measures of processing and operator time, as well as products such as graphics and statistics (Clarke, 1990). The user interface and system documentation can also be assessed. This is an excellent method of choosing between systems, and more effective than viewing vendors' demonstrations and listening to presentations on the experiences of other users. In a benchmarking situation, some of the problems of data entry and conversion, as well as the limitations of analysis operations, may be uncovered. However, the whole exercise requires a fair level of expertise and awareness on the part of the potential buyer, particularly with respect to the identification of the tasks to be performed by the vendors.

Pilot systems

A pilot project or system may be defined as the limited-term use of GIS, using data for a small geographical area to test the planned application and demonstrate the capabilities (Huxhold, 1991). It is a test run on a small-scale system that may involve an investment of 5–10 per cent of the total system costs. The system may have been leased from the vendor, with an option to purchase should the pilot project go well. Such a project provides the opportunity for an organization to test each potential application, test user acceptance, demonstrate facilities to improve awareness and acceptance, and improve cost–benefit analysis. The pilot system may be selected after benchmarking. Overall, a pilot system will decrease the financial risks involved in a large-scale GIS project since it does not require the full commitment of a new full system initially. However, time and resources are still needed to ensure that the pilot system will be a successful test of the GIS application.

There are many examples of pilot projects in the GIS literature. For instance, Grimshaw (1994) reports the findings of pilot GIS studies carried out by the Woolwich Building Society, a financial organization in the UK. Its study raised the question of whether a single system to meet all needs, or several systems to deal with specific tasks, would be appropriate, as well as questions about the data to be used. The pilot study was an important step in the design and acquisition of a system that would meet all users' requirements. Bernhardsen (1992) outlines the conclusions of a pilot project undertaken in Oslo for municipal management. This project involved co-operation between a wide range of user groups. The benefits to these user groups were projected forward over 25 years, and the findings led to a change to the originally suggested collective system. A distributed system was developed, and since the pilot project showed that no one system would meet the needs of all users, a total of four or five different software packages were included in the final GIS configuration.

In practice, most GIS users do not appear to use methods such as pilot projects and benchmarking to help with their choice of system. Medyckyj-Scott and Cornelius (1991) found that the most common methods adopted to help with the choice of a new system amongst UK users were visits to and from suppliers, visits to other GIS users and discussions with other GIS users. More formal methods such as feasibility studies, pilot periods, the use of systems analysis techniques, benchmark tests and cost–benefit analysis were far less common. In the London Ambulance Service case study described in Box 11.6, there appears to have been limited use of any of these methodologies, and little attempt to learn from the experiences of other similar users. It remains a matter for speculation whether any of these more formal methods would have helped to prevent the problems that were experienced with the system.

Implementation strategies

As Campbell and Masser (1995) recognize, even if there is nothing technically wrong with a GIS system chosen by an organization, problems and failure may occur. One of the causes of problems can be the way in which the GIS is implemented. Eason (1994) reviews the methods for implementing GIS, and Stern and Stern (1993) discuss similar approaches for IT more generally. Implementation methods fall into four main categories:

- *direct conversion* from the old system to the new;
- *parallel conversion*, where both old and new systems run alongside one another for a short time period;
- *phased conversion*, where some of thefunctions of the old system are implemented first, then others follow; and
- *trial and dissemination*, for example the running of a pilot system, then appropriate conversion to a new system.

Eason (1994) reflects on the level of user involvement with each of these approaches. With a direct conversion, also known as a 'big bang' conversion, the old system is switched off one day, and immediately replaced by the new. Users must adapt immediately. This was the approach used in the case of the London Ambulance Service, and the London Stock Market computer system, which also crashed in 1986. In such an approach the speed of change is rapid, leaving little time for users to learn and adjust to new working practices and methods. However, such an implementation can be successful if managed well and prepared for properly. Parallel running of an old and a new system requires extra resources, but decreases the risks of failure (Grimshaw, 1994). Extra demands are placed on users, who need to operate both systems. An incremental implementation system, with limited conversion to new methods at any one time, gradually results in a fully functional new system. Here the rate of change is much reduced, giving users more time to learn and adapt to the new situation.

Grimshaw (1994) considers organizational characteristics that can help the choice of implementation. If the system needs a 'large critical mass', then a big bang or parallel running approach would be appropriate. If normal business would be placed at risk during the implementation, parallel running or phased implementation would be appropriate. If users are willing and quick to learn a new system, a big bang may be possible, and if the needs of specific groups of users must be considered, pilot projects may be necessary.

The method of implementation is one factor that can affect the success or failure of a GIS project. Gilfoyle (1997) considers that other factors identified by the Department of the Environment (1987) are still applicable a decade later. These suggested that GIS would succeed if:

- the organization can afford some experimental work and trials;
- there is a corporate approach to geographic information and a tradition of sharing and exchanging information;
- there is strong leadership and enthusiasm from the top of the organization with a group of enthusiasts at the working level; and
- there is experience of and commitment to IT and the use of digital data.

Others researchers have examined success factors in particular application areas. Campbell (1994) studied UK local government. She identified success factors as simple applications, an awareness

BOX 11.7

Keys to successful GIS in the retail sector

Following research in retail organizations in Canada, the Netherlands and the UK, it has been possible to identify factors that are critical to the success of GIS in this sector during the planning, use and development of systems:

Planning

- use GIS intelligently to gain competitive advantage;
- identify business issues and then look for specific GIS solutions, not vice versa;
- involve end-users and make sure that everyone who may be affected by the system can make some input in the early stages;
- get support from someone at executive level who will 'sign-off' on the investment;
- quantify possible benefits and costs at the planning stage, but recognize that some will be unknown;
- keep the project manageable, and place it within broader IT and data strategy framework;

Use

- keep initial applications simple;
- allocate resources to training and education;
- secure early and visible quick results in critical area;

Development

- get the GIS champion to promote and market the technology within the organization;
- accept that GIS is likely to be seen as just a part of a complex process – whilst you may view it as central to decision-making, others may not;
- maintain momentum. Plan development to involve more datasets and exploit new analytical methods.

(*Sources*: Hernandez and Verrips, 2000; Hernandez *et al.*, 1999)

of resource limitations, and innovative and stable organizations or those with an ability to cope with change. The importance of these factors has been recognized in other areas. A study of a US Army GIS by Peuquet and Bacastow (1991) revealed slow development with a two-and-a-half-year delay. The reasons for this were felt to be the vague requirements and newness of GIS technology but also the inability of the organization to deal with change, and the non-involvement of the whole organization.

It remains for research whether the factors identified by Campbell and Masser (1995) apply to all other sectors of GIS users. Some areas are already under study, for example international retailing (Box 11.7). It also remains to be investigated whether GIS is actually any different from other areas of IT in this respect. For example, are there additional problems associated with the implementation of GIS as opposed to a large command and control system, or a stock management system?

Organizational changes due to GIS

In most organizations the introduction of a GIS brings with it a period of change. This is in common with other IT applications and other innovations that change working practices. A new GIS system can have a number of impacts: on the organization, on relationships with external suppliers and bodies, and on individuals. For example, a new system for the Happy Valley retailer SkiSupplies Inc. may:

- require the organization to undergo some internal restructuring (for example, a new business analysis department may be created to house the GIS);
- require redeployment and retraining of staff (staff may be redeployed to the new department and retrained in IT, or digitizing, or GIS);
- require revisions to administration arrangements and working practices (delivery vehicles may be supplied with printed maps in addition to written instructions of routes to be followed); and
- require implementation of new data sharing and exchange policies along with appropriate standards and guidelines (the distribution department will need to use the same base map data as the marketing and location analysis department; all address data held will need to be updated to a common format).

Betak and Vaidya (1993) summarize the impacts of GIS, and in particular increased data sharing, at Conrail, the American Consolidated Rail Corporation. These included changes in the way employees related to customers, competitors and suppliers. At a more general level, changes included the consolidation of business functions, a reduction in departmental barriers and the delayering of the management structure.

Changes that influence the success of a GIS project may also be initiated outside the organization. External factors that affect GIS could include the stability of the GIS supplier organizations, political and institutional restructuring, and changes to law and policy in associated areas.

As in any business sector, there are long-established GIS companies that have a broad customer base and stability. Others might be subject to takeover, rapid staff turnover or even closure. Even within a long-established company, particular product lines may be withdrawn or phased out, leaving customers unsupported or unable to upgrade. GIS users often use equipment and software from a range of vendors – the GIS software from one company, the computer hardware from a second, the printer from a third and the digitizer from a fourth – and this can compound problems associated with the stability of suppliers.

Political and institutional restructuring can affect any business, but can cause particular problems when imposed on GIS users. An example is British local government, where many GIS applications have been established and then restructuring of the local government system has been imposed. This has caused job insecurity and uncertainty about the future, and has resulted in the abolition of some local government organizations. The future of the GIS projects was not a major consideration for the decision makers in this process. Some of the newly created local government organizations have been left with two GIS systems, which will need to be rationalized and integrated.

Changes to law and policy can also affect GIS, in both a positive and negative manner. For instance, legislation in the UK to ensure that the register of waste disposal sites was made available to the public resulted in a host of vendors offering GIS-style applications for the management and communication of the waste disposal register. Other policy changes, such as the adoption of standards for data, can also be beneficial in the long term, but disruptive and time-consuming in the short term.

In short, the introduction of a GIS is not simply a matter of choosing the product, ordering the equipment and getting it running, but a complex interplay of technical and human and organizational factors that may reflect change and uncertainty. During and after implementation there are likely to be considerable organizational changes due to the introduction of a GIS. Different users and stakeholders will be affected in different ways. Individuals' jobs and roles may be affected, relationships between departments may change and the nature of the services provided to outsiders may alter. It is worth addressing issues of organizational change from the outset. For example, what impact will a new GIS have on jobs? What difference will it make to the roles of individuals using it? What changes will be required to training programmes? How will the way information flows between departments change? Will the structure of the organization have to change to accommodate the GIS? Consideration of these issues may facilitate any changes required, identify areas of potential conflict and avoid problems of resistance and under-use.

Conclusions

The motor car has led to profound changes in society. Increasing the mobility of the population has influenced the work and social habits of individuals and the widespread use of vehicles has led to changes in the way society is structured, and the way settlements are planned and constructed. Veregin (in Pickles, 1995) considers that the most significant impact of technology tends to occur when the technology becomes indistinguishable from everyday life. Cars, for most of us in Europe, North America and other parts of the world, are indistinguishable from everyday life. In some areas, however, the motor car is less common, and life can be very different. The same is true for GIS. Some users are now at the stage where they do not even know they are using a GIS. GIS has become part of their everyday work, and is seamlessly integrated with other IT in their organizational information strategies. In other cases GIS has had a profound impact on the way individuals work and departments interact. For some organizations, GIS is still a technology they intend to adopt when it can be justified, and for many more, the benefits of adopting a system still need to be evaluated.

This chapter has attempted to ask questions that need to be answered during the implementation of a GIS. Some answers have been proposed, but the best methods of introducing systems, and assessing the impacts of GIS, are areas requiring more research for clear generic recommendations appropriate to all organizations to be made. Perhaps it will be impossible to develop generic guidelines, since all organizations differ in their nature, in the scope of their GIS applications and the range of users. However, many authors have attempted to generalize methodologies for IT implementations, and the question remains whether GIS is really any different from other IT implementations.

REVISION QUESTIONS

- Discuss the nature of GIS users. Why is a user-centred approach to GIS required for successful implementation?
- What factors can be considered when selecting a GIS for an organization? Which of these do you consider to be the most important, and why?
- How can techniques such as cost–benefit analysis, benchmarking and pilot projects be incorporated in the GIS procurement process?
- Describe the strategies available for GIS implementation and discuss their relative advantages and disadvantages.
- Outline the problems which organizations implementing GIS may face. Are these problems any different from those associated with other large IT applications? What makes GIS special?

Further study

A good starting point for reading in this area is Chapter 9, 'Choosing a GIS', in Burrough (1986). This chapter emphasizes the need to match the system chosen with user requirements, and offers a procedure to follow when setting up a new system. However, as Burrough himself notes, this is now a little dated, and should be supplemented with more up-to-date sources.

For an overview of GIS in particular applications areas it is necessary to consult a wide range of sources. There have been some published surveys of users, their applications and difficulties faced when adopting GIS. Examples include Gould (1992), who surveyed UK health authorities and considered their use of GIS and computer-assisted cartography; Campbell and Masser (1992, 1995) and Campbell (1994), who review the applications and nature of GIS in UK local government; and Medyckyj-Scott and Cornelius (1991, 1994), who conducted a more general review of GIS in the UK, looking at the nature of systems being used and problems faced when setting up a GIS. More up-to-date surveys are commonly published in brief in the GIS magazines such as *GIS Europe* and *GIS World*. One interesting example was a survey of utilities in the USA (Battista, 1995).

Medyckyj-Scott and Hearnshaw (1994) offer some interesting chapters on various aspects of GIS in organizations. The chapter by Eason (1994) entitled 'Planning for change: introducing a GIS' is particularly recommended. Eason (1988) is also an interesting book, offering a more general picture of IT in organizations, but with many ideas that are directly applicable to GIS.

An entire section in Longley *et al.* (1999) addresses GIS management issues. The chapter by Bernhardsen (1999) offers comments on choosing a GIS that includes consideration of hardware and software issues. Longley *et al.* (2001) includes a useful chapter on GIS software that expands on the types of systems and provides examples of products in each category.

Obermeyer (1999) takes an in-depth look at cost–benefit analysis for GIS. Campbell (1999) examines the institutional consequences of the use of GIS and discusses different managerial approaches to GIS. Further information on the research into the retail sector by Hernandez can be found in Hernandez *et al.* (1999) and Hernandez and Verrips (2000).

Battista C (1995) GIS fares well in utilities survey. *GIS World* 8 (12): 66–9

Bernhardsen T (1999) Choosing a GIS. In: Longley P A, Goodchild M F, Maguire D J, Rhind D W (eds) *Geographical Information Systems,* 2nd edn. Wiley, New York, 589–600

Burrough P A (1986) *Principles of Geographical Information Systems for Land Resources Assessment.* Clarendon Press, Oxford

Campbell H J (1994) How effective are GIS in practice? A case study of British local government. *International Journal of Geographical Information Systems* 8 (3): 309–26

Campbell H J (1999) Institutional consequences of the use of GIS. In: Longley P A, Goodchild M F, Maguire D J, Rhind D W (eds) *Geographical Information Systems.* Wiley, New York, 621–31

Campbell H J, Masser I (1992) GIS in local government: some findings from Great Britain. *International Journal of Geographical Information Systems* 6 (6): 529–46

Campbell H J, Masser I (1995) *GIS and Organizations: How Effective are GIS in Practice?* Taylor and Francis, London

Eason K D (1988) *Information Technology and Organizational Change.* Taylor and Francis, London

Eason K D (1994) Planning for change: introducing a Geographical Information System. In: Medyckyj-Scott D J, Hearnshaw H M (eds) *Human Factors in Geographical Information Systems.* Belhaven, London, pp. 199–209

Gould M I (1992) The use of GIS and CAC by health authorities: results from a postal questionnaire. *Area* 24 (4): 391–401

Hernandez T, Scholten H J, Bennison D, Biasiotto M, Cornelius S, van der Beek M (1999) Explaining Retail GIS: the adoption, use and development of GIS by retail organizations in the Netherlands, the UK and Canada. *Netherlands Geographical Studies* 258, Utrecht

Hernandez T, Verrips A (2000) Retail GIS: more than just pretty maps. *GeoEurope* 9 (4): 16-18

Longley P A, Goodchild M F, Maguire D J, Rhind D W (eds) (1999) *Geographical Information Systems.* Wiley, New York

Longley P A, Goodchild M F, Maguire D J, Rhind D W (2001) *Geographic Information Systems and Science.* Wiley, Chichester

Medyckyj-Scott D J, Cornelius S C (1991) A move to GIS: some empirical evidence of users' experiences. *Proceedings of Mapping Awareness 1991*, conference held in London, February, Blenheim Online, London

Medyckyj-Scott D J, Cornelius S C (1994) User viewpoint: a survey. In: Hart T, Tulip A (eds) *Geographic Information Systems Report.* Unicom Seminars, Uxbridge, UK

Medyckyj-Scott D J, Hearnshaw H M (eds) (1994) *Human factors in Geographical Information Systems.* Belhaven, London

Obermeyer N J (1999) Measuring the benefits and costs of GIS. In: Longley P A, Goodchild M F, Maguire D J, Rhind D W (eds) *Geographical Information Systems.* Wiley, New York, 601–10

12 GIS project design and management

KEY QUESTIONS AND ISSUES

- How can a GIS project be designed and managed?
- What is a *rich picture*?
- What is the difference between a physical and a conceptual data model?
- What is an analysis scheme?
- What is cartographic modelling?
- What approaches are available for managing the implementation of a GIS project?
- What are the problems that may be encountered when implementing a GIS?

Introduction

The theoretical and technical knowledge you have gained from earlier chapters provides a solid foundation from which to start your GIS projects. This chapter aims to build on this foundation and provide a framework for the development of your own GIS applications. The emphasis is on the practical aspects of designing and managing a GIS application. Design techniques help to identify the nature and scope of a problem, define the system to be built, quantify the amount and type of data necessary and indicate the data model needed and the analysis required. Management techniques help a project to be delivered on time and ensure quality work. Good project design and management are essential to produce a useful and effective GIS application. The Martian is now ready to use his car to go on his first real journey. Before he sets out he must plan, or design, his excursion to ensure that he has all the resources he will need – fuel, food and maps. He must also manage his journey to make sure that he will arrive on time, and in the right place.

The project design and management approach outlined in this chapter is suitable for small-scale GIS projects – the type of project which may be required by a GIS course or as part of a research project. The approach does not embrace any specific design methodology or management philosophy, but it is an integration of many ideas. Various elements of the approach, when scaled up, could provide a methodology for the implementation of larger projects. It should be remembered that there is no generic blueprint for GIS success. Any design and management approach adopted should be adapted to meet the needs of the application, the available technology, the users of the system and the organizational culture in which the GIS must reside.

This chapter starts by considering how the character of the problem for which a GIS solution is being sought can be identified. Two methods are introduced: the *rich picture* and *root definition*. A method for constructing a GIS data model is then discussed. A distinction is made between the conceptual data model and the physical implementation of this model in the computer. Cartographic modelling is then considered, as an approach for structuring the GIS analysis required by an application. A review of several project management approaches and techniques and the tools available for the implementation of a GIS project follow. Next implementation problems and project evaluation are considered. To conclude, a checklist is provided to help with the design and implementation of a GIS project. The house-hunting case study is used to illustrate the approach throughout the chapter.

Problem identification

Before developing a GIS application the problem that the GIS will address must be identified. There are two techniques that can be used to assist problem identification: creating a *rich picture* (a schematic view of the problem being addressed), or developing a *root definition* (a statement of an individual's or group's perspective on the problem). Both these techniques are drawn from the *soft systems approach* to system design (Box 12.1).

The rich picture

A rich picture is a schematic view of the problem a project will address. It presents the main components of the problem, as well as any interactions that exist. The rich picture for the house-hunting GIS (Figure 12.1) adopts the conventions of other authors, in particular Avison and Wood-Harper (1991), Reeve (1996) and Skidmore and Wroe (1988). These include the use of:

- *Crossed swords.* A crossed swords symbol expresses conflict. It is used to indicate the differences between the home buyers and the estate agents. There is conflict since the motives of the two groups for system development are different. The home buyer wishes to find the house that best suits their needs whereas the estate agent wishes to sell a house to gain commission.
- *Eyes.* Eyes are used to represent external observers. Property developers interested in identifying new areas for housing development may be external observers.
- *Speech bubbles.* Personal or group opinions are indicated in speech bubbles. The different priorities home buyers see for the system may be included in the rich picture in this way.

Drawing the rich picture records thoughts on paper and helps to organize ideas. For a small-scale project, the rich picture may be drawn by one individual. A rich picture drawn by a project team will represent a consensus view of a problem reached by all the project participants. A single

BOX 12.1

The soft systems approach

The original soft systems ideas were developed by Checkland (1981) and have been added to more recently by other researchers (Wood-Harper *et al.*, 1995). The soft systems approach to problem identification provides a method for addressing unstructured problems (Skidmore and Wroe, 1988). This is useful in a GIS context because many GIS problems are unstructured and often difficult to define. To understand the difference between structured and unstructured problems it is helpful to revisit the house-hunting case study.

- *Structured problem.* Identifying the location of the properties an estate agent has for sale is an example of a structured problem. To develop a GIS to address this problem, the system developer needs to know where houses are and whether or not they are for sale.
- *Unstructured problem.* Choosing a neighbourhood in which to live is an example of an unstructured problem. In this situation all home buyers will have different opinions on where they want to live and different priorities influencing their choice of property. They may use some common data sets such as proximity to public transport or work place, but how they use these data to make an informed decision will vary. Developing a GIS to meet their needs is much more problematic than developing a GIS that just shows the location of houses currently for sale.

To formulate a problem users should appreciate the context, or *world view,* from which the problem is being considered. This is the key to the soft systems approach. For example, the house-hunting GIS was constructed from the perspective of a home buyer searching for a place to live. An alternative perspective would be that of the estate agent trying to sell property. From the soft systems perspective it is not models of real-world activities which are created, but models of people's perception of an activity. How people feel about and view the activity are included (Checkland, 1988). Therefore, soft systems models are abstract logical models that help with our understanding and structuring of a problem.

composite rich picture can be achieved by asking all members of the team to draw their own rich pictures. These are then discussed and combined to create a single picture that reflects the views of all parties. Skidmore and Wroe (1988) suggest that rich pictures are particularly useful when considering the design of computer systems within organizations because:

- they focus attention on important issues;
- they help individuals to visualize and discuss the roles they have in the organization;
- they establish exactly which aspects of the information flows within the organization are going to be covered by the system; and
- they allow individuals to express worries, conflicts and responsibilities.

The development of a rich picture should not be rushed, particularly if it is trying to reflect an unstructured problem. A poorly defined rich picture may translate into a poor GIS application. An additional check to ensure that the problem is well understood is to develop a root definition.

The root definition

Like rich picture, the term root definition also comes from the soft systems approach (Box 12.1). The root definition is a view of a problem from a specific perspective. Different users have different views of a problem. In the house-hunting GIS, the views of groups involved in the design process might be quite different and lead to a degree of conflict. For example, home buyers may see the GIS as 'a system to help identify and rank possible homes', whereas the estate agents may see it as 'a system to help maximize house sales'. These two statements are the root definitions of these particular groups. The system developer must get these two groups to agree on a common root definition, for example, 'a system that identifies properties for sale which meet the requirements of individual home buyers'.

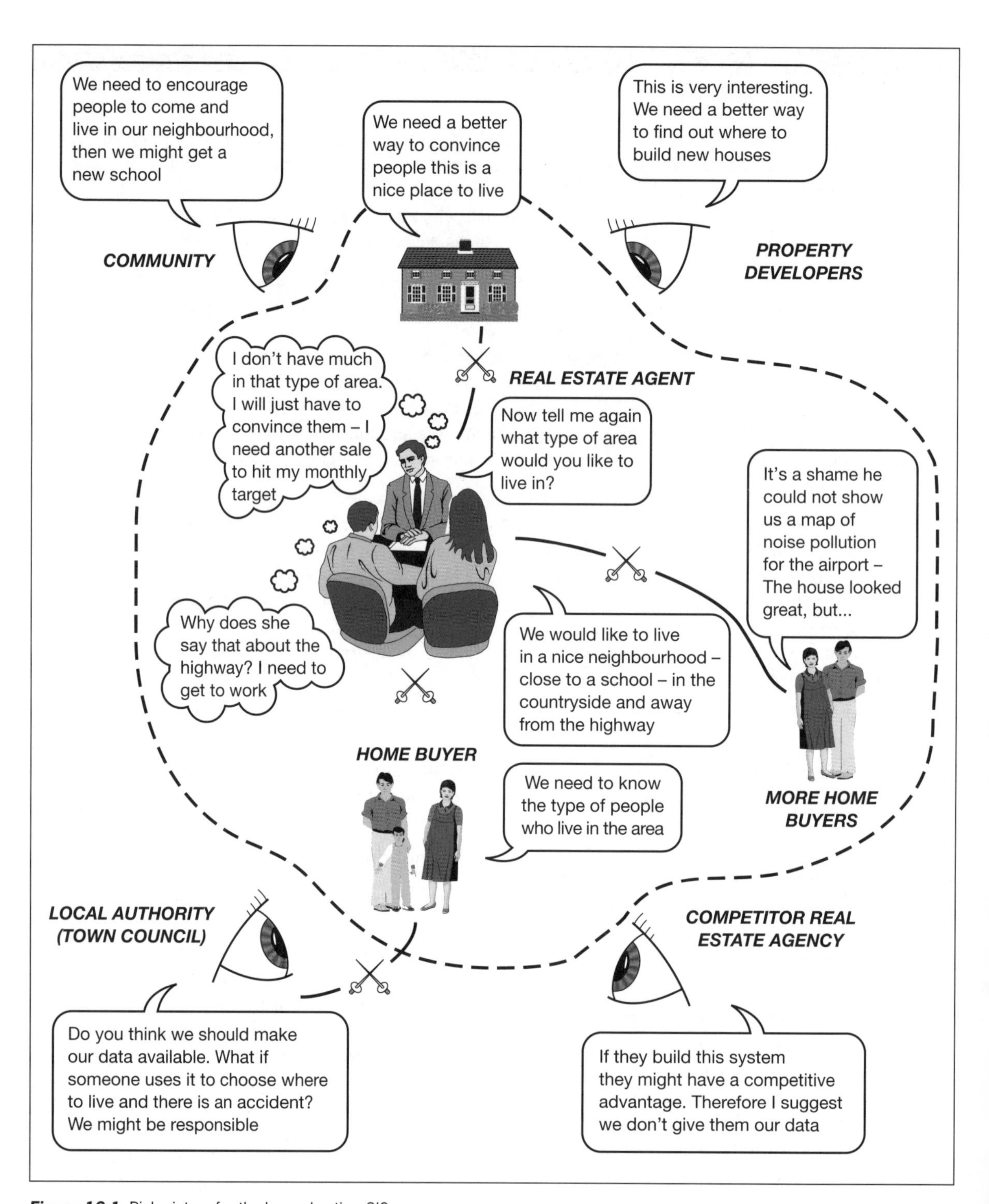

Figure 12.1 Rich picture for the house-hunting GIS

Establishing a common root definition for a problem will help others to evaluate and understand why a GIS has been constructed in a particular way. Likewise, understanding that others may view a problem from a different perspective will ensure a GIS application is designed to address a range of needs. If a single root definition can be agreed upon then there is a greater chance the GIS will meet the requirements of all concerned. Once rich picture and root definition exist the main aims and objectives for a project can be identified and a GIS data model can be created.

If it proves difficult to draw a rich picture or formulate a root definition then the problem being addressed may be unstructured (Box 12.1). Unstructured problems are the most difficult to address with GIS. However, the rich picture method can still be used; typically it will start with only a few elements of the problem clearly defined. Additional elements are added after talking to potential users of the GIS, consulting the literature and discussing the project with others working in a similar field.

As the rich picture is developed and the root definition formulated the resources available to the project must also be considered. In some cases the only resource will be one individual's time and commitment. In larger projects there may be access to several members of staff and a budget. It is important to consider, given the resources available, whether it is possible to address the whole problem that is unfolding, or whether it will be necessary to break the problem down into smaller parts. Breaking the problem down into more manageable pieces may allow quicker results, which may be important where the GIS activities are taking place in a large organization. Small but useful results, produced relatively quickly, will gain recognition and respect for a project. This may result in further support and resources being allocated to the project. Therefore, many system designers use pilot projects to produce results quickly. These results can be disseminated widely throughout the organization to encourage support for the GIS. Box 12.2 offers key questions to be addressed at this stage of the GIS project design process.

BOX 12.2

Problem identification: questions

1 What is the problem you wish to address with GIS? Is the problem structured or unstructured?
2 Can the problem be summarized in a rich picture? Does the rich picture include the views of all those who will use the GIS and all those who will be affected by it?
3 Can a single root definition be defined for the problem? Is this root definition acceptable to all potential users and those who will be affected by the GIS?
4 Can the problem be tackled within the resources available? Would it be better to break the problem into smaller component parts and tackle one of these first?

Designing a data model

The rich picture and root definitions that define a problem must be turned into a GIS data model. Data model is a term that has different meanings in different contexts. In Chapter 3 it was used to describe the method for representing spatial entities in the computer. Peuquet (1984) and Frank and Mark (1991), however, use data model as a collective term for the process of identifying all the design elements used in the construction of a GIS. Worboys (1995) offers a useful solution to this confusion by distinguishing between conceptual and physical data models. The conceptual data model is a high-level view that is independent of the computer system. This is the user's view of a problem and its elements. This is close to the way that Peuquet (1984) and Frank and Mark (1991) use the term. The physical data model, on the other hand, describes the organization of data in the computer. This is how the term has been used in Chapter 3.

Therefore, from the project design and management perspective, it is useful to think of the GIS data model as consisting of two parts: a conceptual model and a physical model. The conceptual data model adds spatial detail to the rich picture by including elements of spatial form and spatial process. The physical data model is concerned with how to represent the conceptual model within the computer. Details about the spatial data model (raster or vector), the appropriate data structure and the analysis scheme are included in the physical data model.

Unfortunately, the data modelling stage is frequently neglected in the design and development of GIS projects, often with disastrous consequences. Insufficient attention to data modelling may lead to the failure of the GIS to meet the expectations of users.

Creating a conceptual and physical data model

One way to create a conceptual data model is to borrow heavily from the ideas of hard systems analysis. Hard systems analysis advocates the clear identification of the elements of the data model: the entities, their states and their relationships to each other (Box 12.3). One method of presenting this is using a flowchart. In systems analysis, flowcharts use a range of symbols to communicate different aspects of the model. This is illustrated in Figures 1.1–1.3 in Chapter 1, where flowcharts represent conceptual models of the three case studies. A box has been used to identify entities and their states, a diamond represents decisions, a hexagon indicates procedures and lines with arrows show relationships and linkages. The aim of the flowcharts is to illustrate all the stages involved in modelling the problem from identifying the data requirements, through defining the analysis, to determining the output requirements. Figures 1.1–1.3 give an idea of the stages that must be gone through to use a GIS to address the problems of siting a radioactive waste dump, identifying areas for nature conservation or locating a neighbourhood in which to live.

GIS terminology can be avoided when constructing the conceptual data model. This is a good idea as the resulting flowchart will then explain what it is the GIS application will do in a way that is clear to all interested parties. It will give those with little GIS experience the opportunity to provide feedback on the approach. Moreover, if the conceptual model is jargon-free it can be given to GIS programmers with different software backgrounds. This may be an advantage in large-scale GIS projects where an organization wishes to compare how well different software products can address a task.

Bell and Wood-Harper (1992) provide a useful checklist for the development of a conceptual model:

1. Develop a *rich picture and root definition.* Everyone associated with the problem should agree upon these. They are used to focus the aims and direction for the project.
2. *Create a list of actions the system must be able to perform.* In the house-hunting example these actions may include permitting users to select neighbourhood characteristics such as proximity to schools, railway stations and shops, and allowing users to weight these characteristics in terms of their relative importance. These actions are known as activities.
3. *Identify a list of system inputs and outputs.* In GIS terms system inputs are data sources and outputs are products such as maps. In the house-hunting example, the data sources would include street networks, public transport routes and the location of properties for sale. Outputs might be a list, or map, of properties meeting home buyers' criteria.
4. *Group activities, inputs and outputs into a logical, chronological order.* Arrows symbolizing some form of action are used to join activities together. For example, in the house-hunting GIS the combination of data from different sources could be effectively represented in this way.

The physical data model requires additional detail that describes how to model the spatial entities, their associated attributes and the relationships between entities in the computer. Chapters 3 and 4 examined the detail of spatial and attribute data models. Therefore the emphasis here is on developing a model of the relationships between entities. This is frequently referred to as an analysis scheme. There are a number of different techniques for designing an analysis scheme that can be used; here we concentrate on an approach known as cartographic modelling.

Cartographic modelling

The origins of cartographic modelling are difficult to establish since, as Tomlin (1991) states, they are

BOX 12.3

Hard systems analysis and GIS

Hard systems analysis advocates trying to understand reality by rebuilding part of it. The link to GIS is clear, as GIS data models attempt to reconstruct parts of reality for specific purposes. During the 1970s and early 1980s the hard systems approach was the dominant methodology used for the design of computer systems. It is possible that the early developers of GIS software used a hard systems approach to design.

There are four phases in hard systems analysis. These are outlined below (after Huggett, 1980). However, before discussing these phases in detail, it is necessary to explain some of the terminology of the hard systems approach. Three key terms are entities, states and relationships. The entities, or elements of a system, are either physical objects or concepts. In GIS terms entities are points, lines, areas, surfaces and networks (Chapter 2). Entities also possess properties known in hard systems terms as states. The states associated with an entity give its character. In GIS terms states are attributes (Chapter 4). In addition, relationships exist between entities. In GIS this relationship could be the topological links between features (Chapter 3).

The four stages in hard systems analysis, in a GIS context, are:

1. *The lexical phase.* The objectives of the lexical phase are:
   - to define the problem;
   - to define the boundaries of the problem;
   - to choose the entities that define the components of the problem; and
   - to establish the states of these entities.

   In GIS this involves:
   - identifying the nature of the application;
   - selecting the study area;
   - defining the real-world features of interest; and
   - identifying associated attributes.
2. *The parsing phase.* In the parsing phase the relationship between entities and groups of entities are defined. The entities and knowledge about their states are used to create a computer model (Chapter 3).
3. *The modelling phase.* In this phase the GIS is used to address the problems identified during the lexical phase. The way in which entities and their states will interact and respond under differing situations is expressed. This may involve linking GIS software to other software.
4. *The analysis phase.* This phase is the validation of the modelling phase. Testing occurs to find out how closely the GIS model (of both form and process) fits what is observed in reality.

derived from a collection of old ideas that have been organized, augmented and expressed in terms amenable to digital processing. However, it is the work described by Tomlin (1983) as 'Map Algebra' and Berry (1987) as 'Mapematics' that established cartographic modelling as an accepted methodology for the processing of spatial information.

Cartographic modelling, at its simplest, is a generic way of expressing and organizing the methods by which spatial variables, and spatial operations, are selected and used to develop a GIS data model. Tomlin (1991) considers that the fundamental conventions of cartographic modelling are not those of any particular GIS. On the contrary, they are generalized conventions intended to relate to as many systems as possible. The truth in this statement is illustrated by the number of GIS software products that use the concepts of cartographic modelling in their approach to spatial analysis.

The concepts that underpin cartographic modelling borrow heavily from mathematics. Cartographic modelling is a geographic data processing methodology that views maps (or any spatial data layer) as variables in algebraic equations. In algebra, real values are represented by symbols, such as x, y and z. In map algebra these symbols may represent numeric attributes of map elements (for example, pH values associated with a given soil type) or even whole maps. Numbers assigned to symbols in an equation interact to generate new numbers using mathematical operators such as add, subtract, multiply and divide. In the same way, in map algebra, maps are transformed or combined into new maps by the use of specific spatial operations. Figure 12.2 shows how a GIS overlay procedure could be represented as a cartographic model and as an algebraic expression.

There are four stages in the development of a cartographic model (after Burrough, 1986):

1 Identify the map layers or spatial data sets required.
2 Use natural language to explain the process of moving from the data available to a solution.
3 Draw a flowchart to represent graphically the process in step 2. In the context of map algebra this flowchart represents a series of equations you must solve in order to produce the answer to your spatial query.
4 Annotate this flowchart with the commands necessary to perform these operations within the GIS you are using.

To explore these stages it is helpful to consider the house-hunting case study. In Chapter 1 we completed stage 1 of the cartographic modelling process by identifying the different data layers to be used in the model (Figure 1.3). Figure 1.3 also shows how natural language was used to describe what happens at various stages in the model (stage 2). The use of natural language in the development of a cartographic model is important because a user who is able to express in words the actions that he wishes to perform on the geographical data should be able to express that action in similar terms to the computer (Burrough, 1986). Tomlin (1983) recognized the role for natural language to express the logic of spatial analysis when developing the analogy between cartographic modelling and algebra. The result was GIS software with a natural language interface: the Map Analysis Package. In this package each spatial operation is a verb which acts on subject nouns, representing map layers, to create object nouns, or new map layers. Table 12.1 provides a selection of key verbs of spatial operations used in Tomlin's approach.

Stage 3 of the cartographic modelling approach involves adding detail to the flowchart to represent the logic of the analysis required by the GIS project. Figure 12.3 provides an example of how this might be done for part of the house-hunting GIS. Table 12.2 presents four of the equations required by the analysis scheme shown in Figure 12.3. The equation numbers refer to those spatial operations identified by a number in Figure 12.3.

The final stage in the cartographic modelling process is to annotate the flowchart with the appropriate commands from the GIS package in which it is intended to perform the analysis.

From the above example, it should be apparent that the power of cartographic modelling lies in the structure it provides to the GIS designer. This structure allows the designer to tackle a complex spatial problem by breaking it down into its components. Simple statements, sections in a flowchart or solvable equations can then express these. However, it may not be possible or sensible to do all your analysis in the GIS. In certain cases it may be necessary to couple the GIS with other applications to obtain results (Chapter 7).

If care has been taken over the construction of the rich picture and root definition and their subsequent translation into a conceptual and physical data model, then, at least in theory, the computer implementation of a GIS data model should be relatively straightforward. In reality, it is likely that building an application will start while detail is still being added to the GIS data model. There is nothing wrong with this approach as long as the

Equation $x + y = z$

where x = Road map
y = Rail map
z = Communication map
+ = The spatial overlay operation 'union'

Graphical representation

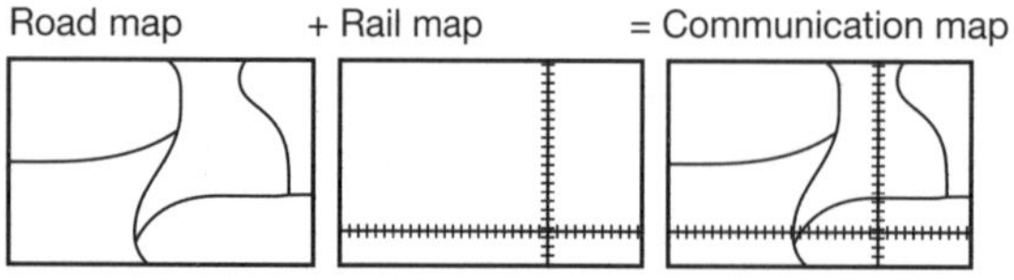

Road map + Rail map = Communication map

Figure 12.2 A simple map algebra equation

Table 12.1 Selected examples of natural language keywords

Keyword	*Operation*	*Description*	*Example of use*
SPREAD	To create a corridor from a linear data set, or zone of influence around a point	Calculate the distance of all geographical positions in the data set from a given starting point or line	To create buffer zones representing proximity to schools, main roads or stations
OVERLAY	To find the intersection of two different sets of area entities covering the same geographical region	Lay two different sets of area entities over each other to produce a new, more complex set of areas	To create a map of accessibility from an overlay of proximity to stations and proximity to main roads
EXTRACT	To extract a new data set from an existing data set	Select specified values and/ or ranges of values from one overlay to make a new overlay	To create a new data layer showing only those houses that are for sale within the area defined by the accessibility map

Source: adapted from Tomlin, 1991

Table 12.2 Simple equations for Figure 12.3

Equation 1. From land use 'extract' countryside	*Equation 2. From road map 'spread' road by distance*
$a - b = c$ where: a = land use map b = urban land use map c = countryside	$(d - e) + f = g$ where: d = road map e = portion of map which is not road f = zone of spread for a specified distance either side of the road g = proximity to road map
Equation 3. From house status 'extract' houses for sale	*Equation 4. 'Overlay' houses for sale in countryside away from roads*
$h - i = j$ where: h = house status map i = houses not for sale j = houses for sale	$j + k = l$ where: j = houses for sale k = countryside away from roads l = houses for sale and countryside away from roads

enthusiasm for implementation does not take over, leaving missing details forgotten. Trial and error with systems development is an accepted approach, but often individuals working alone may spend many hours developing a solution only to find that they have not documented their work and are therefore unable to explain to others how the result was reached. How you implement a physical data model will depend upon the nature of your problem and the organizational setting in which you are working. However, good project management will help ensure that goals are met. Box 12.4 suggests questions that should be addressed at this stage of the data modelling process.

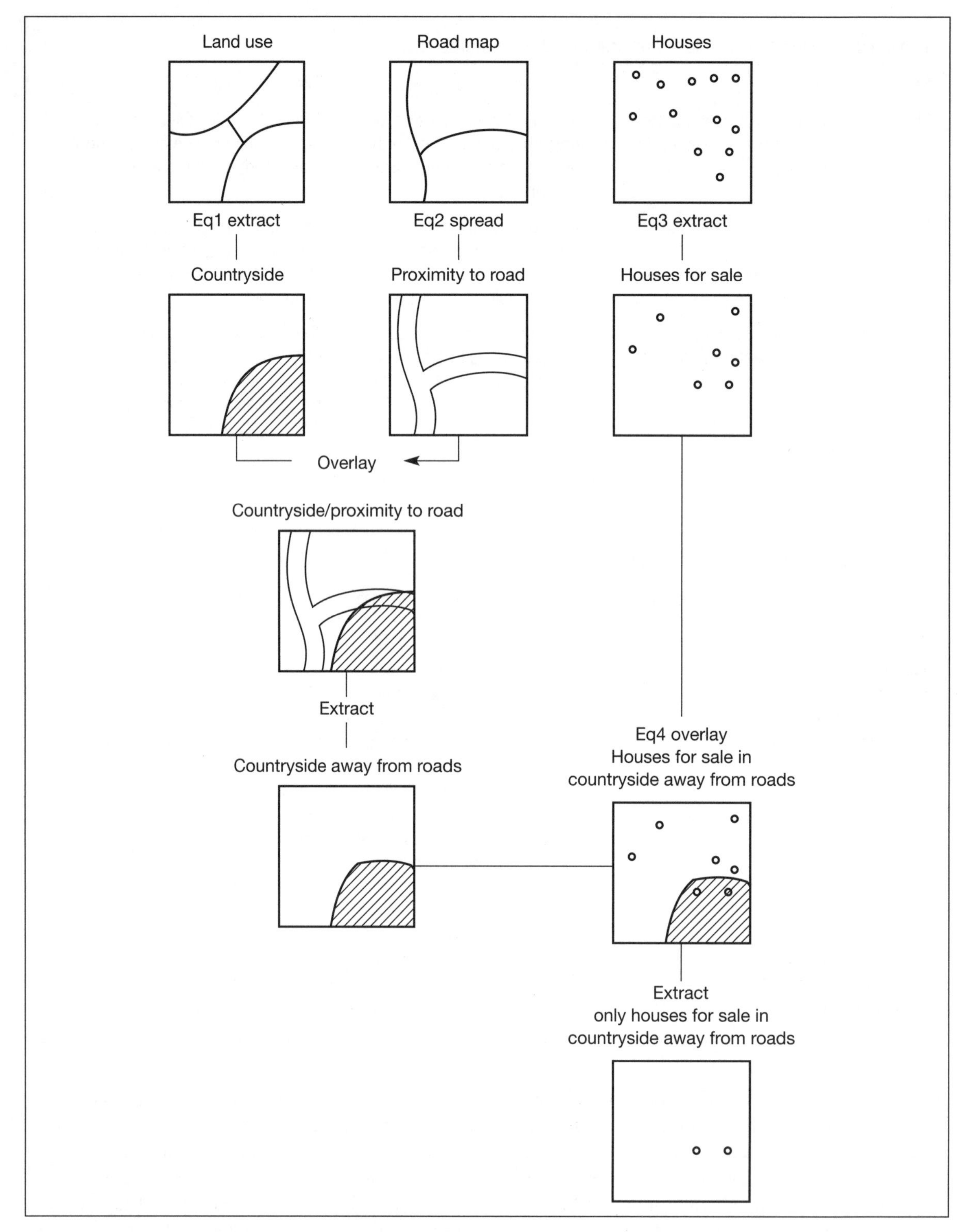

Figure 12.3 Example of part of the analysis required by the house-hunting GIS

BOX 12.4

Designing a data model: questions

1. What are the elements of your conceptual data model: entities, attributes and relationships?
2. What are the four stages in the design of a conceptual model? Can they be represented in a flowchart?
3. What spatial and attribute data models and structures are required?
4. What approach to designing an analysis scheme is appropriate? Can the ideas of cartographic modelling be applied?
5. Can GIS offer all the functions required for the analysis?

Project management

This chapter has considered techniques for identifying the character and extent of a spatial problem and techniques for helping with the design of GIS data models. However, as discussed in Chapter 11, accurate problem definition and good data model design are only part of the story. Once the data model is constructed the GIS must be implemented and in many cases integrated into the wider information strategy of an organization. To help this process good project management is an essential prerequisite for success.

There are many different approaches to managing information technology (IT) projects. Two approaches commonly used by GIS designers are the system life cycle and prototyping.

Systems life cycle approach

The systems life cycle (SLC) approach advocates a linear approach to managing the development and implementation of an IT system. It is also referred to as the 'waterfall model' (Skidmore and Wroe, 1988). The waterfall analogy is used because the outputs from the first stage of the process inform the second phase, and the outputs from the second phase affect the third phase, and so on. There are many variations on the general approach (Figure 12.4). The house-hunting case study provides an example of how a project could be managed using SLC.

1 *Feasibility study*. This would involve asking the real estate agents and home buyers questions about whether they would make use of the system being proposed for development and what the costs and benefits of developing a GIS would be. If the feasibility study is positive then the project moves to the second phase.
2 *System investigation and system analysis*. The GIS designer would try to establish the current way in which home buyers and real estate agents interact to identify houses for sale in appropriate neighbourhoods. This would include identifying the data and analysis requirements as well as the preferred output types. A soft systems approach could be used to help with this phase.
3 *System design*. The GIS data model is constructed using information collected in the previous phase. In the house-hunting example cartographic modelling techniques might be used to help structure the analysis requirements of the GIS.
4 *Implementation, review and maintenance*. Now the house-hunting GIS is built and provided to users. This may be the first opportunity for users to comment on, or interact with, the system since their

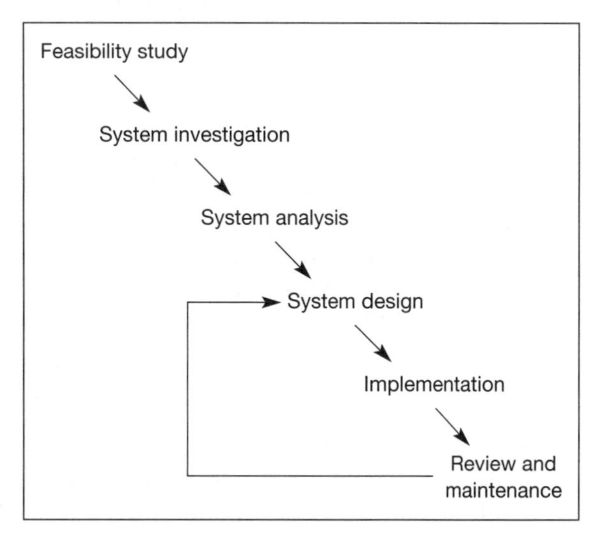

Figure 12.4 The system life cycle 'waterfall' model

involvement in the feasibility study. Users' experiences inevitably require changes to the system. These may include the addition of new data layers, new analysis techniques or new ways of visualizing the output.

The main advantage of the SLC approach is that it provides a very structured framework for the management of a GIS project. This can be extremely important when good time management is an essential aspect of the project. In addition, it is often easier to budget for the resources required by an SLC approach because the requirements of the system are established at an early stage in the project.

Despite its popularity as a project management tool for IT projects there are a number of problems with the SLC approach:

- Designers who use the SLC approach often fail to address the context of the business for which the system is being developed. The approach encourages the designer to focus on only a part of the information problem of an organization. In the case of the house-hunting example an SLC approach might prevent some of the external factors that influence where people choose to live being included in the model.
- The timescale and linear nature of the SLC process do not allow for change in the scope and character of the problem. By the time the system is implemented it may be out of date. In the house-hunting case there may be a government policy review which affects the way houses are bought and sold between the feasibility study and implementation.
- The SLC approach does not put the user at the centre of the system design. It emphasizes the identification of flows of information rather than understanding why they are required. This creates problems because it only allows a system that reflects the current way of doing things. This may be a problem for GIS design as a new system may radically change the way information is managed.
- Finally, the SLC approach is often considered to favour hierarchical and centralized systems of information provision. It offers a very technocentric view of system development.

The prototyping approach

The prototyping approach to IT project management developed as a response to the criticisms of the SLC approach, particularly in response to the lack of consideration of users. Figure 12.5 shows the main stages in the method. The user first defines the basic requirements of the system. This could be achieved by using the rich picture and root definition techniques described earlier in this chapter. The system designer takes these basic ideas to construct a prototype system to meet the needs identified by the user. In GIS projects such systems are often described as demonstrators. The users who identified the original requirements for the system then experiment with the demonstration system to see if it is what they expected. Other potential users of the final system may be brought in at this stage to see if the system is of wider value. The system designer uses their recommendations to improve the system.

The prototyping approach has a number of advantages over the SLC method:

- Users have a more direct and regular involvement in the design of the system.
- It is easier to adapt the system in the face of changing circumstances which were not identified at the outset of the project.
- The system can be abandoned altogether after the first prototype if it fails to meet the needs of users. This reduces the cost of developing full systems.
- If money and time are available a number of prototypes can be built until the user is satisfied.

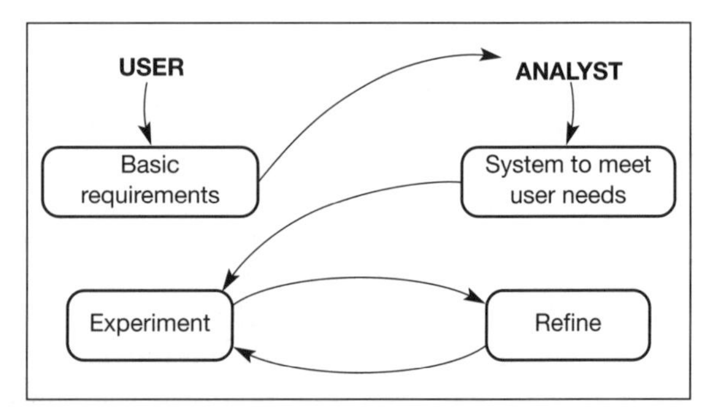

Figure 12.5 The prototyping approach

Table 12.3 Project management techniques

Technique	*Purpose*
SWOT analysis (strengths, weaknesses, opportunities and threats)	This technique is used to establish the strengths, weaknesses, opportunities and threats associated with the development of the GIS. It is often used as part of the feasibility study in the SLC approach
Rich pictures and root definitions	These techniques come from a methodology known as 'soft systems' and are used to help system designers determine the scope of a problem. They can be used both in the feasibility phases of the SLC approach and to help users identify the basic requirements of a system in the prototyping approach
Demonstration systems	These are demonstration GIS applications, which are designed to help users evaluate whether or not the full system being proposed by the system designer will work. They are most frequently used when a prototyping approach to project management is used
Interviews and data audits	These techniques are used to help problem definition and to establish the current information and analysis needs of an organization. They are more structured than the rich picture and root definition techniques described above. Data audits can be particularly valuable in a GIS context, as many organizations are unaware of the spatial data to which they have access
Organization charts, system flowcharts and decision trees	These three techniques are all variations on the flowcharting theme. The organization chart maps out the flows of information within the organization. The system flowchart describes how the system will model these information flows. The decision tree shows the problem from a decision-making perspective and focuses on showing how different decisions cause information to be used in different ways within the organization. The technique used will depend on the experience of the system designer and the character of the problem
Data flow diagrams and data flow dictionaries	These techniques are drawn from hard systems analysis and represent a much more structured approach to system design than some of the techniques described above. They can be of immense value in GIS for tracking what happens to a data layer through the analysis process. This is extremely valuable in monitoring data quality and providing lineage information
Cartographic models and entity relationship diagrams	These techniques are of most value in structuring the analysis schemes used in GIS. In this respect they help plan the functional requirements of each stage of the analysis

The prototyping approach, however, does have problems:

- Prototyping can be difficult to manage. There may be large numbers of users with large numbers of ideas and opinions.
- The resource implications may change following the development of the first prototype.
- Knowing when to stop development can also be a problem. However, some GIS designers argue that this is a positive aspect of the approach since few, if any, GIS systems are ever finished.

The SLC and prototyping approaches are just two of many that can be adopted for the management of a GIS project. There are also many variations on the basic approaches outlined above. It is also possible to pick and mix aspects of the two approaches to develop a management style that suits the development environment. In addition, there are a wide range of project management techniques and tools which can be used to help with various phases in the SLC and prototyping approaches. Several of these – rich pictures, root definitions and flowcharts – have already been considered in this chapter. These are now summarized alongside some new techniques in Tables 12.3 and 12.4. General questions about project management that need to be addressed in relation to a GIS project are presented in Box 12.5.

Implementation problems

There will always be problems for GIS design and development which no amount of prior planning can prepare for. Three of the most common are:

- data in the wrong format for the GIS software;
- a lack of GIS knowledge imposing technical and conceptual constraints on a project; and
- users of the GIS frequently changing their mind about what they want the GIS to do.

BOX 12.5

Project management: questions

1. Which management approach is appropriate for your application – system life cycle or prototyping?
2. What techniques will you use to ensure that your project is adequately documented?
3. Will it be possible for others to recreate how the application has been constructed if things go wrong?
4. Can someone else use the documentation you are producing as part of your system development to tackle the project?

In many GIS projects the data required are unavailable in a format compatible with the GIS software or analysis needs. If this is the case there are two options: to look elsewhere for a supplier, or to convert the data into the desired format. Both these approaches have been dealt with in Chapter 5. However, in the case of the latter, errors may creep in as data are changed from one format to another. Alternatively, the conceptual data model could be revisited to assess the importance of the data, and evaluate alternative data options.

It is inevitable, at some point, that applications will be limited by users' technical or conceptual knowledge about which spatial data model, data structure or analytical operation is most appropriate for the task required. Much can be learnt from other applications and other users. Colleagues working in similar areas, other organizations, or the Internet can all be sources of help. For many organizations, the solution is to employ an independent expert to undertake application development or specific analysis.

The dynamic nature of the GIS design process is such that the information needs of users are often in a constant state of flux. By the time a GIS data model is implemented, the needs of the users and the scope of the problem may have moved away from the original defined by the rich picture. This

Table 12.4 Project management tools

Project management tool	*Description*
GANTT charts	GANTT charts are time management tools that establish when and where, over the life of the project, a particular task will take place. Figure 12.6 shows a GANTT chart for the house-hunting GIS
PERT (program evaluation review techniques)	These are graphical tools for managing a project by showing how a task depends on the completion of others before it can be undertaken. One of the main values of a PERT chart is that it shows which tasks can be undertaken in parallel (Figure 12.7)
CASE (computer-assisted software engineering) tools	These are software tools that help reduce program development time. They can be used in a number of ways. They may help with project management, for example in the construction of GANTT and PERT charts. They can be used to assist software development and the construction of conceptual models and analysis flowcharts. And in the later stages of a project they may be used, like 'wizards' that allow automatic construction of documents in a word processor, to help create prototypes

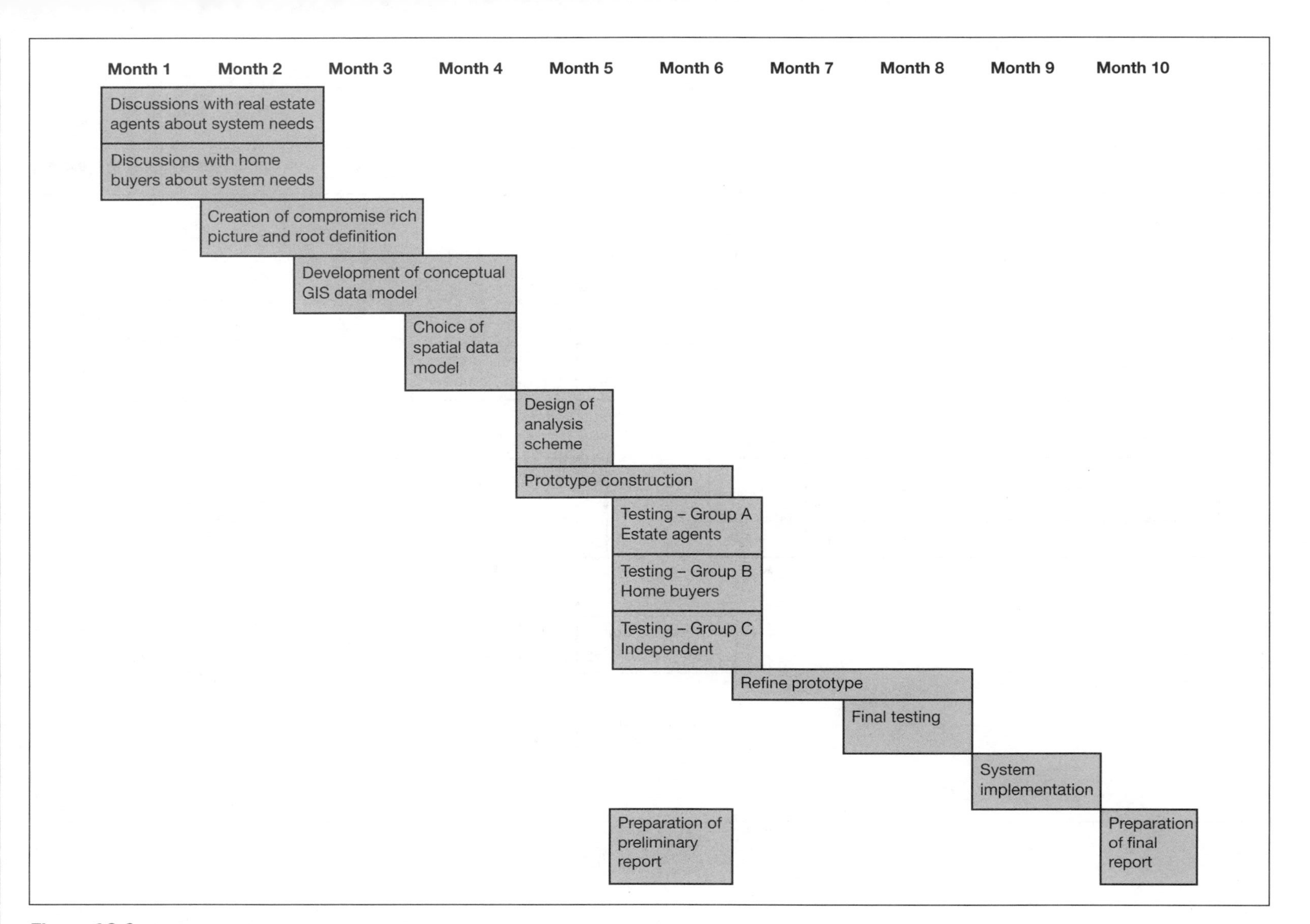

Figure 12.6 GANTT chart for the house-hunting GIS

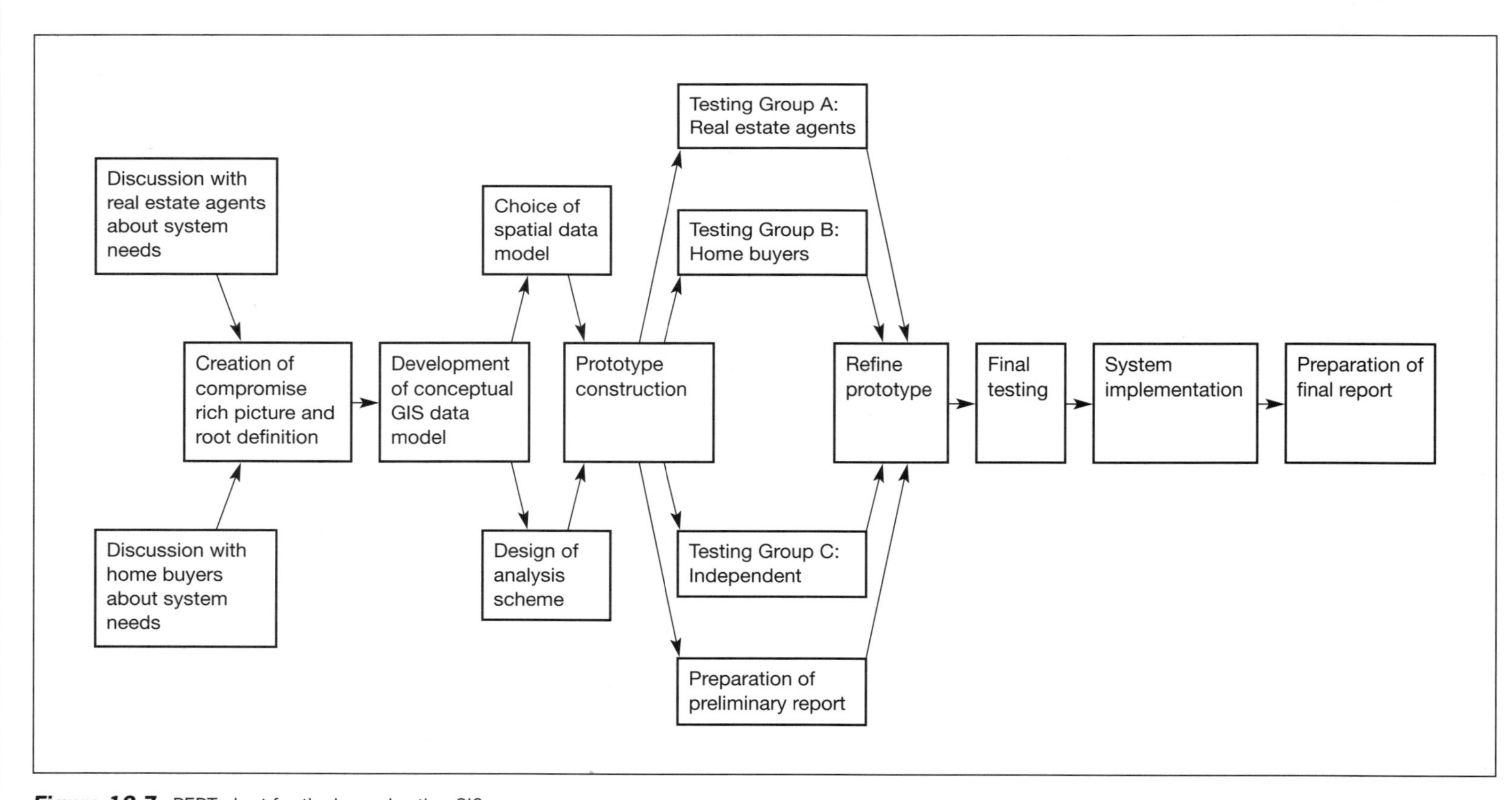

Figure 12.7 PERT chart for the house-hunting GIS

is a major issue in the development of GIS applications for larger organizations, where applications development may take considerable time, and the awareness of key players about GIS may increase in the meantime. The solution is to gain frequent feedback from the individuals who will be the end-users of the GIS. They should reconfirm that the scope of the project has not changed, or allows changes to be brought into the design process. Even in a small-scale GIS project, as the knowledge of the GIS analyst grows there may be changes in the aims and scope of a project. Box 12.6 presents some of the questions that should be addressed during the implementation of a GIS.

BOX 12.6

Implementation problems: questions

1. Are data available? What format are they in? Can they be entered into the GIS?
2. Can an appropriate attribute database be created?
3. Is appropriate software available? Does the software impose restrictions on the analysis that can be conducted? If so, can the analysis scheme be revised or the software changed?
4. Will the output produced still meet users' needs? Have the users' needs changed? Has the root definition changed in the light of increased knowledge and awareness of GIS and the problem to be addressed?

Project evaluation

After a GIS application has been constructed, some problems may be just about to start. It is important that the output produced by the system is usable, valid and meets the goals set at the beginning of the project. Validation of results is often difficult to achieve, particularly if results are in the form of predictions. However, if you are working with an organization, testing the GIS and validating output will be a crucial part of the design process. In many cases, this may well result in adjustments to the rich picture and the GIS data model. In extreme cases the GIS may have to be abandoned, and the project restarted. This feedback process can be very costly and often explains why many organizations adopt the prototyping approach to project management described earlier in this chapter. Prototyping should prevent a system from being inappropriate, as frequent testing and evaluation should be taking place.

There are three tests that can be used to check whether a GIS application meets the goals set for it at the start of the design process. First, all the parties involved in the design and development of the GIS can be asked if they are using the application for the purpose for which it was designed. If they are not, or have even reverted to using the old methods, it is a sure sign that something has gone wrong somewhere. The goals originally used to guide them in helping to identify the scope of the problem may have changed as time and work priorities have altered. Alternatively, users may find the application difficult to use, or without a key feature. In such cases, further training, or adaptations to software, may be all that is required to ensure that the GIS are effectively used.

Second, the GIS output can be checked against reality. This type of test would be appropriate for the avalanche prediction model as the location of avalanches could be predicted before the start of a season and then compared against actual avalanches at the end of the season. For the house-hunting GIS the best check would be to see if home buyers found a house that met all their requirements.

Third, the adaptations and changes that had to be made when moving from the rich picture through the GIS data model to the GIS implementation can be evaluated. Whether these were due to knowledge deficiencies, to poor problem definition or to system adaptations because the software or data would not permit implementa-

BOX 12.7

Project evaluation: questions

1. Does the GIS help to address the problem identified at the outset?
2. Has the project been implemented as planned?
3. Have there been many system adaptations or modifications?
4. Are the users of the GIS satisfied with the product?

tion of the model as planned, can be assessed. If the system adaptations have dominated the development of the application then it may be that a technical solution has been provided that has little resemblance to the reality of the initial problem.

Conclusions

This chapter has reviewed the stages involved in the design and management of a GIS application. Details of a methodology that will encourage successful design, development and implementation of a GIS have been presented. The methodology described draws heavily on the ideas of both hard and soft systems analysis, as well as ideas from other system design methodologies including the system life cycle and prototyping approaches. It does not favour one specific model over another but takes elements from several. Different design methods have advantages in different situations depending upon the type, scale and nature of the GIS application being developed.

A key element of the methodology is the time and importance attached to problem identification, the design of the GIS data model, the development of an appropriate analysis schema and careful project management. In addition, the method stresses the need for all individuals concerned to be involved in the design and validation of the GIS.

While this chapter has presented a theoretical framework for the design and implementation of a GIS project, in practice the design issues need to be addressed concurrently with a whole set of practical matters including:

- data availability;
- data encoding;

BOX 12.8

Checklist of details to be included in a GIS project log

1 GIS project title
2 Aim and scope of the project
 2.1 Root definition
 2.2 Rich picture
3 Geographical area concerned
 3.1 Dimensions of the area concerned
 3.2 Resolution required
4 Numbers of users of the GIS
 4.1 Who are the users?
 4.2 What kind of knowledge do they have?
5 List of digital thematic maps required
 5.1 Feature
 5.2 Entity type (point, line or area)
 5.3 Raster or vector?
 5.4 Accuracy
 5.5 Source
6 List of attribute data requirements
 6.1 Feature
 6.2 Description
 6.3 Character or numeric?
 6.4 Accuracy
 6.5 Source
7 Database requirements
 7.1 Database queries required
 7.2 Frequency of updating
8 Analysis requirements
 8.1 List of analysis functions required
 8.2 Simple language flowchart of the GIS analysis model
 8.3 System specific flowchart of the GIS analysis model
9 Description or picture of interfaces required for user interaction
10 Choice of GIS software
 10.1 Principal requirements
 10.2 PC/workstation/network and number of installations
 10.3 Risks
11 Choice of hardware
 11.1 PCs/workstations
 11.2 Networking
 11.3 Peripherals
12 GANTT chart
13 Resources and costs
 13.1 Fixed investments (PCs, workstations, other hardware, software, data)
 13.2 Continuous costs (manpower, GIS specialists, system manager, project manager, maintenance, data updating

- analysis methods available;
- output requirements;
- software availability;
- time available; and
- resources available.

Some of these issues have been touched upon, and in reality these may dictate the stages in the design and development of a project. For example, in the Zdarske Vrchy case study a lack of remotely sensed data for a part of the study area meant that this area had to be excluded from some of the analysis, despite it being included in the project design. Thus, results of analysis were unavailable for this region. However, good project design and management mean that less will be left to chance and a GIS application will be more likely to meet the needs of its users. Box 12.8 offers a checklist to facilitate implementation of the ideas in this chapter. This checklist, together with the questions from boxes earlier in the chapter, provides a useful framework for project management.

REVISION QUESTIONS

- What is a rich picture and how is it used to assist GIS project design?
- What are the differences between conceptual and physical data models?
- What is cartographic modelling? How can cartographic modelling be used to develop a GIS analysis scheme?
- Review the main problems that might occur during the implementation of a GIS application.
- Compare and contrast the system life cycle and prototyping approaches to managing a GIS project.
- Discuss the role of flowcharts in the management and design of GIS projects.
- How can a GIS application be evaluated?

Further study

There are several sources of further reading which provide details of designing and managing the development of a GIS application. DeMers (1997) offers a comprehensive overview of the software engineering approach which draws heavily on the ideas of hard systems analysis. Reeve and Petch (1998) put an emphasis on soft systems and prototyping methodologies. Chrisman (1997) discusses several ideas for assessing data quality and defining the analysis requirements of a GIS project. Perhaps one of the most comprehensive overviews of GIS project management is provided by Huxhold (1991) in his chapter titled 'The model urban GIS project'. This chapter provides a good overview of the practicalities of developing a large-scale GIS project, and also considers how organizational factors influence the process.

From a technical perspective, the tutorial material accompanying specific GIS software may provide a good starting point and many of the tips presented for one system often have generic application. In addition, software-specific books for systems such as ARC/INFO (Zeiler, 1994), MapInfo (Daniel *et al.*, 1996) and MicroStation (Sahai, 1996) provide useful information for designing and managing a GIS project using these software products.

Outside the GIS literature there is an extensive literature on IT project management and design. Although a little dated, the books by Skidmore and Wroe (1988) and Avison and Fitzgerald (1988) still provide good introductions to the subject for non-computer scientists. If you are unable to find these two books virtually any text on IT project management will cover the range of techniques and tools introduced in this chapter.

Avison D E, Fitzgerald G (1988) *Information Systems Development: Methodologies, Techniques and Tools*. Blackwell Scientific, Oxford

Chrisman N R (1997) *Exploring Geographic Information Systems*. Wiley, New York

Daniel L, Paula L, Whitener A (1996) *Inside MapInfo Professional*. OnWord Press, USA

DeMers M N (1997) *Fundamentals of Geographic Information Systems*. Wiley, New York

Huxhold W E (1991) *An Introduction to Urban Geographic Information Systems*. Oxford University Press, New York

Reeve D E, Petch J R (1998) *GIS Organizations and People: a Socio-technical Approach*. Taylor and Francis, London

Sahai R (1996) *Inside MicroStation 95*. OnWord Press, USA

Skidmore S, Wroe B (1988) *Introducing Systems Analysis*. NCC Publications, Manchester

Zeiler M (1994) *Inside ARC/INFO*. OnWord Press, USA

13 The future of GIS

KEY QUESTIONS AND ISSUES

- What are the problems with current GIS?
- What technological developments will occur over the next few years?
- Will the relationship between GIS and other technologies have an impact on future developments?
- How will GIS developments affect human and organizational issues?
- What can we expect GIS to be in 10, 20 or even 50 years time?
- Will GIS survive long into the twenty-first century?

Introduction

Making predictions for a field changing and developing as rapidly as GIS is very difficult. In spite of this, there are many authors who have been prepared to predict the future of GIS. Sometimes they look forward as an intellectual exercise, using their knowledge and imagination to try to anticipate the world in a few years or decades time. They also look forward to assist businesses that depend on staying one step ahead and meeting the next, perhaps as yet unidentified, customer need. Managers and decision makers contemplating spending thousands or even millions of pounds on a GIS also need to look to the future. They need to 'future proof' a new system, and make sure that it will last, or at least be upgradeable when the next set of technological developments comes along. Some of the well-known techniques for justifying investment in new technology, such as cost–benefit analysis, involve an element of crystal ball gazing. Costs and benefits are predicted to justify a project, despite the fact that the benefits may never actually be achieved. Strategic planning exercises, such as writing development or research plans for a business or company, also require a look into the future.

Imagine that the Martian is visiting Earth on a research project. He has to produce a report that assesses the current state and future potential of the motor car. What does he need to consider? His adventures in previous chapters have helped to establish many of the technical and practical issues of car ownership and use. Now he must consider the future. First he may examine the predictions of others for the next 20 years. If these are not available he can make his own predictions. To do this he must understand the nature of the current industry, and identify strengths and weaknesses. For example, perhaps he would be concerned that current cars use too much fuel, create too much pollution, or that the roads are too congested for efficient car use. Having identified and evaluated current problems, he could suggest areas where more research is necessary and begin to make predictions for the future.

This chapter presents some of our ideas for the future of GIS. First, the GIS of the 1990s is reviewed and some of the predictions for GIS in the 1990s are revisited. The problems and limitations of current GIS are considered, and from this research and development issues for the future are identified. The predictions of other authors and the impact of GIS on our daily lives are considered. Finally, the question of whether GIS will still exist in a few years time is addressed.

GIS in the 1990s

In the early 1990s GIS were general-purpose software packages. Automated computer cartography products were widely used, and many users were beginning the long task of establishing digital databases for large applications in the utilities, local government and environmental research.

With many applications in the process of being established, data encoding was an important topic. Data encoding was a drawn-out and expensive process because of the limited availability of digital data. In the UK, for example, the national mapping agency (Ordnance Survey) had digitized only part of the national base data that are essential for many applications. Data for many areas, particularly rural areas, were not available. Widespread use of manual digitizing techniques made the data-conversion process slow. Large volumes of data were generated, but storage devices were still relatively limited. PCs were widely available, but CD-ROMs, disks with storage capacities of over a few hundred megabytes and the use of electronic communications for data transfer and exchange were relatively rare. Large applications needed serious computing power – either workstations or mainframe computers. Data sets were being digitized by many different organizations, but it was difficult to find out if someone else had already digitized the data that you needed. If you could find digital data, there were questions of format, price, copyright and confidentiality, which all made data exchange a difficult process.

At a more conceptual level, the layer model of the world was dominant in GIS. Some packages offered a vector view of the world, and others offered only raster-based functionality. Analysis was restricted to one data model or the other, with little or no integration of raster and vector data sets. Even the inclusion of a raster background map was relatively unusual. Modelling was often conducted outside the GIS package, with the results being imported for map display. Alternatively, GIS was used to generate variables for models run elsewhere.

Issues of software usability were being discussed. The GIS of the early 1990s was not Windows-based – most packages were command-line driven. Programs on PCs were DOS-based; those on mainframes used other non-intuitive operating systems. Users had to become expert at remembering complex commands. These commands had to be typed in, without errors, to operate software. If data transfer or editing were required, GIS users would write programs, using programming languages such as FORTRAN, to undertake reformatting, as spreadsheets and database software were unable to perform many of the tasks required. A high level of skill was required, and generally GIS expertise was at a premium. Training and awareness of GIS were important, and GIS courses were becoming widely available. In the commercial world many organizations that could benefit from GIS were unaware that it existed, and in others an individual 'champion' had promoted the case for GIS to get it into their work place. The case for GIS generally was hampered by a lack of good case studies of successes and problems, and lack of any evaluated cost–benefit analysis to justify investment in systems. On the other hand, a huge amount of hype was beginning to emerge, with large-scale conferences, vendor offers to education and the emergence of new products. Very little, if anything, was really known about the impact of GIS on organizations, although the potential of GIS to add value to an organization's data was clear, and research had begun to improve data issues and help organizations with the implementation of systems.

BOX 13.1

Issues for GIS in the 1990s

Practical and technical issues

- How can the conversion of map data to digital format be made more efficient and cost-effective?
- How can large geographical data sets be stored when file sizes may exceed machine capacity? How can large data sets be accessed efficiently?
- How can data be commodified? How can data sets be identified and obtained? What about issues of pricing, liability, copyright, ownership and access? What will be the role of national mapping agencies in the supply and generation of data?
- How will new technologies affect the traditional methods of geographical data collection and compilation?

Conceptual GIS issues

- How can data models represent the complexity of the real world more accurately?
- How can GIS cope with different types of data about our world, for example data that cannot be represented by co-ordinates or pixels (sound, photographs, video and free-form text)? How can GIS complement other technologies and techniques for handling these data?
- How can GIS database functions be linked with more advanced modelling capabilities, such as those available in other software?
- How can uncertainty in data sets and in analysis operations be dealt with?

Implementation and user issues

- How can GIS be made easier to use? What makes an interface effective and appropriate?
- What are the costs and benefits of introducing GIS? How can these be assessed?
- What is the impact of GIS on an organization? What can be learned from successful and unsuccessful applications?
- How can GIS awareness be raised, and what education and training is necessary for organizations? How can training in particular systems be improved?
- How will the rigorous, objective perspective of GIS be adapted to the imprecise, subjective world of human reasoning and decision making?

(adapted from Maguire, 1991; Rhind *et al.*, 1991; and Rhind, 1992)

At the beginning of the 1990s Maguire *et al.* (1991) considered that GIS had not found their full potential. Instead of being used for decision support, and for exploration and analysis of our world, the systems in use were often limited to mapping, inventory and information management type applications. Maguire *et al.* (1991) went on to identify the issues they considered were facing the GIS community in the early 1990s. These issues – data problems, conceptual, analysis and user issues – together with others suggested by Rhind *et al.* (1991) and Rhind (1992), are summarized in Box 13.1.

Rhind (1992) offered two contrasting views of the future of GIS. First, he suggested that GIS might fragment and disappear. More positively, and more accurately, he predicted a convergence in GIS that would result in:

- better tools for the handling and exchange of information (faster computers and improved networks);
- a better-served GIS community (with conferences, textbooks and magazines);
- new players in the GIS industry;
- the development of standards for data issues such as exchange and structures; and
- the development of a core set of ideas to inform teaching and training in GIS.

These are just some of the predictions that Rhind made. Others are given in Box 13.2 and the progress made in these areas is reviewed below.

Progress on conceptual and technical issues

Current data models for GIS are limited. The third and fourth dimensions are poorly catered for, and there is an overemphasis on the map and image presentations of spatial data. The object-oriented data model has undergone significant development in the 1990s and commercial systems using this model are now available. However, most GIS are still firmly fixed in the raster and vector views of the world, and there are no widely used GIS with established methods for handling three-dimensional or temporal data, despite advances in related fields such as geological modelling. The lack of tools for handling temporal data restricts the integration of process models with GIS, an issue important in many environmental applications (Goodchild, 1992).

There has been widespread development of analysis functionality since the early 1990s. In addition, links to other software are often provided in GIS products, and model and menu builders exist in some packages to allow the integration of analysis from elsewhere, or the addition of new functions programmed by the user. Visualization of both 3D and time data is also an area of research and development (Openshaw and Perree, 1996; Dorling and Openshaw, 1992; Dorling, 1993).

Data lineage tracking functions in some software have assisted quality control; however, there has been less progress on the widespread integration of error-handling tools. Apart from Monte Carlo simulation techniques, it seems that to most software developers the inclusion of error-handling tools would be an admission that their software is faulty! This is disappointing, since interest in error was clear while the USA's National Center for Geographic Information and Analysis (NCGIA) initiative in this area was in progress. This ended in the mid-1990s. In other areas, quality issues are still being considered. Data quality is often examined and documented by the data supplier, and there are data standards for supply and transfer of data. Beyond this, GIS users are often left on their own.

Multimedia has been an area of general developments in computing in the 1990s, and GIS has not missed out. Many applications now include scanned images (for instance, pictures of property and assets), but sound and video are less common. There are some examples of advances; for example, Shiffer (1995a) has illustrated the development of multimedia GIS from the video disks of the late 1980s to recent 'object-based spatial–nodal videos on the World Wide Web', which are designed to assist participatory planning and mediation.

The Global Positioning System (GPS) has been a major growth area of the 1990s. There has been a proliferation of products from 'cheap and cheerful' navigation devices to expensive surveying equipment. GPS can be used to collect field data, and conversion routines and standard data formats allow direct input into GIS packages. The integration of GIS and GPS has progressed further than Rhind perhaps expected. GPS receivers can

BOX 13.2

Twenty predictions for GIS

A. Conceptual and technical predictions

1 Data models will be developed to handle 3D and time characteristics, and complex interactions between objects.
2 There will be new analysis and support functions to allow, for example, the tracking of data lineage or visual interaction with the stages in an analysis process.
3 There will be support for quality assurance and quality control.
4 Support for multimedia will become common.
5 GIS and GPS developments will occur. GIS and GPS will be integrated for data collection and compilation, and a decline in mass digitizing will be accompanied by a growth in use of GPS.
6 Tools for visualizing 3D and time-dependent data will be developed.
7 There will be a convergence in general-purpose GIS, with most systems running under Unix and functionality becoming more similar.
8 Sector-specific products will probably appear.

B. Market predictions

9 Vendors will attempt to differentiate products by efficiency of coding, friendship schemes and other means.
10 The bulk of the GIS market will be outside the USA. Europe and Japan will challenge American supremacy, and political factors will ensure that software and system creation will be carried out in multiple locations.
11 Users will want local customization of global products, for example translation into their own language.
12 There will be a growth in 'value-added services'.

C. Data predictions

13 Digitizing will be done where it is cheapest. The peak of mass digitizing will be over in the USA and UK by 2000.
14 The data volume problem will disappear in some applications (due to, for instance, CD-ROM).
15 Ownership of data will become an issue, and data issues will continue to be affected by government policies.
16 Concern over privacy and confidentiality of data will be an issue in socio-economic applications.

D. Human and organizational issues

17 Improved techniques will become available for the GIS acquisition and project management process.
18 GIS will become part of wider management information systems in many organizations.
19 The skills and professionalism of individuals involved will influence the success of applications.
20 Education and training will have to concentrate on setting and demonstrating standards as well as curriculum content.

(adapted from Rhind, 1992; Rhind *et al.*, 1991)

frequently collect attribute as well as positional data, and field-based computing systems combining GIS and GPS technology allow direct plotting of data onto a map base whilst in the field. Muller (1993) also foresaw GPS as having a major impact on GIS. He was concerned that duplication of data collection should be avoided, and that standards would be necessary. As GPS becomes routinely used, duplication is a problem, and standards and quality are difficult to ensure in a field where complex equipment collect data which are difficult to interpret. This is an area in which there is now a wide published literature with research papers and books such as Cornelius *et al.* (1994), Hoffman-Wellenhof *et al.* (1994) and Kennedy (1996).

Rhind predicted a convergence in general-purpose GIS with most systems running under Unix and functionality becoming more similar. However, this has not been seen. In fact there has been a resurgence of interest in PC-based products under the Windows operating system. Functionality has become more similar in the sense that most systems can now handle rasters and vectors, and data exchange is standard. However, the development of new products for map viewing and querying, and for specific market sectors, has meant that functionality between products has diverged. Some of the companies producing 'high-end' multi-purpose GIS for large-scale organizational applications have even failed. The market has moved away from this type of product. Mangold (1997) suggests that it is time for a change of emphasis in GIS products in the commercial world, from top-down consultant-designed systems to bottom-up small-scale user- and problem-led solutions. This probably ties in with the near saturation of the market for systems in organizations such as utilities, local government and large commercial organizations, and the predicted expansion of markets in service and business sectors, as foreseen for Europe *by Mapping Awareness* magazine (Anon., 1997a).

Progress on products and markets

With the development of PC-based systems there have been attempts to develop products for different market sectors. These handle specific applications and offer targeted functionality. The software may arrive pre-loaded with relevant data. The functionality offered may include data query and presentation tools, and operations to meet the needs of specific users from different backgrounds. For example, hybrid GIS/CAD systems have emerged. Other software products, including databases, spreadsheets and statistical analysis packages, have also been given mapping extensions.

In the light of these changes, as Rhind predicted, the vendors need to differentiate their products to promote them to potential customers. Wilson (1997) has identified four trends amongst GIS vendors:

- the development of strategic alliances amongst companies of equal status and the signing of co-operative agreements;
- mergers and acquisitions to allow companies to extend their capabilities;
- the establishment of third-party developer programs which allow independent developers to deliver value-added product enhancements; and
- the forging of partnerships with customers to promote successful use of their products to other potential clients.

The customers for GIS are also changing. Rhind predicted that the bulk of the market would be from outside the USA. The market outside the USA is certainly expanding, as witnessed by the presence of major vendors in countries such as the Netherlands, the UK and Australia. Another indication of the growth of markets elsewhere is the translation of products into languages such as Russian and Chinese, and the growth of user communities in Eastern Europe, Latin America, Africa and the Far East. However, there are still problems for the development of GIS in large parts of the developing world. Limited availability and poor quality of data can be a major obstacle, for example in Latin America (Maskrey, 1997). However, developments in GPS and remote sensing will help ease these data problems, and data availability is being assisted by the provision of some data sets freely over the Internet.

In addition to changes in the GIS software market there has been, as Rhind predicted, a growth in 'value-added services'. Consultant services to meet one-off applications have been an area of particular expansion, with some consultants also producing custom software. Market analysis is an area where this has been particularly prevalent (Birkin *et al.*, 1996).

Progress on data issues

Rhind made several predictions relating to data. First, he foresaw a decline in mass digitizing. Mass digitizing is certainly largely over in the USA and UK, as the national mapping agencies have completed the initial encoding of their resources, and many of the major utilities are completing their data capture programmes. The new challenges are in updating and revising these data. In a new twist, the digitizing of UK population census boundaries was put out to tender and subcontracted to more than one organization to ensure quality and competitiveness. Perhaps Rhind's vision of mass digitizing moving to the Far East will not become a reality, as the development of automatic digitizing tools and the increased resolution and availability of good aerial photography and satellite imagery offer alternatives. However, there is still a great deal of data capture to be completed in many areas of the world, and in some cases the production of a useful digital base topographic data set is hampered by the limitations and incompleteness of available paper maps (Maskrey, 1997).

The data produced by digitizing need to be stored, and many of the digital data sets that have been created are huge. Despite the decreasing cost and increased capacity of data storage media there are still some constraints on the size of GIS applications. The size of individual images and attribute files can make processing difficult, and the amount of memory demanded can exceed that available. And, as the authors can testify, most PC users always seem to be working in the last 10 per cent of disk space on their machines, despite routinely increasing their disk size exponentially! Thus, data handling is still a challenge for the new breed of PC products.

In the USA, spatial information is regarded as a public good (Muller, 1993), with access available to every citizen for a nominal charge. The situation in other countries can be very different, often due to government policy. Rhind's predictions for the supply of data seem valid – there are still some countries where government policies have a profound effect on data supply. In the UK the Ordnance Survey (OS) is required to recover costs, so some users are excluded from GIS due to their need for digital base data from the OS, and their inability to afford them. Rhind also had concerns about privacy and the confidentiality of data. These are certainly still issues for users.

Progress on human and organizational issues

Rhind predicted that improved techniques would become available to assist the GIS acquisition and project management process. As the number of business applications of GIS has expanded, there has been a renewal of interest in making the business case for GIS. However, there is still limited published information, although there are some useful case studies to be found (for example, in the proceedings of the European GIS in Business Conference, GeoInformation International, 1995). Research work has been undertaken on the impacts and diffusion of GIS, and attempts made to try to identify 'critical factors' for success (Campbell and Masser, 1995; Masser *et al.*, 1996). The prediction that GIS will become part of wider management information systems in many organizations is certainly true for commercial applications. Even where GIS was originally set up as a separate system, the utility of GIS tools across organizations has frequently led to later inclusion in corporate information systems plans. As GIS becomes more closely integrated with other technologies, such as CAD, or even office software, this trend is likely to continue.

The importance of trained staff for the success of an application was highlighted in Rhind's predictions. Certainly training and educational opportunities have expanded. In the UK, GIS has appeared in the National Curriculum. In the USA, the NCGIA's core curriculum has been widely adopted in universities and colleges, and vendors have been addressing the needs of the K-12 (kindergarten to year 12) GIS curriculum. Undergraduate, postgraduate and professional development programmes have expanded over the last decade, to the point where there are courses available at technician level, analyst level and manager level; by traditional study or distance learning; or even over the Internet. GIS still has to find a professional home, and accreditation by professional bodies is missing. Resources for those wishing to update their skills independently are also myriad and include CD-ROMs, computer-based learning material and instruction books for different software products. The range of texts and resources has mushroomed since the time of Rhind's predictions, although many still aim at their local market, rather than the global GIS community. In Europe, attempts to standardize a GIS curriculum have been made by Kemp and Frank (1996), whilst in the USA the updating of the NCGIA's influential core curriculum in GIS is attempting to provide resources for all trainers and learners of GIS.

In summary, progress has been made in most of the areas of Rhind's predictions. However, there are some notable and important areas still to be addressed, which will be considered in the next section.

Where next for GIS?

What is GIS? Is it a computer system, a method of handling geographic data for spatial problem solving or a way of thinking? Definitions can change rapidly and they vary according to who is giving them. The definitions of the early 1990s were functional in nature. They focused on what a GIS could do and what it would produce. Now definitions tend to address issues like 'information strategies' and adding value to data. The range of views of GIS has increased over the last few years, in response to the expansion of the market for GIS and the applications to which it has been applied. Is there a danger, as with the term 'management information systems', that 'GIS' could become a meaningless term, meaning different things to different people (Grimshaw, 1994)? Possibly not, because in the case of GIS, it is the G, for the geographic or spatial data, which is arguably most important.

Goodchild (1992) reviews what is unique about spatial data and gives GIS a flavour different from other types of information systems. It is these unique features which make special data handling tools necessary. First, geographical data can be accessed using a spatial reference. These make up a continuous referencing system in two dimensions, so that any location can be visited or studied, and attribute data can be retrieved for any location. Second, locations close together are likely to have similar characteristics. This is Tobler's first law of geography, and justifies, for example, the assumption that all the people with the same postcode have similar income levels, professional status and family characteristics. Third, geographical data are distributed continuously, in three dimensions, all over the surface of the Earth. To date most analysis has been performed on a projected set of these data. There are still few possibilities for the analysis of geographical data in their correct spherical configuration.

As long as we continue to have a requirement for the analysis and management of geographical data, tools that can cope with the special characteristics of these data will be essential. In fact, Goodchild (1995) suggests that this is a 'philosophy' on which GIS is based. The range of geographical data, and the fact that they can be described, structured and handled in similar ways, gives a justification for GIS.

Goodchild has also attempted a review of the successes and limitations of current GIS (Goodchild, 1995), which prompts some questions for the future of GIS. He considers that GIS are:

- two-dimensional with limited abilities to handle the third dimension;
- static, with limited abilities to cope with temporal data;
- good at capturing the physical positions of objects, their attributes and their spatial relationships, but with very limited capabilities for representing other forms of interaction between objects;
- offering a diverse and confusing set of data models; and
- still dominated by the idea of a map, or the view of a spatial database as a collection of digital maps.

These observations prompt some questions. Can GIS be truly three-dimensional? Can real-time GIS become a reality? Can computers be used to model the interactions between features? Is there a better way of modelling geographical data than the current raster/vector/object-oriented paradigms? Are there other ways of representing space that do not rely on maps? These questions give us topics to address to consider the future of GIS. The following section considers four aspects of the future – possible applications, the future of the technology, the future of geographic data and the future for users and organizations.

GIS applications

Maguire (1991), perhaps rather rashly, suggested that by the end of the century everyone in the developed world would use GIS every day for routine operations. It is true that GIS is infiltrating other areas of IT, such as spreadsheet, database and graphics packages; that maps and mapping functions are available over the Internet to anyone who has a modem; and that GIS is appearing in everyday use in other types of applications. Soon all spreadsheets will include mapping functions, all cars will have a navigation aid and maps will be queried and retrieved over the Internet. Your mapping package will be as easy to use as your word processor, your in-car navigation aid as easy as your car radio, and the Internet (or whatever follows it) will be operated on your TV screen with a remote-control handset.

There are other exciting developments afoot which include elements of GIS, and certainly include the handling of geographic information. Gelernter (1992) suggested that you will look into your computer screen and be able to see reality: a 'mirror world'. The idea of creating a mirror of reality is already being addressed. For geographers, virtual field trips are being developed as learning aids (Fisher *et al.*, 1997), while more generally computer games offer exciting simulations of the real world with which users interact.

GIS are also being used in new ways. Beyond the original decision-support and problem-solving role suggested by early definitions and applications, GIS are being mooted as tools for generating ideas, and for improving the participation of all players in the decision-making process. So, from a situation where a planning authority would present a set of authoritative computer-produced maps to gain support for its actions, now the local community can be involved in the setting of important criteria, the modelling of alternative scenarios and in decision-making that will affect them (Heywood and Carver, 1994).

Aside from specific new applications, Birkin *et al.* (1996) see developments in business GIS applications taking two lines. First, GIS will be a tool similar to a database or spreadsheet for middle managers. Generally available, inexpensive systems will be used for mapping to support day-to-day activities. Second, more sophisticated, intelligent GIS will provide inputs to the highest levels within organizations. In this second role, GIS is used for strategic advice, and the systems required are customized, expensive products that will be provided by niche players in the GIS marketplace.

In the face of these developments will GIS continue to exist? What will be the implications of this increased access to GIS technology? What are the implications for society, for organizations and for individuals, and for the field of GIS as a whole? GIS is an integrative discipline involving those who produce the computer tools, collect and prepare the data, and interpret the output. Computer scientists, surveyors, photogrammetrists, remote sensors, spatial analysts, database technicians, cartographers, planners, utility managers and decision makers are all involved, and new applications for the technology will be found in all these disciplines, and in areas involving more than one of these disciplines. The integrative nature of GIS has led Goodchild (1992) to suggest that the abbreviation GIS should actually stand for Geographic Information *Science*, to reflect the breaking down of barriers between some of these traditionally isolated and insular disciplines.

GIS technology – how will we handle spatial data?

Barr (1995a) suggests that GIS as software products will disappear from sight. In the marketplace today GIS is sold, not so much as a versatile 'toolkit' as it was in the past, but as straightforward applications for solving single problems. There are, for instance, routing packages and geodemographic data analysis packages. GIS as a recognizable and separate category of software products may fade from sight.

In the face of these expected disappearances, of GIS the field and GIS the product, there will remain a need, as suggested earlier, for software and analysis tools for the handling, management and analysis of geographical data. In fact, these data are being collected at an increasing rate. Data are one commodity we are not going to go short of in coming years, in fact we are more likely to drown in it, as suggested in Chapter 4. So, the need for tools to handle geographical data effectively and reliably is growing ever greater (Barr, 1995a).

The tools we use to handle geographic data in the future are likely to be different. Barr (1995b) speaks for many users when he suggests that we accept odd quirks and tantrums in our current software, since it is much better than it was a few years ago. However, with increased exposure to information technology, future users of GIS will be confident with computers and impatient with time-consuming, clumsy and unreliable software tools. They will be used for seamless access to information – multimedia, CD encyclopedias, TV pictures and computer games. They may ask why their computer games can model the world in a more exciting and interactive way than their GIS software. Future users of GIS will expect systems with tools for multimedia, for real-time modelling of complex phenomena and for better representation of the real world.

Openshaw is one author who has reflected on the nature of future tools. He considers that analysis and decision-support algorithms will be of prime importance. Thus, products such as his proposed 'geographical analysis machine' (GAM) might become more generally available (Openshaw *et al.*, 1987). These would combine GIS and artificial intelligence techniques to search for patterns in data and inform decision-making. In effect, if current GIS can turn data into information, Openshaw's GAM idea would help to go the next step towards turning information into knowledge. The need for this is clear. At present GIS can provide, for example, a range of options for the siting of a new ski piste, based on data provided and criteria chosen and applied by the end-user. Information is output – the routes of a set of possible ski pistes. However, human interpretation is required to assess the appropriateness of the options presented, and to evaluate if there are additional criteria that need to be applied to the problem. GAMs and their equivalent might be able to go the next step – to sift through options to suggest the best and to identify areas where further data are required.

While the GAM is an excellent idea, developments such as this 'black box' toolkit using complex statistical techniques must always be placed back into the cultural context within which they will operate. Decision making has always been the focal application of GIS. However, it could be argued that decisions are never made on the basis of GIS output, and should not in fact be made in this way. In many businesses, it is the case that GIS results will be used to confirm or authenticate decisions that have been made in other ways. In the retail sector, for example, the selection of sites for new stores has, in the past, often been done in an unscientific manner, using 'gut feel' (Hernandez and Bennison, 1996). Now, however, GIS output is used to justify these, still essentially 'gut feel', siting decisions. Retailers and other business users may be unwilling to trust techniques that they either do not understand or cannot use to validate the decisions they have already made by other methods.

One of the major trends in the late 1990s in the technology sphere is the emergence of co-operation between vendors. A number of major players have, for example, come together to produce common standards to allow the development of open systems that will support real-time access to geographic data across diverse systems and the use of the most appropriate software tool for a particular task (Anon., 1997b). Further development of open systems will be a major theme in coming years.

Emphasizing the information

Barr (1996a) echoes Openshaw's call for a change in emphasis from data to information. The major data entry projects of the 1990s have been completed. National utilities and national mapping agencies have their data digitized and ready to use. So, a shift is needed from tools that help with the input, correction and structuring of data to those that add value to data and provide information. We can now start on the exciting application of GIS to improve our knowledge and understanding of the world. We can start to develop and implement tools that will help with the understanding of geographical phenomena and the solution of practical problems.

Most of the data problems have been overcome, but not all. None of the applications and technical developments will be possible without access to appropriate data. Data access issues have been a subject of debate for well over a decade. In the UK, the Chorley Report (Department of the Environment, 1987) highlighted barriers to the development of GIS, which focused in the main on data issues such as referencing, standards and confidentiality. These issues are still, in part, unresolved (Heywood, 1997). Despite major strides forward, including access to data over the Internet, the development of commercial data providers and the development of standards for address referencing, data formats and data exchange, there is still plenty of work to be done. Barr (1995c) focuses on the problems over standardizing the European Address system. He considers that the address is rapidly becoming the single most important reference for GIS. This is because every individual knows their address whereas they do not necessarily know the grid reference of their home, or the name of the electoral district they live in. In the USA, the Census Bureau keeps the TIGER addressing system up to data, including referencing, assisted by appropriate legislation. However, in Europe no such system exists. In the UK, a standard for addresses has been developed (BS7666), though there is a long way to go before a 'master address file' is created. And then, even if it is created, it needs to be maintained. In the UK, it is clear that the utilities and credit card companies are already using a central address file. Standardization of address formats at an international level has yet to be considered.

Using the huge amounts of data we are now producing presents some exciting opportunities and challenges for the future. Barr (1996b) cites the example of an Internet site that shows real-time maps of traffic conditions. He considers that the maps on display could be downloaded and analysed to help traffic congestion and routing of freight. Add to that data available in real time from CCTV pictures, car monitoring, store cards and other automatic data capture techniques and we have huge new resources of data to exploit to improve our understanding of the world.

The data mountain continues to grow, with data being collected at every opportunity. More and more data have a spatial reference. If you speed past a police camera, your photograph may be taken, and the location of your misdemeanour recorded. If you shop in a supermarket with your store loyalty card, information about your purchases will be added to a database already containing your address and location details. If you telephone a company for information on its services you will be asked for your postcode so that the company database can be updated with your location. Thus, the data mountain continues to grow, and there is a need for new techniques to use this resource. Data issues are still foremost among problems and challenges for the future.

People and GIS

Human and organizational issues are not going to go away. Consider the training and updating of users. Even now, if you take a holiday from the office, you may have to catch up with new releases of software, new Internet sites and a computer mail box overflowing with e-mail messages. Barr (1996c) suggests that it has been estimated that a quarter of your IT knowledge can become obsolete in a year. This ever-increasing pace of change and development is exciting, but also frightening at an individual level. How can you remain an expert? Lifelong learning, retraining and professional development will become necessary for everyone in a society where change is the norm rather than the exception. Tools and aids to learning and professional updating will be required. Software agents, which search the Internet on topics in which you are interested, are being developed (Hughes, 1997). Other tools are needed. Individuals in the developed world are already adapting to a world where jobs for life are rare, where working from home is becoming more common and where major career changes are frequent. Organizations too are adapting and changing.

Two areas of skill and knowledge shortage in GIS were identified by Marble (1997). These are in the basic understanding of geographical phenomena – spatial thinking and high-level technical skills. Skill shortages in these areas are critical. A lack of understanding of spatial issues and topics such as geodesy and cartography will prevent GIS from being used effectively. Users will be unable to recognize the potential and the possible applications of GIS. At the other end of the skill spectrum, a lack of highly skilled technical personnel with skills in programming, software research and product development will limit the development of the tools and programs used to handle geographical data.

Wilson (1997) suggests that the users are the winners in current developments in GIS. The technology is finally catching up with demands for easy-to-use corporate interfaces and seamless data integration, so the focus can move from the development of software tools to adding value to geographic information. However, at the same time, the important issues of training and education must be considered.

Conclusions

What will the Martian report back on the future of the car industry? There are problems to be overcome, challenges to be addressed, but still a huge potential as there are vast regions of the world with limited car ownership. From a purely business perspective there are opportunities for making money. From an environmental perspective there are opportunities for polluting our planet, and challenges for those who want to 'save' it. From an individual perspective there are opportunities for personal mobility and freedom that will come up against problems of traffic congestion and the changing status of the car in society. Some of the problems and challenges for the car industry have been around for some time; others are the result of increased car ownership and increased reliance on the car for everyday life. The same is true of GIS, and our predictions listed in Box 13.3 consist of issues that have been around for some time, and are not yet solved, and new challenges resulting from more recent technological and market developments. Technological developments will continue to drive GIS, and as always technology will continue to be the easy part.

These are our predictions. You may disagree, or know of evidence that some of the issues have already been tackled, or that others will be more important. There are other sources that may give a different view. The National Research Council (1997), Harding and Wilkinson (1997) and Heywood (1997) are three examples, giving views of the future from an American, European and UK perspective, respectively.

BOX 13.3

Predictions for the twenty-first century

- New and improved data models will be developed, especially for true 3D modelling, global data sets and temporal modelling.
- Developments in related fields will influence developments in GIS (for example, artificial intelligence, neural computing, geocomputation, spatial statistics, multimedia, virtual reality, the Internet and office packages).
- There will be a continued repackaging of products to meet sector-specific needs, and, with continued dominance of the PC, further expansion of the desktop market and a reduction in the workstation/mainframe market.
- There will be further additions to GIS functionality, for example much-needed exploratory data analysis and error analysis.
- Data collection and exchange will continue to be influenced by GPS development, coupled with future developments in satellite communications and data collection methods.
- Peripherals such as high-quality printers will become widely available, possibly through on-line Internet service providers. In fact, the need for hard copy will be reduced, as new forms of output, such as animation, fly-through and 3D models, can be easily exchanged over communications networks and more effectively presented in digital format.
- Data storage will cease to be an issue, although the development of real-time systems will challenge even the largest storage devices!
- The ideas of interoperability, allowing the use of the most appropriate software tool for the solution of a particular problem, will extend to influence other areas of GIS, including data and education.
- New education and training products will be developed to address skill shortages for both potential users and the research and development needs of the GIS industry.

As we said at the outset, predicting the future is a dangerous occupation. What is clear is that if GIS is to be more widely used there are still challenges to be addressed. On the other hand there is still a huge potential for the development of tools to handle geographical data. Consider what might happen in the case study situations. The radioactive waste siting GIS could become a system allowing widespread public participation in the decision-making process. Using the Internet, or via their TV, members of the public could influence the siting criteria from their armchairs. The Internet may also have a role to play in the house-hunting example. An on-line house sales service could offer multimedia presentations of houses for sale, virtual reality displays of houses yet to be built and automatic selection of houses meeting buyers' criteria. Printouts of maps and details would help them find the properties. Direct links to land registry systems (as are already being developed) could then speed up the house-buying process. In the Zdarske Vrchy case study there are opportunities for automatic data collection and the generation of new data to help model building and predictions. GPS and satellite data collection facilities could be coupled with bar-coded visitor passes to find out about the resources of the area and the pressures on them.

And in Happy Valley the possibilities are endless! Perhaps the valley will have its own real-time management information system. This could be used to enhance services, target potential clients, provide effective avalanche prediction, and for real-time management of resources. Information on avalanche risk, weather forecasts and special events could be continuously provided in multiple languages on personal communicators issued to all resort guests, and on hand-held computers used by emergency services and service providers.

These are possible scenarios for the future. At the heart of them all are a number of challenges. The first is data. Automatic data collection, data exchange and real-time modelling using a range of data types all need to be tackled. Another challenge is the technology. It is likely that the technological challenges, as in the past, will be easier to address than the organizational and human issues, and many new opportunities will come from areas outside the field of GIS. A third challenge is how to present geographical data and the results of geographical modelling to a whole new audience. New ways of visualizing the results of modelling and analysis are needed to help decision-making and resource management. New interfaces are needed to make it possible for people to interact with geographical data and decision-support systems. And finally, there is the challenge for education and training. A new audience of users must be prepared to work and live with geographical information.

REVISION QUESTIONS

- Have the predictions of GIS writers over the last 10 years been realized? What are the characteristics of GIS in the early twenty-first century?
- What are the problems facing future developments of GIS?
- How will developments in related fields, such as hardware, communications and multimedia, influence the future of GIS?
- Give examples of the use of GIS-type products in daily life, and suggest other areas where GIS use might become more widespread among the general population.

Further study

For further details of the predictions of others as to the character of GIS today, consult the full text of articles by Rhind *et al.* (1991), Rhind (1992), Maguire *et al.* (1991) and Goodchild (1992). For predictions for the future of GIS, you need to read up-to-date literature. GIS magazines often contain the predications of key individuals, in annual reviews of the state of the GIS industry. Barr has provided a lively and thought-provoking column for the last few years in *GIS Europe*, 'Dangling segment', which has speculated on many future issues for GIS based on the problems and issues faced by users (see, for example, Barr, 1996a, 1996b, 1996c). Other ideas for developments come from the research literature, from reading up-to-date articles by innovative researchers. The proceedings of research conferences such as GIS Research UK, which are published in a series of books titled *Innovations in GIS* edited by Fisher (1995), Parker (1996), Kemp (1997) and Carver (1998), give a feel for current research directions. Reading in other areas will also give you ideas for the development of GIS. One example is an interesting journal, *Personal Technologies*, which focuses on developments in hand-held and mobile information appliances.

Three interesting reports (National Research Council, 1997; Harding and Wilkinson, 1997; Heywood, 1997) give an insight into the views of the American Mapping Science Committee, the European Union and the UK's Association for Geographic Information, respectively.

Barr R (1996a) Data, information and knowledge in GIS. *GIS Europe* 5 (3): 14–15

Barr R (1996b) Look out! Someone is watching you. *GIS Europe* 5 (7): 12–13

Barr R (1996c) Desperately seeing solutions. *GIS Europe* 5 (8): 14–15

Carver S J (ed.) (1998) *Innovations in GIS* 5. Taylor and Francis, London

Fisher P (ed.) (1995) *Innovations in GIS* 2. Taylor and Francis, London

Goodchild M F (1992) Geographical Information Science. *International Journal of Geographical Information Systems* 6 (1): 31–45

Harding S M, Wilkinson G G (1997) *A strategic view of GIS research and technology development for Europe*. Report of the expert panel convened at the Joint Research Centre, Ispra, 19–20 November 1996. EUR 17313 EN, European Commission, Brussels

Heywood D I (1997) *Beyond Chorley: Current Geographic Information Issues*. AGI, London

Kemp Z (ed.) (1997) *Innovations in GIS* 4. Taylor and Francis, London

Maguire D J, Goodchild M F, Rhind D W (eds)(1991) *Geographical Information Systems: Principles and Applications*. Longman, London

National Research Council (1997) *The Future of Spatial Data and Society*. National Academy Press, Washington DC

Parker D (ed.) (1996) *Innovations in GIS* 3. Taylor and Francis, London

Rhind D (1992) The next generation of Geographical Information Systems and the context in which they will operate. *Computers, Environment and Urban Systems* 16: 256–68

Rhind DW, Goodchild M F, Maguire D J (1991) Epilogue. In: Maguire D J, Goodchild M F, Rhind DW (eds) *Geographical Information Systems: Principles and Applications*. Longman, London, Vol. 2, pp. 313–27

References

Ackoff R L (1971) Towards a system of systems concepts. *Management Science* **17** (11): 661–71

Andersson S (1987) The Swedish land data bank. *International Journal of Geographical Information Systems* **1** (3): 253–63

Anon. (1993) Survey of choice of system. *GIS World* **6**

Anon. (1994) Government is the largest GIS user. *GIS World* 7 (9): 12

Anon. (1995) *GIS World* interview: Ian McHarg reflects on the past, present and future of GIS. *GIS World* 8 (10): 46–9

Anon. (1996) Platform preference varies with size of GIS user base. *GIS World* **9** (1): 16

Anon. (1996) *GIS World* Interview: Roger Tomlinson the father of GIS. http://www.geoplace.com/gw/1996/0496/0496 feat2.asp

Anon. (1997a) GIS software survey. *Mapping Awareness* **11**(2): 29–36

Anon. (1997b) Competitors unite to set standards. *Mapping Awareness* **11**(2):

Anon. (2001) *GIS World* Interview: Ian McHarg reflects on the past, present and future of GIS. http://www.geoplace.com/gw/2001/0601/0601mem.asp

Anon. (no date) Wigwam publicity flier, unpublished

Aronoff S (1989) *Geographic Information Systems: A Management Perspective.* WDL Publications, Ottawa

Arthur C (1992) Ambulance computer system was too complicated. *New Scientist* **136** (1847): 7

Arthur C (1993) Pressurized managers blamed for ambulance failure. *New Scientist* **137** (1863): 5

Avison D E, Fitzgerald G (1988) *Information Systems Development: Methodologies, Techniques and Tools*. Blackwell Scientific, Oxford

Avison D E, Wood-Harper A T (1991) Information systems development research: an exploration of ideas in practice. *Computer Journal* **34** (2): 98–112

Barnard M E (1992) The Global Positioning System. *IEE Review* (March) 99–102

Barr R (1995a) Is GIS a Cheshire cat? *GIS Europe* **4** (10): 20–1

Barr R (1995b) Mind the generation gap. *GIS Europe* **4** (7): 18–19

Barr R (1995c) Addressing the issues. *GIS Europe* **4** (9): 18–19

Barr R (1996a) Data, information and knowledge in GIS. *GIS Europe* **5** (3):14–15

Barr R (1996b) Look out! Someone is watching you. *GIS Europe* **5** (7): 12–13

Barr R (1996c) Desperately seeing solutions. *GIS Europe 5* (8): 14–15

Battista C (1995) GIS fares well in utilities survey. *GIS World* **8** (12): 66–9

Batty M, Xie Y (1994a) Modelling inside GIS: Part 1. Model structures, exploratory spatial data analysis and aggregation. *International Journal of Geographical Information Systems* 8 (4): 291–307

Batty M, Xie Y (1994b) Modelling inside GIS: Part 2. Selecting and calibrating urban models using ARC/INFO. *International Journal of Geographical Information Systems* **8** (5): 451–70

Batty P (1990) Exploiting relational database technology in GIS. *Mapping Awareness* **4** (6): 25–32

Beale H (1987) The assessment of potentially suitable repository sites. In: *The Management and Disposal of Intermediate and Low Level Radioactive Waste*. Mechanical Engineering Publications, London, pp. 11–18

Beck M B, Jakeman A J, McAleer M J (1995) Construction and evaluation of models of environmental systems. In: Jakeman A J, Beck M B, McAleer M J (eds) *Modelling Change in Environmental Systems*. Wiley, Chichester, pp. 3–36

Bell S, Wood-Harper T (1992) *Rapid Information Systems Development: A Non Specialist's Guide to Analysis and Design in an Imperfect World*. McGraw-Hill, London

Berger P, Meysembourg P, Sales J, Johnston C (1996) Toward a virtual reality interface for landscape visualization. *Proceedings of the 3rd International Conference/Workshop on Integrating GIS and Environmental Modeling* CD-ROM

Bernhardsen T (1992) *Geographic Information Systems*. VIAK IT, Norway

Bernhardsen T (1999) Choosing a GIS. In: Longley P A, Goodchild M F, Maguire D J and Rhind D W (eds) *Geographical Information Systems*. Wiley, New York, 589-600

Berry J K (1987) Fundamental operations in computer-assisted map analysis. *International Journal of Geographical Information Systems* **1**: 119–36

Berry J K (1993) *Beyond Mapping: Concepts, Algorithms and Issues in GIS*. GIS World Inc., Colorado

Betak J, Vaidya A (1993) GIS: a change agent in a transportation company. *Proceedings of the GIS in business '93* conference, 7–10 March, Boston, Mass. GIS World Books, Fort Collins, Colorado, pp. 275–82

Bickmore D P, Shaw M A (1963) *Atlas of Great Britain and Northern Ireland*. Clarendon Press, Oxford

Birkin M, Clarke G, Clarke M, Wilson A (1996) *Intelligent GIS: Location Decisions and Strategic Planning*. GeoInformation International, Cambridge, UK

Blakemore M J (1984) Generalization and error in spatial databases. *Cartographica* **21**: 131–9

Bolstad P V, Gessler P, Lillesand T M (1990) Positional uncertainty in manually digitized map data. *International Journal of Geographical Information Systems* **4** (4): 399–412

Bonham-Carter G F (1991) Integration of geoscientific data using GIS. In: Maguire D J, Goodchild M F, Rhind D W (eds) *Geographical Information Systems: Principles and Applications*. Longman, London, Vol. 1, pp. 171–84

Bonham-Carter G F (1995) *Geographic Information Systems for Geoscientists: Modelling with GIS*. Pergamon Press, New York

Boulding K E (1956) On systems theory and analysis in urban planning. *Papers in Planning* **11**. UWIST, Cardiff

Brown C (1989) *Implementing a GIS: Common Elements of Successful Sites*. Presented at Conference of URISA, Boston, MA.

Brown S J, Schreirer H E, Woods G, Hall S (1994) A GIS analysis of forestry/caribou conflicts in the transboundary region of Mount Revelstoke and Glacier National Parks, Canada. In: Price M F, Heywood D I (eds) *Mountain Environments and Geographical Information Systems*. Taylor and Francis, London, pp. 235–48

Brunsdon C, Carver S, Charlton M, Openshaw S (1990) A review of methods for handling error propagation in GIS. *Proceedings of 1st European Conference on Geographical Information Systems*, Amsterdam, April, pp. 106–16

Brusegard D, Menger G (1989) Real data and real problems: dealing with large spatial databases. In: Goodchild M F, Gopal S (eds) *The Accuracy of Spatial Databases*. Taylor and Francis, London, pp. 177–86

Buchanan H J (1992) NTF – an introduction. *Mapping Awareness* **6** (5): 38–41

Burgan R E, Rothermel R C (1984) BEHAVE: *fire behaviour and prediction fuel modelling*

system – fuel subsystem. General technical report INT–167, US Department of Agriculture, Ogden, Utah

Burrough P A (1986) *Principles of Geographical Information Systems for Land Resources Assessment*. Clarendon Press, Oxford

Burrough P A, McDonnell R (1998) *Principles of Geographical Information Systems*. Oxford University Press, Oxford

Buttenfield B P, Mackaness W A (1991) Visualization. In: Maguire D J, Goodchild M F, Rhind D W (eds) *Geographical Information Systems: Principles and Applications*. Longman, London, Vol. 1, pp. 427–43

Campbell H J (1994) How effective are GIS in practice? A case study of British local government. *International Journal of Geographical Information Systems* 8 (3): 309–26

Campbell H J (1999) Institutional consequences of the use of GIS. In: Longley P A, Goodchild M F, Maguire D J and Rhind D W (eds) *Geographical Information Systems*. Wiley, New York, 621–631

Campbell H J, Masser I (1992) GIS in local government: some findings from Great Britain. *International Journal of Geographical Information Systems* 6 (6): 529–46

Campbell H J, Masser I (1995) *GIS and Organizations: How Effective are GIS in Practice?* Taylor and Francis, London

Carver S J (1991a) Error modelling in GIS: who cares? In: Cadoux-Hudson J, Heywood D I (eds) *The Association for Geographic Information Yearbook 1991*. Taylor and Francis/Miles Arnold, London, pp. 229–34

Carver S J (1991b) Integrating multi-criteria evaluation with geographical information systems. *International Journal of Geographical Information Systems* 5 (3): 321–39

Carver S J (ed.) (1998) *Innovations in GIS 5*. Taylor and Francis, London

Carver S, Brunsdon C (1994) Vector to raster conversion error and feature complexity: an empirical study using simulated data. *International Journal of Geographical Information Systems* 8 (3): 261–72

Carver S J, Heywood D I, Cornelius S C (1995) Evaluating field-based GIS for environmental characterization, modelling and decision support. *International Journal of Geographical Information Systems*. 9 (4): 475–86

Carver S, Blake M, Turton I, Duke-Williams O (1997) Open spatial decision-making: evaluating the potential of the World Wide Web. In: Kemp Z (ed.)) *Innovations in GIS 4*. Taylor and Francis, London, pp. 267–78

Cassettari S (1993) *Introduction to Integrated Geo-information Management*. Chapman & Hall, London

Charlesworth F R, Gronow W S (1967) A summary of experience in the practical application of siting policy in the UK. In: *Containment and Siting of Nuclear Power Plants*. IAEA, Vienna, pp. 143–70

Checkland P B (1981) *Systems Thinking – Systems Practice*. John Wiley, Chichester

Checkland P B (1988) Information systems and systems thinking: time to unite? *International Journal of Information Management* 8: 239–48

Chen P P-S (1976) The entity-relationship model – towards a unified view of data. *Association for Computing Machinery Transactions on Database Systems* **1** (1): 9–36

Chen Z, Guevara J A (1987) Systematic selection of very important points (VIP) from digital terrain models for construction of triangular networks. *Proceedings of AutoCarto 8*. ASPRS/ACSM, Falls Church, Virginia, pp. 50–6

Chou Y-H (1996) *Exploring Spatial Analysis in Geographic Information Systems*. OnWord Press, USA

Chrisman N R (1987) Efficient digitizing through the combination of appropriate hardware and software for error detection and editing. *International Journal of Geographical Information Systems*. **1**: 265–77

Chrisman N R (1989) Modelling error in overlaid categorical maps. In: Goodchild M F, Gopal S (eds) T*he Accuracy of Spatial Databases*. Taylor and Francis, London, pp. 21–34

Chrisman N R (1997) *Exploring Geographic Information Systems*. Wiley, New York

Churchman C W (1968) *The Systems Approach*. Dell, New York

Clarke A L (1991) GIS specification, evaluation and implementation. In: Maguire D J, Goodchild M F, Rhind D W (eds) *Geographical Information Systems: Principles and Applications*. Longman, London, Vol. 1, pp. 477–88

Clarke K C (1990) *Analytical and Computer Cartography*. Prentice-Hall, Englewood Cliffs, New Jersey

Clayton K (1995) The land from space. In: O'Riordan T (ed.) *Environmental Science for Environmental Management*. Longman, London, pp. 198–222

Codd E (1970) A relational model for large shared data banks. *Communications of the Association for Computing Machinery* **13** (6): 377–87

Collins (1981) *Pocket English Dictionary*. Collins, Glasgow

Coppock J T (1988) The analogue to digital revolution: a view from an unreconstructed geographer. *American Cartographer* **15** (3): 263–75

Coppock J T, Rhind D W (1991) The history of GIS. In: Maguire D J, Goodchild M F, Rhind D W (eds) *Geographical Information Systems: Principles and Applications*. Longman, London, Vol. 1, pp. 21–43

Cornelius S C, Medyckyj-Scott D (1991) 'If only someone had said' Human and organizational barriers to GIS success. *Mapping Awareness* **5** (7): 42–5

Cornelius S C, Sear D A, Carver S J, Heywood D I (1994) GPS, GIS and geomorphological field work. *Earth Surface Processes and Landforms* **19**: 777–87

Crain I K (1985) *Environmental Information Systems: An Overview*. Environment Canada, Ottawa

Cresswell P (1995) Customized and proprietary GIS: past, present and future. In: Longley P, Clarke G (eds) *GIS for Business and Service Planning*. Geoinformation International, Cambridge, UK, pp. 192–226

Curran P (1989) *Principles of Remote Sensing*. Longman, London

Dale P F, McLaughlin J D (1988) *Land Information Management: An Introduction with Special Reference to Cadastral Problems in Third World Countries*. Clarendon Press, Oxford

Dangermond J (1991) The commercial setting of GIS. In: Maguire D J, Goodchild M F, Rhind D W (eds) *Geographical Information Systems: Principles and Applications*. Longman, London, Vol. 1, pp. 55–65

Daniel L, Paula L, Whitener A (1996) *Inside MapInfo Professional*. OnWord Press, USA

Date C J (1986) *An Introduction to Database systems*, 2nd edn. Addison-Wesley, Reading,Mass.

Davies C, Medyckyj-Scott D (1996) GIS users observed. *International Journal of Geographical Information Systems* **10** (4): 363–84

Davis D (1999) *GIS for Everyone*. ESRI Press, Redlands

Davis J C (1986) *Statistics and Data Analysis in Geology*. Wiley, New York

DeFloriani L, Fakidieno B, Nagy G, Pienovi C (1984) A hierarchical structure for surface approximation. *Computer Graphics* **32**: 127–40

DeMers M N (1997) *Fundamentals of Geographic Information Systems*. Wiley, New York

Densham P J (1991) Spatial decision support systems. In: Maguire D J, Goodchild M F, Rhind D W (eds) *Geographical Information Systems: Principles and Applications*. Longman, London, Vol. 1, pp. 403–12

Department of the Environment (1985) *Disposal facilities on land for low and intermediate level radioactive wastes: principles for the protection of the human environment*. HMSO, London

Department of the Environment (1987) *Handling Geographic Information*. Report of the Committee of Enquiry chaired by Lord Chorley. HMSO, London

Dorling D (1993) From computer cartography to spatial visualisation: a new cartogram algorithm. *AutoCarto* **11**: 208–17

Dorling D (1995) Visualizing changing social structure from a census. *Environment and Planning A* **26** (3): 353–78

Dorling D, Openshaw S (1992) Using computer animation to visualise space–time patterns. *Environment and Planning B* **19**: 639–50

Douglas D H, Peucker T K (1973) Algorithms for the reduction of the number of points required to represent a digitized line or its caricature. *Canadian Cartographer* **10** (4): 110–22

Downey I, Heywood D I, Petch J R (1991) Design and construction of a Geographical Information System (GIS) for Zdarske Vrchy landscape protected area, Czechoslovakia. In: Jackson M C, Mansell G J, Flood R L, Blackham R B, Probert S V E (eds) *Systems Thinking in Europe*. Plenum, New York, pp. 151–158

Downey I, Pauknerova E, Petch J R, Brokes P, Corlyon A (1992) Habitat analysis and modelling for endangered species. In: Willison *et al.* (eds) *Science and the Management of Protected Areas*. Elsevier, Amsterdam, pp. 271–7

Dunn R, Harrison A R, White J C (1990) Positional accuracy and measurement error in digital databases of land use: an empirical study. *International Journal of Geographical Information Systems* **4** (4): 385–98

Eason K D (1988) *Information Technology and Organizational Change*. Taylor and Francis, London

Eason K D (1994) Planning for change: introducing a Geographical Information System. In: Medyckyj-Scott D J, Hearnshaw H M, *Human Factors in Geographical Information Systems*. Belhaven, London, pp. 199–209

Eastman R, Kyem P, Toledano J, Jin W (1993) *GIS and Decision Making*. Explorations in Geographic Information Systems technology. Volume 4. United Nations Institute for Training and Research, Switzerland

Ellisor E (1992) What's ahead in data access and management? *GIS World* (April), 50–3

Elmasri R, Navathe S B (1994) *Fundamentals of Database Systems*, 2nd edn. The Benjamin/Cummings, Calif.

ESRI (1995) ESRI builds national GIS for the real estate industry. *ARC News* **17** (4): 1–2

ESRI (1996) http://www.esri.com

ESRI (2001) ESRI Timeline http://www.esri.com/company/about/timeline/flash/index.html

Evangelatos T V (1991) Digital geographic interchange standards. In: Taylor D R F (ed.) *Geographic Information Systems: The Microcomputer and Modern Cartography*. Pergamon Press, Oxford, London and New York, pp. 151–66

Evans I S (1977) The selection of class intervals. *Transactions of the Institute of British Geographers* **2** (1): 98–124

Faust N L (1995) The virtual reality of GIS. *Environment and Planning B: Planning and Design*, **22**: 257–68

Fisher P (1991) Modelling soil map-unit inclusions by Monte Carlo simulation. *International Journal of Geographical Information Systems* **5** (2): 193–208

Fisher P (ed.) (1995) *Innovations in GIS 2*. Taylor and Francis, London

Fisher P, Dykes J, Moore K, Wood J, McCarthy T, Raper J, Unwin D, Williams N, Jenkins A (1997) The virtual field course: an educational application of GIS. In: Carver S (ed.) *Proceedings of GIS Research UK, 5th National Conference*, University of Leeds. pp. 37–40

Fishwick A, Clayson J (1995) *GIS Development Project: Final Report to the Countryside Commission*. Lake District National Park Authority, Kendal, UK

Foresman T (ed.) (1997) *The History of GIS*. Prentice Hall, New Jersey

Fotheringam A S, Brunsdon C, Charlton M (2000) *Quantitative Geography: Perspectives on Spatial Data Analysis*. Sage, London

Fotheringham A S, O'Kelly M E (1989) *Spatial Interaction Models: Formulations and Applications*. Kluwer Academic, Dordrecht

Fotheringham A S, Rogerson P (eds) (1994) *Spatial Analysis and GIS*. Taylor and Francis, London

Fowler R J, Little, J J (1979) Automatic extraction of irregular network digital terrain models. *Computer Graphics* **13**: 199–207

Frank A U, Mark D M (1991) Language issues for GIS. In: Maguire D J, Goodchild M F, Rhind D W (eds) *Geographical Information Systems: Principles and Applications*. Longman, London, Vol. 1, pp. 147–63

Gatrell A C (1991) Concepts of space and geographical data. In: Maguire D J, Goodchild M F, Rhind D W (eds)) *Geographical Information Systems: Principles and Applications*. Longman, London, Vol. 1, pp. 119–43

Gelernter D (1992) *Mirror World or the Day Software Puts the University in a Shoebox...How it will Happen and What it will Mean.* Oxford University Press, New York

Geoffrion A M (1983) Can OR/MS evolve fast enough? *Interfaces* **13** (10)

GeoInformation International (1995) *GIS in Business: Discovering the Missing Piece in your Business Strategy.* GeoInformation International, Banbury, UK

Gilfoyle I (1997) The Chorley Report: philosophy and recommendations. What Chorley got right. Presented at: *The Future for Geographic Information: 10 Years after Chorley*, Association for Geographic Information, London, 1 May

GISNews (2001) Guide to the new Ordnance Survey. Supplement issued with Vol. 2 no. 5

Goodchild M (1978) Statistical aspects of the polygon overlay problem. In: Dutton G H (ed.) *Harvard Papers on Geographical Information Systems.* Addison-Wesley, Reading, Mass. pp. 1–21

Goodchild M F (1988) Stepping over the line the technological constraints and the new cartography. *American Cartographer* **15** (3): 277–89

Goodchild M F (1990) Spatial information science. *Proceedings of 4th International Symposium on Spatial Data Handling.* International Geographical Union, Columbus Ohio. Vol. 1, pp. 3–12

Goodchild M F (1991a) Geographic Information Systems. *Progress in Human Geography* **15** (2): 194–200

Goodchild M F (1991b) The technological setting of GIS. In: Maguire D J, Goodchild M F, Rhind D W (eds) *Geographical Information Systems: Principles and Applications.* Longman, London, Vol. 1, pp. 45–54

Goodchild M F (1992) Geographical Information Science. *International Journal of Geographical Information Systems* **6** (1): 31–45

Goodchild M F (1995) GIS and geographic research. In: Pickles J (ed.) *Ground Truth: The Social Implications of Geographic Information Systems.* Guilford Press, New York, pp. 31–50

Goodchild M F (1997) What is Geographic Information Science? NCGIA Core Curriculum in GIScience. http//www. ncgia.ncsb.edu/giscc/units/u002/ u002.html, posted October 7, 1997.

Goodchild M F, Gopal S (eds) (1989) *The Accuracy of Spatial Databases.* Taylor and Francis, London

Goodchild M F, Kemp K (1990) Developing a curriculum in GIS: the NCGIA core curriculum project. In: Unwin D J (ed.) *GIS Education and Training*, Collected Papers of a Conference, University of Leicester, 20–21 March 1990, Midlands Regional Research Laboratory, Leicester

Goodchild M F, Parks B O, Steyaert L T (eds) (1993) *Environmental Modelling with GIS.* Oxford University Press, Oxford and New York

Goodchild M F, Rhind D W (1990) The US National Center for Geographic Information and Analysis: some comparisons with the Regional Research Laboratories. In: Foster M F, Shand P J (eds) *The Association for Geographic Information Yearbook 1990.* Taylor and Francis, London, pp. 226–32

Goodchild M F, Yang S (1989) A hierarchical spatial data structure for global geographic information systems. *NCGIA Technical Paper 89–5.* National Center for Geographic Information and Analysis (NCGIA), University of California, Santa Barbara

Goodwin T (1995) GIS – getting started. In: *GIS in Business: Discovering the Missing Piece in your Business Strategy.* GeoInformation International, Banbury, UK, pp. 144–8

Gould M I (1992) The use of GIS and CAC by health authorities: results from a postal questionnaire. *Area* **24** (4): 391–40

Green D R, Rix D, Corbin C (1997) *The AGI Source Book for Geographic Information Systems 1997.* Association for Geographic Information, London

Griffith D A, Amrhein C G (1991) *Statistical Analysis for Geographers.* Prentice-Hall, Englewood Cliffs, New Jersey

Grimshaw D J (1994) *Bringing Geographical Information Systems into Business.* Longman, London

Gunn J, Hunting C, Cornelius S, Watson R (1994) The proposed Cuilcagh National History Park, County Fermanagh: a locally based conservation initiative. In: O'Halloran D, Green C, Harley M, Stanley M, Knill J (eds) *Geological and Landscape Conservation*. Geological Society, London, pp. 337–41

Haggett P, Chorley R J (1967) Models, paradigms and the new geography. In: Chorley R J, Hagget P (eds) *Models in Geography*. Methuen, London, pp. 19–42

Haines-Young R, Green R D, Cousins S H (eds) (1994) *Landscape, Ecology and GIS*. Taylor and Francis, London

Halpin P N (1994) GIS analysis of the potential impacts of climate change in mountain ecosystems and protected areas. In: Price M F, Heywood D I (eds) *Mountain Environments and Geographical Information Systems*. Taylor and Francis, London, pp. 281–301

Hanna K C (1999) *GIS for Landscape Architects*. ESRI Press, Redlands

Hanold T (1972), An executive view of MIS. *Datamation* **18** (11): 66

Harder C (1997) *Arcview GIS Means Business: Geographic Information System Solutions for Business*. ESRI, California

Harder C (1998) *Serving Maps on the Internet: Geographic Information and the World Wide Web*. ESRI Press, Redlands, California

Harding S M, Wilkinson G G (1997) *A strategic view of GIS research and technology development for Europe*. Report of the expert panel convened at the Joint Research Centre, Ispra, 19–20 November 1996. EUR 17313 EN, European Commission, Brussels

Hardisty J, Taylor D M, Metcalfe S E (1993) *Computerized Environmental Modelling: a Practical Introduction Using Excel*. Wiley, New York

Harley J B (1975) *Ordnance Survey Maps: a Descriptive Manual*. OS, Southampton

Healey R G (1991) Database management systems. In: Maguire D J, Goodchild M F, Rhind D W (eds) *Geographical Information Systems: Principles and Applications*. Longman, London, Vol. 1, pp. 251–67

Hearnshaw H M, Unwin D J (1994) *Visualization in Geographical Information Systems*. Wiley, New York

Heit M, Shortreid A (1991) *GIS Applications in Natural Resources*. GIS World Inc., Colorado

Henry B, Pugh J (1997) CAD and GIS software find common ground: a review of the market leaders. *GIS World* **10** (5): 32–7

Hernandez T, Bennison D (1996) Selling GIS to retailers: organisational issues in the adoption and use of Geographical Information Systems. *Proceedings of the AGI/GIS 96 Conference*. 22–24 September, Birmingham, UK

Hernandez T, Scholten H J, Bennison D, Biasiotto M, Cornelius S, van der Beek M (1999) Explaining retail GIS: the adoption, use and development of GIS by retail organizations in the Netherlands, the UK and Canada. *Netherlands Geographical Studies* **258** Utrecht

Hernandez T, Verrips A (2000) Retail GIS: more than just pretty maps. *GeoEurope* **9** (4): 16–18

Heuvelink G B, Burrough P A, Stein A (1990) Propagation of errors in spatial modelling with GIS. *International Journal of Geographical Information Systems* **3**: 303–22

Hewitson B D, Crane R G (1994) *Neural Nets: Applications in Geography*. Kluwer Academic, Dordrecht

Heywood D I (1997) *Beyond Chorley: Current Geographic Information Issues*. AGI, London

Heywood D I, Blaschke T, Carlisle B (1997) Integrating geographic information technology and multimedia: the educational opportunities. In: Hodgson S, Rumor M, Harts J J (eds) *Proceedings of the Joint European Conference and Exhibition on Geographical Information Systems*, Vienna, pp. 1281–90

Heywood D I, Oliver J, Tomlinson S (1995) Building an exploratory multi-criteria modelling environment for spatial decision support. In: Fisher P (ed.) *Innovations in GIS 2*. Taylor and Francis, London, pp. 127–36

Heywood I (1990) Monitoring for change: a Canadian perspective on the environmental role for GIS. *Mapping Awareness* **4** (9): 24–6

Heywood I, Carver S (1994) Decision support or idea generation: the role for GIS in policy formulation. *Proceedings Symposium für Angewante Geographische Informationsverarbeitung (AGIT '94)*, Salzburg, Austria, July, pp. 259–66

Hobby J (1992) This is an emergency. The *Guardian*, 5 November

Hoffman-Wellenhof B, Lichtenegger H, Collins J (1994) *Global Positioning Systems: Theory and Practice*. Springer-Verlag

Hohl P (ed) (1998) *GIS Data Conversion: Strategies, Techniques, Management*. Onword Press, Santa Fe

Holland D (2001) Delivering the digital national framework in GML. *GeoEurope* August 2001: 29–30.

Huggett R (1980) *Systems Analysis in Geography*. Clarendon Press, Oxford

Hughes T (1997) The new seekers. *Compuserve Magazine* **2**: 18–20

Huxhold W E (1991) *An Introduction to Urban Geographic Information Systems*. Oxford University Press, New York

IAEA (1983) *Disposal of Low and Intermediate Level Solid Radioactive Wastes in Rock Cavities*. Safety Series No. 59, IAEA, Vienna

Jackson M J, Woodsford P A (1991) GIS data capture hardware and software. In: Maguire D J, Goodchild M F, Rhind D W (eds) *Geographical Information Systems: Principles and Applications*. Longman, London, Vol. 1, pp. 239–49

Jacobsen F, Barr R, Theakstone W (1993) Exploring glaciers with GIS: the Svartisen and Okstindan project. *GIS Europe* **2** (4): 34–6

Jakeman A J, Beck M B, McAleer M J (eds) (1995) *Modelling Change in Environmental Systems*. Wiley, Chichester

Janssen R (1992) *Multi-objective Decision Support for Environmental Management*. Kluwer Academic, Dordrecht

Janssen R, Rievelt P (1990) Multicriteria analysis and GIS: an application to agricultural land use in the Netherlands. In: Scholten H J, Stillwell J C H (eds) *Geographical Information Systems for Urban and Regional Planning*. Kluwer Academic, Dordrecht

Jenks G F (1981) Lines, computer and human frailties. *Annals of the Association of American Geographers* **71** (1): 142–7

Keates J S (1982) *Understanding Maps*. Longman, London

Keefer B K, Smith J L, Gregoire T G (1988) Simulating manual digitizing error with statistical models. *Proceedings of GIS/LIS '88*. ACSM, ASPRS, AAG and URISA, San Antonio, Texas, pp. 475–83

Kellaway G P (1949) *Map Projections*. Methuen, London

Kemp K, Frank A (1996) Toward consensus on a European GIS curriculum: the international post-graduate course on GIS. *International Journal of Geographical Information Systems* **10** (4): 447–97

Kemp Z (ed.) (1997) *Innovations in GIS 4*. Taylor and Francis, London

Kennedy M (1996) *The Global Positioning System and GIS: an Introduction*. Ann Arbor Press, Ann Arbor, Michigan

Kingston R, Carver S, Evans A, Turton I (2000) Web-based public participation Geographical Information Systems: an aid to local environmental decision-making. *Computers, Environment and Urban Systems*, **24** (2): 109–25

Korte G B (1993) *The GIS Book*, 3rd edn. OnWord Press, USA

Korte G B (2000) *The GIS Book*. 5th edn. OnWord Press, USA

Kostblade J T (1981) Mapping frequency distributions by the box and whisker. *Professional Geographer* **33** (4): 413–18

Kraak M-J, Brown A (eds) (2000) *Web Cartography: Developments and Prospects*. Taylor and Francis, London

Kubo S (1987) The development of GIS in Japan. *International Journal of Geographical Information Systems* **1** (3): 243–52

Lam N S (1983) Spatial interpolation methods: a review. *American Cartographer* **10**: 129–49

Lang L (1998) *Managing Natural Resources with GIS*. ESRI Press, Redlands

Langran G (1992) *Time in Geographical Information Systems*. Taylor and Francis, London

Laurini R, Thompson D (1992) *Fundamentals of Spatial Information Systems*. Academic Press, London

Lee J (1991) Comparison of existing methods for building triangular irregular network models of terrain from grid elevation models. *International Journal of Geographical Information Systems* **5** (3): 267–85

Lewis O M, Ware J A (1997) The use of census data in the appraisal of residential properties within the United Kingdom: a neural network approach. In: Hodgson S, Rumor M, Harts J J (eds) *Proceedings of the Joint European Conference and Exhibition on Geographical Information Systems*, Vienna, pp. 226–33

Limp F (1999) Mapping hits warp speed on the world wide web! *GeoWorld* Sept 1999 available online at http://www.geoplace.com/gw/1999/0999/999tec.asp

Longley P, Clarke G (eds) (1995) *GIS for Business and Service Planning*. GeoInformation International, Cambridge, UK

Longley P A, Goodchild M F, Maguire D J, Rhind D W (eds) (1999) *Geographical Information Systems: Principles, Techniques, Management and Applications*. John Wiley, New York

Longley P A, Goodchild M F, Maguire D J, Rhind D W (2001) *Geographical Information Systems and Science*. Wiley, Chichester

MacEahren A M (1994) *Some Truth with Maps*. Association of American Geographers, Washington DC

MacEahren A M, Taylor D R F (1994) *Visualization in Modern Cartography*. Pergamon Press, New York

Maffini G (1987) Raster versus vector data encoding and handling: a commentary. *Photogrammetric Engineering and Remote Sensing* **53** (10): 1397–8

Maguire D J (1989) *Computers in Geography*. Longman, London

Maguire D J (1991) An overview and definition of GIS. In: Maguire D J, Goodchild M F, Rhind D W (eds) *Geographical Information Systems: Principles and Applications*. Longman, London, Vol. 1, pp. 9–20

Maguire D J (1995) Implementing spatial analysis and GIS applications for business and service planning. In: Longley P, Clarke G (eds) *GIS for Business and Service Planning*. GeoInformation International, Cambridge, UK, pp. 171–270

Maguire D J, Goodchild M F, Rhind D W (eds) (1991) *Geographical Information Systems: Principles and Applications*. Longman, London

Maguire D J, Worboys M F, Hearnshaw H M (1990) An introduction to object oriented GIS. *Mapping Awareness* **4** (2): 36–9

Mahoney R P (1991) GIS and utilities. In: Maguire D J, Goodchild M F, Rhind D W (eds) *Geographical Information Systems: Principles and Applications*. Longman, London, Vol. 2, pp. 101–14

Malczewski J (1999) *GIS and Multi-criteria Decision Analysis*. Wiley, New York

Mangold R (1997) Are the walls of GIS finally falling down? *Earth Observation Magazine* **6** (2): 4

Marble D (1997) Rebuilding the top of the pyramid: providing technical GIS education to support both GIS development and geographic research. Presentation to GIS in Higher Education 97 conference, 30 Oct–2 Nov, NCGIA and Towson University, Virginia, USA

Mark D M, Chrisman N, Frank A U, McHaffie P H, Pickels J (No date) The GIS History project: http://www.geog.buffalo.edu/ncgia/gishist/bar_harbor.html

Martin D (1991) *Geographical Information Systems and their Socio-economic Applications*. Routledge, London

Martin D (1996) *Geographical Information Systems and their Socio-economic Applications*, 2nd edn. Routledge, London

Martin D, Bracken I (1991) Techniques for modelling population related raster databases. *Environment and Planning A* **23**: 1065–79

Maskrey A (1997) *The design of GIS for risk analysis in Latin America*. Unpublished MSc thesis, Department of Environmental and Geographical Sciences, Manchester Metropolitan University

Masser I (1990) The Regional Research Laboratory initiative: an update. In: Foster M J, Shand P J (eds) *The Association for Geographic Information Yearbook 1990*. Taylor and Francis/Miles Arnold, London, pp. 259–63

Masser I, Campbell H, Craglia M (1996) *GIS Diffusion: The Adoption and Use of*

Geographical Information Systems in Local Government in Europe. GISDATA 3, Taylor and Francis, London

Mather P M (1991) *Computer Applications in Geography*. Wiley, New York

Mather P M (ed.) (1993) *Geographical Information Handling – Research and Applications*. Wiley, Chichester

McAlpine J R, Cook B G (1971) Data reliability from map overlay. *Proceedings of the Australian and New Zealand Association for the Advancement of Science 43rd Congress, Section 21 – Geographical Sciences*. Australian and New Zealand Association for the Advancement of Science, Brisbane, May 1971

McDonnell R, Kemp K (1998) *International GIS Dictionary*, 2nd edn. GeoInformation International, London

McHarg I L (1969) *Design with Nature*. Doubleday, New York

McLaren R A (1990) Establishing a corporate GIS from component datasets – the database issues. *Mapping Awareness* **4** (2): 52–8

McMaster R B, Shea K S (1992) *Generalization in Digital Cartography*. Association of American Geographers, Washington DC

Medyckyj-Scott D J (1989) Users and organisational acceptance of GIS: the route to success or failure. *Proceedings of 1st National Conference for Geographic Information Systems 'GIS – a corporate resource'*. AGI, London, pp. 4.3.1–4.3.8

Medyckyj-Scott D J (1991) What have I done now? In: Shand P J, Moore R V (eds) *The Association for Geographic Information Yearbook 1991*. Taylor and Francis/Miles Arnold, London

Medyckyj-Scott D J, Cornelius S C (1991) A move to GIS: some empirical evidence of users' experiences. *Proceedings of Mapping Awareness 1991*, conference held in London, February, Blenheim Online, London

Medyckyj-Scott D J, Cornelius S C (1994) User viewpoint: a survey. In: Hart T, Tulip A (eds) *Geographic Information Systems Report*. Unicom Seminars, Uxbridge, UK

Medyckyj-Scott D J, Hearnshaw H M (1994) *Human Factors in Geographical Information Systems*. Belhaven, London

Mitchell A (1997) *Zeroing in: Geographic Information Systems at Work in the Community*. ESRI, California

Monmonier M (1972) Contiguity-biased class-interval selection: a method for simplifying patterns on statistical maps. *Geographical Review* **62** (2): 203–28

Monmonier M (1993) *Mapping it Out*. University of Chicago Press, Chicago

Monmonier M (1995) *Drawing the Line*. Henry Holt, New York.

Monmonier M (1996) *How to Lie with Maps*. University of Chicago Press, Chicago

Mounsey H (1991a) Eurodata – myth or reality? In Cadoux-Hudson J, Heywood D I (eds) *The Association for Geographic Information Yearbook 1991*. Taylor and Francis/Miles Arnold, London, 59–64

Mounsey H (1991b) Multisource, multinational environmental GIS: lessons learnt from CORINE. In: Maguire D J, Goodchild M F, Rhind D W (eds) *Geographical Information Systems: Principles and Applications*. Longman, London, Vol. 2, pp. 185–200

Mounsey H (1995) GIS: Justification in the face of organizational change. In: *GIS in Business: Discovering the Missing Piece in your Business Strategy*. GeoInformation International, Banbury, UK, pp. 25–8

Muller J-C (1991) Generalization of spatial databases. In: Maguire D J, Goodchild M F, Rhind D W (eds) *Geographical Information Systems: Principles and Applications*. Longman, London, Vol. 1, pp. 457–75

Muller J-C (1993) Latest developments in GIS/LIS. *International Journal of Geographical Information Systems* **7** (4): 293–303

Mullin J (1992) Computer focus for ambulance inquiry. The *Guardian*, 30 October

National Research Council (1997) *The Future of Spatial Data and Society*. National Academic Press, Washington DC

NCGIA (1989) The research plan of the National Center for Geographic Information and Analysis. *International Journal of Geographical Information Systems* **3** (2): 117–36

Newcomer J A, Szajgin J (1984) Accumulation of thematic map errors in digital overlay analysis. *American Cartographer* **11** (1): 58–62

Nijkamp P (1980) *Environmental Policy Analysis: Operational Methods and Models*. Wiley, New York

Nijkamp P, Scholten H J (1993) Spatial information systems: design, modelling and use in planning. *International Journal of Geographical Information Systems* **7** (1): 85–96

NJUG (1986) *Proposed data exchange format for utility map data. NJUG 11*, National Joint Utilities Group, 30 Millbank, London SW1P 4RD

Obermeyer N J (1999) Measuring the benefits and costs of GIS. In: Longley P A, Goodchild M F, Maguire D J, Rhind D W (eds) *Geographical Information Systems*. Wiley, New York, pp. 601–10

O'Callaghan J F, Garner B J (1991) Land and geographical information systems in Australia. In: Maguire D J, Goodchild M F, Rhind D W (eds) *Geographical Information Systems: Principles and Applications*. Longman, London, Vol. 2, pp. 57–70

Odum T H (1971) *Environment, Power and Society*. Wiley, London

OGC (2001) OGC vision, mission, values and goals: http://www.opengis.org/info/mv.htm

Onsrud H J, Pinto J K (1991) Diffusion of geographic information innovations. *International Journal of Geographical Information Systems* **5** (4): 447–67

Openshaw S (1984) The modifiable areal unit problem. *Concepts and Techniques in Modern Geography*, Vol. 38, GeoBooks, Norwich, UK

Openshaw S (1989) Learning to live with errors in spatial databases. In: Goodchild M F, Gopal S (eds) *The Accuracy of Spatial Databases*. Taylor and Francis, London, pp. 263–76

Openshaw S (1990) Spatial referencing for the user in the 1990s. *Mapping Awareness* **4** (2): 24

Openshaw S (1993) GIS crime and criminality *Environment and Planning A* **25** (4):451–8

Openshaw S (ed.) (1995) *The 1991 Census User's Handbook*. Longman GeoInformation, London

Openshaw S, Carver S, Fernie J (1989) *Britain's Nuclear Waste: Safety and Siting*. Belhaven, London

Openshaw S, Charlton M, Carver S (1991) Error propagation: a Monte Carlo simulation. In: Masser I, Blakemore M (eds) *Handling Geographic Information: Methodology and Potential Applications*. Longman, London, pp. 78–101

Openshaw S, Charlton M, Craft A W, Birch J M (1988) An investigation of leukemia clusters by use of a geographical analysis machine. *Lancet* **1:** 272–3

Openshaw S, Charlton M, Wymer C, Craft A (1987) A mark 1 geographical analysis machine for the automated analysis of point data sets. *International Journal of Geographical Information Systems* **1:** 335–58

Openshaw S, Goddard J (1987) Some implications of the commodification of information and the emerging information economy for applied geographical analysis in the United Kingdom. *Environment and Planning A* **19:** 1423–39

Openshaw S, Peree T (1996) User-centred intelligent spatial analysis of point data. In: Parker D (ed.) *Innovations in GIS 3*. Taylor and Francis, London, pp. 119–34

Otawa T (1987) Accuracy of digitizing: overlooked factor in GIS operations. *Proceedings of GIS '87*. ACSM, ASPRS, AAG and URISA, San Fransisco, pp. 295–9

Oxborrow E P (1989) *Databases and Database Systems: Concepts and Issues*, 2nd edn. Chartwell Bratt, Sweden

Parker D (ed.) (1996) *Innovations in GIS 3*. Taylor and Francis, London

Perkal J (1956) On the epsilon length. *Bulletin de L'Académie Polonaise des Sciences* **4** (7): 399–403

Petch J R, Pauknerova E, Heywood D I (1995) GIS in nature conservation: the Zdarske Vrchy project, Czech Republic. *ITC Journal* **2:** 133–42

Peterson M P (1995) *Interactive and Animated Cartography*. Prentice-Hall, Englewood Cliffs, New Jersey

Peuquet D J (1984) A conceptual framework and comparison of spatial data models. *Cartographica* **21** (4): 66–113

Peuquet D J (1990) A conceptual framework and comparison of spatial data models. In:

Peuquet D J, Marble D F (eds) *Introductory Readings in Geographical Information Systems*. Taylor and Francis, London, pp. 209–14

Peuquet D J (1999) Time in GIS and geographical databases. In: Longley PA, Goodchild M F, Maguire D J, Rhind D W (eds) *Geographical Information Systems: Principles and Technical Issues*. Wiley, New York, Vol. 1, pp. 91–103.

Peuquet D J, Bacastow T (1991) Organizational issues in the development of geographical information systems: a case study of US Army topographic information automation. *International Journal of Geographical Information Systems* **5** (3): 303–19

Peuquet D J, Marble D F (1990) *Introductory Readings in Geographical Information Systems*. Taylor and Francis, London

Pickles J (ed.) (1995) *Ground Truth: the Social Implications of Geographic Information Systems*. Guilford Press, New York

Plewe B (1997) *GIS Online: Information Retrieval, Mapping, and the Internet*. OnWord Press, Santa Fe, NM

Poiker T K (1982) Looking at computer cartography. *GeoJournal* **6** (3): 241–49

Price M F, Heywood D I (eds) (1994) *Mountain Environments and Geographic Information Systems*. Taylor and Francis, London

Rana S, Haklay M, Dodge M (2001) GIS Timeline: http://www.casa.ucl.ac.uk/gistimeline

Raper J (1997) Experiences of integrating VRML and GIS. Paper presented at *ESIG97*, Lisbon

Raper J F, Kelk B (1991) Three-dimensional GIS. In: Maguire D J, Goodchild M F, Rhind D W (eds) *Geographical Information Systems: Principles and Applications*. Longman, London, Vol. 1, pp. 299–317

Raper J F, Rhind D W (1990) UGIX(A): the design of a spatial language interface to a topological vector GIS. *Proceedings of the 4th International Conference on Spatial Data Handling*. International Geographical Union, Columbus, Ohio. pp. 405–12

Raper J F, Rhind D W, Shepherd J W (1992) *Postcodes: the New Geography*. Longman, Harlow, UK

Reeve D E (1996) *Module 4: Attribute data*. UNIGIS postgraduate diploma in GIS by distance learning. Course materials, 5th edn. Manchester Metropolitan University

Reeve D E (1997) *Module 10: GIS in organisations*. UNIGIS postgraduate diploma in GIS by distance learning. Course materials, 6th edn. Manchester Metropolitan University

Reeve D E, Petch J R (1998) *GIS Organizations and People: a Socio-technical Approach*. Taylor and Francis, London

Reina P (1997) At long last, AM/FM/GIS enters the IT mainstream. *Information Technologies for Utilities* (Spring): 29–34

Rhind D W (1981) Geographical Information Systems in Britain. In: Wrigley N, Bennett R J (eds) *Quantitative Geography: Retrospect and Prospect*. Routledge and Kegan Paul, London pp. 17–35

Rhind D (1987) Recent developments in GIS in the UK. *International Journal of Geographical Information Systems* **1** (3): 229–41

Rhind D (1988) Personality as a factor in the development of a discipline: The example of computer-assisted cartography. *American Cartographer* **15** (3): 277–89

Rhind D W (1989) Why GIS? *ARC News* (Summer): pp. 28–9

Rhind D (1992) The next generation of Geographical Information Systems and the context in which they will operate. *Computers, Environment and Urban Systems* **16:** 256–68

Rhind D W, Goodchild M F, Maguire D J (1991) Epilogue. In: Maguire D J, Goodchild M F, Rhind D W (eds) *Geographical Information Systems: Principles and Applications*. Longman, London, Vol. 2, pp. 313–27

Rhind D W, Mounsey H W (1989) Research policy and review 29: The Chorley Committee and 'Handling Geographic Information'. *Environment and Planning A* **21:** 571–85

Rhind D W, Wyatt B K, Briggs D J, Wiggins J C (1986) The creation of an environmental information system for the European Community. *Nachrichter aus dem Karten und Veressungsvesen Series 2*, **44:** 147–57

Richards D R, Jones N L, Lin H C (1993) Graphical innovations in surface water flow analysis. In: Goodchild M F, Parks B O, Steyaert L T (eds) *Environmental Modelling*

with GIS. Oxford University Press, Oxford and New York, pp. 188–95

Ripple W J (ed.) (1994) *The GIS Applications Book: Examples in Natural Resources: A Compendium*. American Society for Photogrammetry and Remote Sensing, Maryland

Robertson B (1988) *How to Draw Charts and Diagrams*. North Light, Cincinnati, Ohio

Robinson A H, Morrison J L, Muehrecke P C, Kimerling A J, Guptill S C (1995) *Elements of Cartography*, 6th edn. Wiley, New York

Sahai R (1996) *Inside MicroStation 95*. OnWord Press, USA

Salge F (1996) MEGRIN: a pragmatic approach towards a European Geographic Information infrastructure. In: Peckham R J (ed.) Proceedings of the 2nd EC – GIS Workshop, Genova, Italy, 26–28 June. 16441 EN, European Commission, Brussels

Samet H (1989) *Applications of Spatial Data Structures: Computer Graphics Image Processing and Other Areas*. Addison-Wesley, London

Schimel D S, Burke I C (1993) Spatial interactive models of atmosphere–ecosystem coupling. In: Goodchild M F, Parks B O, Steyaert L T (eds) *Environmental Modelling with GIS*. Oxford University Press, Oxford and New York, pp. 284–9

Scholten H (1997) *Towards a European spatial meta information system: ESMI*. Presentation at 3rd EC – GIS Workshop. 25–27 June 1997. Leuven, Belgium

Seaborn D (1995) *Database Management in GIS – Past, Present, Future: an Enterprise Perspective for Executives*. Seaborn Associates, 475 Cloverdale Road, Ottawa, Canad

See L, Openshaw S (1997) An introduction to the fuzzy logic modelling of spatial interaction. In: Hodgson S, Rumor M, Harts J J (eds) *Proceedings of the Joint European Conference and Exhibition on Geographical Information Systems*, Vienna, Austria, pp. 809–18

Seegar H (1999) Spatial referencing and coordinate systems. In: Longley P A, Goodchild M F, Maguire D J, Rhind D W (eds) *Geographical Information Systems*. Wiley, New York, vol. 1, pp. 427–36

Shand P, Moore R (eds) (1989) *AGI Yearbook 1989*. Taylor and Francis/Miles Arnold, London

Shiffer M J (1995a) Interactive multimedia planning support: moving from stand alone systems to the Web. *Environment and Planning B: Planning and Design*, **22**: 649–64.

Shiffer M J (1995b) Multimedia representational aids in urban planning support system. In: Marchese F (ed.) *Understanding Images*. Springer-Verlag. New York, pp. 77–90

Shupeng C (1987) Geographical data handling and GIS in China. *International Journal of Geographical Information Systems* **1** (3): 219–28

Sinton D F (1992) *Reflections on 25 Years of GIS*. Supplement to *GIS World*

Skidmore S, Wroe B (1988) *Introducing Systems Analysis*. NCC Publications, Manchester

Smith D A, Tomlinson R F (1992) Assessing costs and benefits of geographical information systems: methodological and implementation issues. *International Journal of Geographical Information Systems* **6** (3): 247–56

Smith T R, Menon S, Star J L, Estes J E (1987) Requirements and principles for the implementation of large-scale geographic information systems. *International Journal of Geographical Information Systems* **1** (1): 13–31

Sobel J (1990) Principal components of the Census Bureau's TIGER File. In: Peuquet D J, Marble D F (eds) *Introductory Readings in Geographic Information Systems*. Taylor and Francis, London, pp. 12–19

Stern N, Stern R A (1993) *Computing in the Information Age*. Wiley, New York

Stewart I (1989) *Does God Play Dice? The Mathematics of Chaos*. Basil Blackwell, Oxford

Steyaert L T (1993) A perspective on the state of environmental simulation modelling. In: Goodchild M F, Parks B O, Steyaert L T (eds) *Environmental Modelling with GIS*. Oxford University Press, Oxford and New York, pp. 16–30

Stocks A M, Heywood D I (1994) Terrain modelling for mountains. In: Price M F,

Heywood D I (eds) *Mountain Environments and Geographical Information Systems*. Taylor and Francis, London, pp. 25–40

Symington D F (1968) Land use in Canada: the Canadian land inventory. *Canadian Geographical Journal*

Taylor D R F (1991a) *Geographic Information Systems: the Microcomputer and Modern Cartography*. Pergamon Press, Oxford, London and New York

Taylor D R F (1991b) GIS and developing nations. In: Maguire D J, Goodchild M F, Rhind D W (eds) *Geographical Information Systems: Principles and Applications*. Longman, London, Vol. 2, pp. 71–84

Taylor P J (1990) GKS. *Political Geography Quarterly* **9** (3): 211–12

Thill J-C (1999) *Spatial Multicriteria Decision Making and Analysis: a Geographic Information Systems Approach*. Ashgate, Aldershot

Tobler W A (1970) A computer movie simulating urban growth in the Detroit region. *Economic Geography* **46:** 234–40

Tobler W R (1976) Spatial interaction patterns. *Journal of Environmental Systems* **6:** 271–301

Tomlin C D (1983) *Digital cartographic modelling techniques in environmental planning*. Unpublished PhD dissertation, Yale University, Connecticut

Tomlin C D (1990) *Geographic Information Systems and Cartographic Modelling*. Prentice-Hall, Englewood Cliffs, New Jersey

Tomlin C D (1991) Cartographic modelling. In: Maguire D J, Goodchild M F, Rhind D W (eds) *Geographical Information Systems: Principles and Applications*. Longman, London, Vol. 1, pp. 361–74

Tomlinson R F (1988) The impact of the transition from analogue to digital cartographic representation. *American Cartographer* **15** (3): 249–63

Tomlinson R F (1990) Geographic Information Systems – a new frontier. In: Peuquet D J, Marble D F (eds) *Introductory Readings in Geographical Information Systems*. Taylor and Francis, London, pp. 18–29

Tomlinson R F, Calkins H W, Marble D F (1976) *Computer Handling of Geographic Data*. UNESCO Press, Paris

Trotter C M (1991) Remotely sensed data as an information source for geographical information systems in natural resource management: a review. *International Journal of Geographical Information Systems* **5** (2): 225–39

Tufte E (1983) *The Visual Display of Quantitative Information*. Graphics Press. Cheshire, CT

Tyler C (1989) Climatology: technology's answer to the British weather. *Geographical Magazine* **LXI** (7): 44–7

Unwin D (1981) *Introductory Spatial Analysis*. Methuen, London

Unwin D J (1991) The academic setting of GIS. In: Maguire D J, Goodchild M F, Rhind D W (eds) *Geographical Information Systems: Principles and Applications*. Longman, London, Vol. 1, pp. 81–90

Unwin D J *et al.* (1990) A syllabus for teaching geographical information systems. *International Journal of Geographical Information Systems* **4** (4): 457–65

US Bureau of Census (1990) Technical description of the DIME system In: Peuquet D J, Marble D F (eds) *Introductory Readings in Geographic Information Systems*. Taylor and Francis, London, 100–11

Vasconcelos M J, Pereira J M C, Zeigler B P (1994) Simulation of fire growth in mountain environments. In: Price M F, Heywood D I (eds) *Mountain Environments and Geographical Information Systems*. Taylor and Francis, London, pp. 167–85

Veregin H (1989) Error modelling for the map overlay operation. In: Goodchild M F, Gopal S (eds) *The Accuracy of Spatial Databases*. Taylor and Francis, London, pp. 3–18

Vertex (1997) http://www.vertex.com

Vincent P, Daly R (1990) GIS and large travelling salesman problems. *Mapping Awareness* **4** (1): 19–21

Von Bertalanffy L (1968) *General Systems Theory: Foundation, Developments and Applications*. Penguin, London

Voogd H (1983) *Multicriteria Evaluation for Urban and Regional Planning*. Pion, London

Walsby J (1995) The causes and effects of manual digitizing on error creation in data input to GIS. In: Fisher P (ed.) *Innovations in GIS 2*. Taylor and Francis, London, pp. 113–24

Walsh S J, Lightfoot D R, Butler D R (1987) Recognition and assessment of error in Geographic Information Systems. *Photogrammetric Engineering and Remote Sensing* **53** (10): 1423–30

Ward R C (1975) *Principles of Hydrology*, 2nd edn. McGraw-Hill, Maidenhead, UK

Waters N M (1989) *NCGIA Core Curriculum*. Lectures 40 and 41. University of California, Santa Barbara

Weibel R, Heller M (1991) Digital terrain modelling. In: Maguire D J, Goodchild M F, Rhind D W (eds) *Geographical Information Systems: Principles and Applications*. Longman, London, Vol. 1, pp. 269–97

Wilson A G (1975) Some new forms of spatial interaction model: a review. *Transportation Research* **9**: 167–79

Wilson J D (1997) Technology partnerships spark the industry. *GIS World* **10** (4): 36–42

Wood C H, Keller C P (eds) (1996) *Cartographic Design*. Wiley, Chichester

Wood D (1993) *The Power of Maps*. Routledge, London

Woodcock C E, Sham C H, Shaw B (1990) Comments on selecting a GIS for environmental management. *Environmental Management* **14** (3): 307–15

Wood-Harper A T, Antill L, Avison D E (1995) *Information Systems Definition: the Multiview Approach*. Blackwell Scientific, Oxford

Worboys M F (1995) *GIS: a Computing Perspective*. Taylor and Francis, London.

Worboys M F, Hearnshaw H M, Maguire D J (1991) Object-oriented data modelling for spatial databases. *International Journal of Geographical Information Systems* **4** (4): 369–83

Zeiler M (1994) *Inside ARC/INFO*. OnWord Press, USA

Glossary

This glossary contains the definition of most of the non-standard words used in the text. It is not intended as a comprehensive GIS glossary. Links to several comprehensive GIS glossaries are provided on the book web page.

***A posteriori* models**: Models that are designed to explore an established theory.

***A priori* models**: Models that are used to model processes for which a body of theory has yet to be established.

Aerial photograph: Photograph taken from an aerial platform (usually an aeroplane), either vertically or obliquely.

Aggregation: The process of combining smaller spatial units, and the data they contain, into larger spatial units by dissolving common boundaries and lumping the data together.

AM/FM: Automated mapping and facilities management.

Analysis scheme: The logical linking of spatial operations and procedures to solve a particular application problem.

Analytical hill shading: The process of calculating relief shadows and using these as a visualization technique to enhance the portrayal of relief on a map.

Annotation: Alphanumeric or symbolic information added to a map to improve communication.

Arc: An alternative term for a line feature formed from a connected series of points.

ARC/INFO: A leading GIS software package developed by the Environmental Systems Research Institute (ESRI).

Area: The entity type used to represent enclosed parcels of land with a common description.

Area cartogram: A type of cartogram that is used to depict the size of areas relative to their importance according to some non-spatial variable such as population or per capita income.

Artificial Intelligence (AI): Field of study concerned with producing computer programs capable of learning and processing their own 'thoughts'.

Artificial Neural Network (ANN): See 'Neural networks'.

Aspatial query: The action of questioning a GIS database on the basis of non-spatial attributes.

Aspect: The direction in which a unit of terrain faces. Aspect is usually expressed in degrees from north.

Attributes: Non-graphical descriptors relating to geographical features or entities in a GIS.

Automated cartography: See 'Computer cartography'.

Azimuthal projections: Map projections in which the surface of the globe is projected onto a flat plane.

Basic Spatial Unit (BSU): The smallest spatial entity to which data are encoded.

Block coding: Extension of the run-length encoding method of coding raster data structures to two dimensions by using a series of square blocks to store the data.

Boolean algebra: Operations based on logical combinations of objects (using AND, NOT, OR and XOR).

Boolean overlay: A type of map overlay based on Boolean algebra.

Buffering: The creation of a zone of equal width around a point, line or area feature.

Canadian Geographic Information System (CGIS): An early GIS using data collected for the Land Inventory System that was developed as a result of the requirements of the Canadian Agriculture and Development Act.

Cartesian co-ordinates: The system for locating a point by reference to its distance from axes intersecting at right angles, often represented as a grid on a map.

Cartograms: Maps that show the location of objects relative to a non-Euclidean variable such as population density or relative distance.

Cartographic modelling: A generic methodology for structuring a GIS analysis scheme.

Cartography: The profession of map drawing; the study of maps.

CASE tools: Computer-assisted software engineering. Software tools used to help reduce program development time.

CD-ROM: Method of storing large volumes of data on optical discs.

Census: The collection of data about the entire population of a country or region.

Central point linear cartogram: A form of cartogram that shows space distorted from a central point on the basis of time taken or cost incurred to travel to a series of destination areas.

Chain coding: Method of raster data reduction that works by defining the boundary of the entity.

Chaos theory: Idea that seemingly minor events accumulate to have complex and massive effects on dynamic natural systems.

Chart junk: Unnecessary, often confusing, annotation added to maps and charts.

Choice alternatives: Range of feasible solutions to be evaluated in a decision problem in MCE.

Choice set: The set of choice alternatives in MCE.

Choropleth map: A thematic map that displays a quantitative attribute using ordinal classes. Areas are shaded according to their value and a range of shading classes.

Cleaning: See 'Data editing'.

Computer-aided cartography: See 'Computer cartography'.

Computer-aided design (CAD): Software designed to assist in the process of designing and drafting. This is normally used for architectural and engineering applications, but can also be used for drafting maps as an input to GIS.

Computer-assisted cartography (CAC): Software designed to assist in the process of designing and drawing maps.

Computer-assisted drafting (CAD): See 'Computer-aided design (CAD)'.

Computer cartography: The generation, storage and editing of maps using a computer.

Computer movies: Animated maps showing how a chosen variable changes with time. These are used to spot temporal patterns in spatial data.

Conceptual data model: A model, usually expressed in verbal or graphical form, that attempts to describe in words or pictures quantitative and qualitative interactions between real-world features.

Concordance–discordance analysis: Method of MCE based on lengthy pairwise comparison of outranking and dominance relationships between each choice alternative in the choice set.

Conic projections: Map projections in which the surface of the globe is projected onto a cone (for example, Alber's Equal Area projection).

Constraints: Set of minimum requirements for a decision problem in MCE.

Contour: A line on a topographic map connecting points of equal height and used to represent the shape of the Earth's surface.

Cookie cutting: See 'IDENTITY overlay'.

CORINE: The Co-ordinated Information on the European Environment programme, initiated in 1985 by the European Union to create a database that would encourage the collection and co-ordination of consistent information to aid European Community policy.

Criteria: Attributes by which choice alternatives are evaluated in MCE.

Cylindrical projections: Map projections in which the surface of the globe is projected onto a cylinder (for example, the Mercator projection).

Dangling arc: A dead-end line or arc feature that is connected to other lines in the data layer at one end only.

Data: Observations made from monitoring the real world. Data are collected as facts or evidence, which may be processed to give them meaning and turn them into information.

Data accuracy: The extent to which an estimated data value approaches its true value.

Data bias: The systematic variation of data from reality.

Data conversion: The process of converting data from one format to another. With the many different GIS data formats available this can be a difficult task requiring specialized software.

Data editing: The process of correcting errors in data input into a GIS. This can be carried out manually or automatically.

Data error: The physical difference between the real world and a GIS facsimile.

Data input: The process of converting data into a form that can be used by a GIS.

Data precision: The recorded level of detail of the data.

Data stream: The process of progressing from raw data to integrated GIS database. This includes all stages of data input, transformation, re-projection, editing and manipulation.

Database: A collection of data, usually stored as single or multiple files, associated with a single general category.

Database management system (DBMS): A set of computer programs for organizing information at the core of which will be a database.

Decision support system (DSS): A system, usually computerized, dedicated to supporting decisions regarding a specific problem or set of problems.

Delaunay triangulation: Method of constructing a TIN model such that three points form the corners of a Delaunay triangle only when the circle that passes through them contains no other points.

Deterministic model: A model for which there is only one possible answer for a given set of inputs.

Differential GPS: The process of using two GPS receivers to obtain highly accurate and precise position fixes.

Diffusion: The spatio-temporal process concerned with the movement of objects from one area to another through time. Examples include forest fire development, the movement of pollutants through any medium, and the spread of a disease through a population.

Digital elevation model (DEM): A digital model of height (elevation or altitude) represented as regularly or irregularly spaced point height values.

Digital mapping: See 'Computer cartography'.

Digital terrain model (DTM): A digital model of a topographic surface using information on height, slope, aspect, breaks in slope and other topographic features.

Digitizer: A piece of computer hardware used to convert analogue data into digital format. See 'Digitizing table', 'Scanner' and 'Laser line follower'.

Digitizing: The process of converting data from analogue to digital format.

Digitizing table: A table underlain by a fine mesh of wires and connected to a computer that allows the user to record the location of features on a map using an electronic cursor. These are used to digitize maps and other graphical analogue data sources manually.

Disaggregation: The reverse of 'aggregation'.

Distance decay: A function that represents the way that some entity or its influence decays with distance from its geographical location.

Douglas–Peucker algorithm: A geometric algorithm used to thin out the number of points needed to represent the overall shape of a line feature.

DXF: Data eXchange Format. A file format used for the exchange of GIS data.

Dynamic model: A model in which time is the key variable while all other input variables remain constant. Outputs from the model vary as time progresses.

Ecological fallacy: A problem associated with the 'modifiable areal unit problem' that occurs when it is inferred that data for areas under study can be applied to individuals in those areas.

Edge matching: The process of joining data digitized from adjacent map sheets to ensure a seamless join.

Electronic distance metering (EDM): A theodolite combined with an optical rangefinder for accurate distance measurements.

Enumeration district (ED): The area over which one census enumerator delivers and collects census forms. The size of these units, though small, is enough to ensure anonymity of census respondents in data reported at this level of detail.

Envelope polygon: The external polygon of a vector map inside which all other features are contained.

Epsilon modelling: A method of estimating the effects of positional error in GIS overlay operations. Epsilon modelling is based on the use of buffer zones to account for digitizing error around point, line and area features.

Equilibrium forecast model: A model in which forecasts are made on the basis of change in one or more elements of the system or process while all other factors remain constant.

Error propagation: The generation of errors in a GIS database at various stages of the data stream and during subsequent analyses.

Euclidean distance: A straight-line distance measured as a function of Euclidean geometric space.

Evaluation matrix: A matrix containing criterion scores (rows) for each choice alternative (columns) in the choice set for any MCE problem.

Evapotranspiration: The process by which water vapour re-enters the atmosphere directly through evaporation and from plant transpiration.

Eyeballing: See 'Line threading'.

Feasibility study: An evaluation of the costs and benefits of adopting an IT solution to a problem.

Feature codes: Unique codes describing the feature to which they are attached in the GIS database.

Features: See 'Spatial entities'.

Filter: A cell matrix of varying shape and size used to modify the cell values in a raster map layer through a variety of mathematical procedures such as mean, sum, maximum and minimum. A filter is often used to smooth noisy data.

GANTT charts: Graphical time charts used to assist project management.

Gazetteer: A dictionary or index of geographical names.

General systems theory: A theory based on the idea that to understand the complexity of the real world, we must attempt to model this complexity.

Generalization: The process by which information is selectively removed from a map in order to simplify pattern without distortion of overall content.

Geographical analysis: Any form of analysis using geographical data.

Geomorphology: The study of the Earth's surface and the physical processes acting on it.

Global Positioning System (GPS): A system of orbiting satellites used for navigational purposes and capable of giving highly accurate geographic co-ordinates using hand-held receivers.

GLONASS: Global Orbiting Navigation Satellite System. The Russian equivalent of the US GPS system.

Graphical User Interface (GUI): Computer interface, such as MS Windows, that is used to provide a user-friendly interface between a user and the computer.

Great circle: A circle with the same diameter as the Earth. This is used to describe the shortest route between any two points on the Earth's surface.

Groundwater percolation: The process by which water from the soil enters the underlying geology by slow downward movement.

Hard systems analysis: A set of theory and methods for modelling the complexity of the real world.

Hydrological cycle: A dynamic natural system concerned with the circulation of water through the atmosphere, biosphere and geosphere.

Ideal point analysis: An MCE algorithm based on the evaluation of choice alternatives against a hypothetical ideal solution.

IDENTITY overlay: Polygon-on-polygon overlay corresponding to the Boolean OR and AND overlays. The output map will contain all those polygons from the first map layer and those which fall within these from the second map layer. Also referred to as 'cookie cutting'.

Infiltration rate: The rate at which surface runoff and precipitation soaks into the ground surface.

Internet, The: The global network of computers, originally set up as a means of secure communications for military and intelligence purposes.

Internet protocols: An established language used for the communication between systems on the Internet. HTTP and TCP/IP are examples.

INTERSECT overlay: Polygon-on-polygon overlay corresponding to the Boolean AND overlay. The output map will contain only those polygons that cover areas common to both sets of input polygons.

Intranet: A secure closed network that uses Internet protocols.

Isoline: A line joining areas of equal value.

JAVA Applet: A program that will run within a web page, written in the JAVA language.

Keyboard entry: A method of entering small volumes of data at the computer keyboard.

Key-coding: See 'Keyboard entry'.

Landsat: An orbiting satellite giving regular repeat coverage of the whole of the Earth's surface.

Laser line follower: An automatic digitizer that uses a laser beam to follow and digitize lines on a map.

Latitude: Angular measurement north and south of the equator. Represented on the globe as parallel lines circling the globe perpendicular to the lines of longitude.

Layer-based approach: An approach to organizing spatial data into thematic map layers, wherein each map layer contains information about a particular subject and is stored as a separate file (or series of files) for ease of management and use.

Line: Entity type used to represent linear features using an ordered set of (x,y) co-ordinate points or a chain of grid cells.

Line-in-polygon overlay: The process of overlaying a line map over a polygon map to determine which lines cross which polygons.

Line of true scale: Line on a map along which distances are not distorted due to the effects of the map projection method used.

Line threading: A method of spatial interpolation reliant on estimating isolines through a series of observed values by hand and eye.

Lineage: The record of a data set's origin and all the operations carried out on it.

Linear weighted summation: The simplest MCE algorithm based on the simple addition of weighted criterion scores for each alternative. Under this decision rule the choice alternative with the maximum score is the best.

Link impedance: The cost associated with traversing a network link, stopping, turning or visiting a centre.

Linked display: Method of dynamically linking map and non-map output such as charts and data plots such that changes in one are reflected by changes in the other. Such displays are used to aid exploratory data analysis.

Local area network (LAN): A local network of computers, usually within the same building, which may be used, for instance, to connect a GIS fileserver to a number of GIS access terminals.

Location-allocation modelling: The use of network analysis to allocate the location of resources through the modelling of supply and demand through a network.

Longitude: Angular measurement east and west of the prime or Greenwich meridian. Represented on the globe as a series of great circles intersecting at both poles.

Loose coupling: A method of linking models to GIS in which the user may have to perform transformations on the data output from GIS, spatial analysis or process modelling software before they can be used in another software environment.

Manhattan distance: The method used to calculate distance between cells in a raster data layer by counting the distance along the sides of the raster cells.

Manual digitizing: The process of digitizing maps and aerial photographs using a table digitizer.

Map algebra: A method of combining raster data layers. Mathematical operations are performed on individual cell values from two or more input raster data layers to produce an output value.

Map overlay: The process of combining the data from two or more maps to generate a new data set or a new map. This is a fundamental capability of GIS.

Map projection: A mathematical and/or geometric method used to transfer the spherical surface of the Earth onto a flat surface such as a map. Many different map projections exist, each with its own advantages and disadvantages.

Mapematics: See 'Map algebra'.

Mathematical model: A model that uses one or more of a range of techniques, including deterministic, stochastic and optimization methods.

METEOSAT: A geostationary weather satellite monitoring the whole of the western hemisphere.

Middleware: Technology that breaks down information into packets for exchange between servers and clients.

Modifiable areal unit problem (MAUP): A problem arising from the imposition of artificial units of spatial reporting on continuous geographic phenomena resulting in the generation of artificial spatial patterns.

Monte Carlo simulation: A statistical technique used to simulate the effects of random error in data analyses. Random 'noise' is added to the input data and analyses run repeatedly to estimate the effects on the output data.

Multi-criteria evaluation (MCE): A method of combining several, possibly conflicting, criteria maps to derive suitability maps based on trade-off functions and user-specified criteria preference weights.

Multi-criteria modelling: See 'Multi-criteria evaluation (MCE)'.

Multimedia: Digital media presented in multiple forms, including text, pictures, sound and video.

National Center for Geographic Information and Analysis (NCGIA): An organization set up with funding from the US National Science Foundation to lead US GIS research.

National transfer format (NTF): Standard data transfer format developed and used by the Ordnance Survey in the UK.

Natural analogue model: A model that uses actual events or real-world objects as a basis for model construction.

Natural language: Spoken or written language.

NAVSTAR: American GPS satellite constellation consisting of 24 satellites used for location fixing.

Network: (1) A special type of line entity used to represent interconnected lines that allow for the flow of objects (for example, traffic) or information (for example, telephone calls). (2) Two or more interconnected computers used to facilitate the transfer of information.

Neural networks: Classification systems that look for pattern and order in complex multivariate data sets. A key component of Artificial Intelligence (AI) techniques.

Node: Point entity representing the start or end point of a line or its intersection with another line.

Nuclear Industry Radioactive Waste Executive (NIREX): Organization set up to oversee the disposal of radioactive waste in the UK.

Object-oriented approach: Approach to organizing spatial data as discrete objects in a single map space.

Octree: Three-dimensional modification of the quadtree data structure.

Optical character recognition (OCR): Software used to recognize the shape of alphanumeric characters in a scanned image and convert these to ASCII text files.

Optimization model: A model that is constructed to maximize or minimize some aspect of its output.

Ordnance Survey: UK mapping agency.

PERT: Program evaluation and review technique. A project management tool.

Pilot project: An example software application that shows the potential of a larger application. This is often referred to as a 'demonstrator'. See also 'Prototyping'.

Pixel: A single cell in a raster data model. Short for picture element.

Planar co-ordinates: See 'Cartesian co-ordinates'.

Plotter: Output device used to draw maps and other diagrams from digital data.

Plug-ins: Software that works automatically when particular types of files are encountered by a web browser.

Point: A single (x,y) co-ordinate or pixel used to describe the location of an object too small to be represented as a polygon or area feature within the working scale of the GIS.

Point dictionary: Vector data structure in which all points are numbered sequentially and contain an explicit reference which records which co-ordinates are associated with which polygon.

Point-in-polygon overlay (PIP): The process of laying a point map over a polygon map to determine which points fall within which polygon.

Point mode digitizing: A method of manual digitizing whereby the operator decides where to record each co-ordinate.

Polygon: A multifaceted vector graphic figure used to represent an area and formed from a closed chain of points.

Polygon-on-polygon overlay: Process of overlaying two polygon maps to determine how the two sets of polygons overlap. There are three main types of polygon overlay: UNION, INTERSECT and IDENTITY.

Postal code: Non-geographic co-ordinate system consisting of alphanumeric codes used to increase the efficiency of mail sorting and delivery. Linked to geographic co-ordinates these can be a useful form of spatial reference.

Primary data: Data collected through first-hand observation. Field notes, survey recordings and GPS readings are examples of primary data.

Prototyping: An IT software development and project management methodology that uses 'prototype' software applications to test ideas with potential users. See also 'Pilot project'.

Public land survey system (PLSS): Recursive division of land into quarter sections which can be used as a non-co-ordinate referencing system. Used in the western USA.

Quadtree: A compact raster data structure in which geographical space is subdivided into variably sized homogeneous quarters.

Quaternary triangular mesh (QTM): A referencing system that tries to deal with irregularities on the Earth's surface using a regular mesh of triangles. An alternative to the geographic (latitude and longitude) referencing system.

Raster data model: System of tessellating rectangular cells in which individual cells are used as building blocks to create images of point, line, area, network and surface entities.

Rasterization: The process of converting data from vector to raster format.

Reclassification: The process of reclassifying values in a map layer to produce a new map layer.

Rectangular co-ordinates: See 'Cartesian co-ordinates'.

Regional Research Laboratories (RRL): A network of research laboratories set up by the Economic and Social Research Council to pump-prime GIS research in the UK.

Relational database: A computer database employing an ordered set of attribute values or records known as tuples grouped into two-dimensional tables called relations.

Relations: Two-dimensional tables of data in a relational database.

Relief shadows: Areas calculated to be in the shadow of hills using a DTM. The amount of shadow varies according to the relative position of the Sun as determined by the time of day and date.

Remote sensing: The science of observation without touching. Often used to refer to Earth observation from satellite platforms using electromagnetic sensors.

Resolution: The size of the smallest recording unit or the smallest feature that can be mapped and measured.

Rich picture: A graphical method for expressing the scope of a problem.

Root definition: A clear statement of purpose. In GIS this may be a statement of why the GIS is required and the purpose it is to serve.

Route tracing: Application of network analysis to trace the route of flows through a network from origin to destination. This is particularly useful where flows are unidirectional (as in hydrological networks).

Routed line cartograms: Maps in which routes are plotted as generalized lines showing the sequence of stops and connections rather than the actual route taken. An example is the map of the London Underground system.

Rubber sheeting: The method used to adjust the location of features in a digital map layer in a non-uniform or non-linear manner.

Run-length encoding: The method of reducing the volume of raster data on a row-by-row basis by storing a single value where there are a number of cells of a given type in a group, rather than storing a value for each individual cell.

Satellite image: Graphical image (usually in digital form) taken of the Earth's surface using electromagnetic sensors on board an orbiting satellite or spacecraft.

Scale: The size relationship or ratio between the map document and the area of the Earth's surface that it represents.

Scale analogue model: A model that is a scaled down and generalized replica of reality, such as a topographic map or aerial photograph.

Scale-related generalization: Cartographic simplification of mapped features, the level of which is directly related to the scale of the map being drawn.

Scanner: Raster input device similar to an office photocopier used to convert maps and other analogue data into digital data using the raster data model.

Secondary data: Data collected by another individual or organization for another primary purpose. Many data sources used in GIS, including maps, aerial photographs, census data and meteorological records, are secondary.

Shortest path problem: A classic problem in network analysis which involves identifying the shortest and/or quickest route through a network, taking link and turn impedances as well as distance into account.

Sieve mapping: The consecutive overlay of various maps to find a set of feasible areas that satisfy a given set of criteria.

Simple raster: Simplest raster data structure in which the value of each pixel is stored as a separate value in a matrix.

Sliver polygons: Small, often long and narrow, polygons resulting from the overlay of polygons with a common but separately digitized boundary. Errors in the digitizing process mean two versions of the common boundary are slightly different.

Slope: The steepness or gradient of a unit of terrain, usually measured as an angle in degrees or as a percentage.

Soft systems analysis: A general-purpose methodology for investigating unstructured management problems.

Spaghetti model: Simplest form of vector data structure representing the geographic image as a series of independent (x,y) co-ordinate strings.

Spatial: Anything pertaining to the concepts of space, place and location.

Spatial analysis: See 'Geographical analysis'.

Spatial autocorrelation: Tobler's Law of Geography (1976), which states that points closer together in space are more likely to have similar characteristics than those that are further apart.

Spatial data: Data that have some form of spatial or geographical reference that enables them to be located in two- or-three dimensional space.

Spatial data model: A method by which geographical entities are represented in a computer. Two main methods exist: raster and vector data models.

Spatial data retrieval: The process of selectively retrieving items from a GIS database on the basis of spatial location.

Spatial data structure: Approach used to provide the information that the computer requires to reconstruct the spatial data model in digital form. There are many different data structures in use in GIS.

Spatial decision support system (SDSS): A decision support system (DSS) with a strong spatial component and incorporating spatial data models and spatial analysis to assist the user in arriving at a solution.

Spatial entities: Discrete geographical features (points, lines and areas) represented in a digital data structure.

Spatial interaction models: Models that are used to help understand and predict the location of activities and the movement of materials, people and information.

Spatial interpolation: The procedure of estimating the values of properties at unsampled sites within an area covered by existing observations (Waters, 1989).

Spatial model: A model of the real world, incorporating spatial data and relationships, used to aid understanding of spatial form and process.

Spatial process models: Models that simulate real-world processes which have a significant spatial component. These are used to help evaluate and understand complex spatial systems.

Spatial query: Action of questioning a GIS database on the basis of spatial location. Spatial queries include 'What is here?' and 'Where is...?'

Spatial referencing: The method used to fix the location of geographical features or entities on the Earth's surface.

Splines: Mathematical method of smoothing linear features, often used to interpolate between points digitized along a curved feature.

SPOT: A French remote sensing satellite (Système Pour l'Observation de la Terre) yielding high-resolution (10 m) data.

Spot height: A single height value on a topographic map, usually representing the location and height of a prominent feature between contour lines.

Standardization: The process of transforming data onto a common scale of measurement. This is used in MCE to ensure comparability between criteria measured in different units and different scales of measurement.

Stereoscopic aerial photography: The use of overlapping vertical aerial photographs to determine height based on the parallax effect gained from viewing the same object from slightly different angles.

Stochastic model: A model that recognizes that there could be a range of possible outcomes for a given set of inputs, and expresses the likelihood of each one happening as a probability.

Stream mode digitizing: The method of manually digitizing lines whereby the digitizer automatically records a co-ordinate at a set time or distance interval.

Structured Query Language (SQL): The computer language developed to facilitate the querying of relational databases. Also known as Standard Query Language.

Structured systems analysis and design methodology (SSADM): A method used for the design of GIS projects.

Surface: The entity type used to describe the continuous variation in space of a third dimension, such as terrain.

Surface drape: The draping of an image on top of a 3D view of a terrain model for the purpose of landscape rendering or visualization.

Surface significant points: Points in a TIN model that cannot be closely interpolated from the height values of neighbouring points.

SYMAP: Synagraphic Mapping. An early commercial computer cartography package.

SYMVU: Early 3D computer mapping package developed from SYMAP.

System life cycle: An IT software development and methodology that uses a linear approach to project management.

Temporal data: Data that can be linked to a certain time or period between two moments in time.

Terrain model: Surface model of terrain. See 'digital terrain model (DTM)', 'digital elevation model (DEM)' and 'triangulated irregular network (TIN)'.

Thematic data: Data that relate to a specific theme or subject.

Thematic maps: Maps pertaining to one particular theme or subject.

Theodolite: A survey instrument for measuring horizontal and vertical angles using a rotating and tilting telescope.

Thiessen polygons: An exact method of interpolation that assumes the values of unsampled locations are equal to the value of the nearest sampled point to produce a mesh of irregular convex polygons.

Tight coupling: A method of linking models to GIS in which the link between the GIS and the model is hidden from the user by an application interface and GIS and model share the same database.

Topographic maps: Maps whose primary purpose is to indicate the general lie of the land. These maps generally show terrain, basic land use, transport networks, administrative boundaries, settlements and other man-made features.

Topology: The geometric relationship between objects located in space. Adjacency, containment and connectivity can describe this.

Total station: A theodolite or EDM combined with data logger and automated mapping software.

Transformation: The process of converting data from one co-ordinate system to another.

Travelling salesperson problem: A problem in network analysis where the best route between a series of locations to be visited in one journey must be identified.

Trend surface analysis: A routine that interpolates a complex surface or series of data points to produce a much-simplified surface showing the overall trend in the data.

Triangulated irregular network (TIN): An irregular set of height observations in a vector data model. Lines connect these to produce an irregular mesh of triangles. The faces represent the terrain surface and the vertices represent the terrain features.

Tuples: Individual records (rows) in a relational database.

UNION overlay: Polygon-on-polygon overlay corresponding to the Boolean OR overlay. The output map will contain a composite of all the polygons in both of the input map layers.

Universal soil loss equation (USLE): Method of predicting soil loss over large areas based on relationships between soil loss and determining factors measured empirically using small erosion plots.

Vector data model: A spatial data model using two-dimensional Cartesian (x,y) co-ordinates to store the shape of spatial entities.

Vectorization: The process of converting data from raster to vector format.

Viewshed: A polygon map resulting from a visibility analysis showing all the locations visible from a specified viewpoint.

Virtual reality (VR): The production of realistic-looking computer-generated worlds using advanced computer graphics and simulation models.

Virtual Reality Modelling Language (VRML): A computer language developed to assist VR development on the World Wide Web.

Visibility analysis: An analysis of visible features on a terrain surface using a DTM.

Voronoi polygons: See 'Thiessen polygons'.

Weights: Priorities or preferences attached to criteria in MCE. Usually specified by the user in order to indicate the relative importance of each criterion.

Weird polygons: See 'Sliver polygons'.

World Wide Web (the Web): Interconnected global network of computers and the software interface used to access and exchange digital information and multimedia.

ZIP code: See 'Postal code'.

Index